GWR LOCOMOTIVE ALLOCATIONS

First and Last Sheds 1922–1967

J. W. P. ROWLEDGE C Eng, M I Mech E

DAVID & CHARLES
Newton Abbot London North Pomfret (Vt)

British Library Cataloguing in Publication Data

Rowledge, J. W. P.
GWR locomotive allocations;
first and last sheds 1922–1967.
1. Great Western Railway—
History 2. Locomotives—
England—History—20th century
I. Title
385′.361′0942 TJ603.4.G72G73

ISBN 0-7153-8820-7

Printed in Great Britain
by Redwood Burn Ltd Trowbridge Wilts
for David & Charles Publishers plc
Brunel House Newton Abbot Devon

Published in the United States of America
by David & Charles Inc
North Pomfret Vermont 05053 USA

Contents

INTRODUCTION	5
GWR ENGINE SHEDS	6
NUMBERING OF GWR LOCOMOTIVES 1922	8
GWR LOCOMOTIVE STOCK MAY 1922	29
LOCOMOTIVES ADDED TO STOCK AUGUST 1922	97
LOCOMOTIVES ADDED TO STOCK JANUARY 1923	98
LOCOMOTIVES ADDED TO STOCK JULY 1923 AND JANUARY 1924	99
STEAM RAIL MOTORS	100
NEW LOCOMOTIVES JUNE 1922 - OCTOBER 1956	101
DIESEL RAILCARS	162
SHEDS OF 2-8-0 PURCHASED 1925	163
LOCOMOTIVES LENT TO GWR 1939-47:	
LMS 0-6-0 DIESEL-ELECTRIC SHUNTER	165
LMS 2F 0-6-0	165
LNER J25 0-6-0	166
LMS(WD) 8F 2-8-0	166
LNER O4 2-8-0	167
SR LOCOMOTIVES	168
USA 2-8-0	168
LMS 8F 2-8-0 (Swindon built)	171
LNER Y9 0-4-0ST	173
USA 0-6-0T	173
WD 2-8-0 (Loans 1944-5)	173
WD 2-8-0 (Loans 1946-7)	175

Introduction

In the Public Records Office at Kew the Great Western Railway and many of its activities are well documented. A series of books in the RAIL 254 series is a fascinating record of the location of the company's engines from 1902 to 1963, only 1921 and 1933 not being available; the real thrill on seeing these books for the first time was the fact that engines were shown at sub-sheds, not solely main depots. Thus the motive power of branch lines can be identified and the distribution of any class of locomotive ascertained during the greater part of this century.

This volume deals with GWR locomotives in two parts, the first taking the stock in the summer of 1922, when so many engines were taken over, and gives:-

1. Locoation of engines at the end of May 1922 (or at a slightly later date for the small number taken over later in 1922-4),
2. Last GWR shed,
3. Last British Railways shed.

Similarly the second part deals with engines built after the Grouping (or acquired), giving the first shed and then similar details as in the first part. Post GWR engines, delivered to British Railways, have of course only a first and last shed. Furthermore locomotives on loan to the company are included. In each section the allocation lists are preceded by locomotive building details and former railway numbers where applicable; individual engine dates are given in the main lists.

As the record books fell out of use in 1963 it has been necessary to use other published sources to complete the record to the end of GWR steam. This means of course that for the last few years it has not been possible to indicate a sub-shed as the last working shed. For 1933 it has been necessary to check with other published sources. As no single code was used throughout it has been necessary to record place names in full, but avoid confusion in the few instances where a name was changed by either the GWR or BR the name used in the records in 1922 has been retained throughout (ie Paddington later became Old Oak Common, or in another case Goodwick became Fishguard). It should be noted that towards the end of GWR steam a number of engines finished their days at sheds originally belonging to the other railways.

The author is greatly indebted to the staff of the Public Records Office, Kew, for their kindness in making the appropriate volumes freely available on the many occasions that it was necessary to consult them, both initially and on checking the draft.

If you, the reader as a researcher, general enthusiast or railway modeller, find as much interest and get as much joy as the author on discovering the location of engines, perhaps those seen years ago, or working from favourite locations or perhaps on a branch line happily recalled, then this volume will serve its purpose and its author more than gratified that fellow railway enthusiasts share his feelings for boyhood and later memories of the GREAT WESTERN RAILWAY.

Newton Abbot.

September 1985.

GWR Locomotive Sheds

The following is a list of all engine sheds on the Great Western Railway in 1922 (and those added since), but not every minor location had an allocation (permanent or semi-permanent), so that some places will not be found in the allocation lists.

Some of the sheds listed were jointly owned with other railways, but this fact is not stated in the list as it is not relevant to the information contained in this volume.

At Whitchurch the LNWR/LMS engine shed was used by the Cambrian/GWR.

Some of the above locations did not have covered accommodation.

Cardiff East Dock and Pantyffynon were new GWR sheds, both opened in 1931.

Of necessity, the following place names have been abbreviated in the allocation lists:-

Bath Road	= Bristol Bath Road	Ebbw Junction	= Newport Ebbw Junction
Bristol SPM	= Bristol St. Philips Marsh	Moreton'stead	= Moretonhampstead
Cardiff E Dock	= Cardiff East Dock	N'castle Emlyn	= Newcastle Emlyn
Cardiff E M	= Cardiff East Moors	Plymouth	= Plymouth Millbay‡
Chel High St.	= Cheltenham High Street	Severn Tunnel	= Severn Tunnel Junction
Chipping N	= Chipping Norton	Stafford Road	= Wolverhampton Stafford Road
Ciren' MSWJR	= Cirencester MSWJR	Stratford	= Stratford-on-Avon
Cleobury M	= Cleobury Mortimer	Swansea E Dock	= Swansea E Dock
Dock Street	= Newport Dock Street	Weston-S-Mare	= Weston-Super-Mare

‡ Millbay was used for this shed from soon after the Grouping; although closed in 1924 the shed was used from 1925 to 1931 to give relief to Laira (No. 1365 was shown in the GWR 1947 allocation book as Millbay on 31st December 1947, but in fact was located at the small shed known as Plymouth Dock).

In the 1922 list the following places covered more than one shed:-

Bristol	= Bath Road and St. Philips Marsh
Newport	= Ebbw Junction and Dock Street
Wolverhampton	= Stafford Road and Oxley

Cardiff refers solely to the main GWR shed at Cardiff, latterly known as Cardiff Canton.

Former GWR engines were also withdrawn by British Railways from sheds which were never GWR property, these places being:-

- ex-LMS - Bath, Barrow Road (Bristol), Bromsgrove, Dingwall, Gloucester, Gurnos, Paxton Street (Swansea), Templecombe, Tredegar, Upper Bank (Swansea).
- ex-LNER - Wrexham.
- ex-SR - Exmouth Junction, Nine Elms, Wadebridge, Yeovil Town.

The following abbreviations are used for locomotive manufacturers:-

AB	= Andrew Barclay	KS	= Kerr, Stuart
AE	= Avonside Engine Co.	LNWR	= London & North Western Railway, Crewe
AW	= Armstrong, Whitworth	MW	= Manning, Wardle
BE	= Brush Electrical Engineering	NBL	= North British Locomotive Co.
BER	= Bristol & Exeter Railway	NW	= Nasmyth, Wilson
BP	= Beyer, Peacock	NR	= Neilson, Reid
C	= Cooke Locomotive Co. (America)	P	= Peckett
Cam	= Cambrian Railways	P&J	= Parfitt & Jenkins
CF	= Chapman & Furneaux	RS	= Rbt. Stephenson
D	= Dubs	RSH	= Rbt. Stephenson & Hawthorn
D&M	= Davies & Metcalfe	RWH	= R.& W.Hawthorn
FB	= Franco Belge	SAB	= S.A.Belge
FJ	= Fletcher Jennings	SG	= Slaughter, Grunning
FW	= Fox, Walker	SS	= Sharp, Stewart
GWRS	= Great Western Railway, Swindon	TVR	= Taff Vale Railway
GWRW	= Great Western Railway, Wolverhampton	VF	= Vulcan Foundry
HC	= Hudswell, Clarke	WE	= Worcester Engine Co.
HL	= Hawthorn, Leslie	WGB	= W.G.Bagnall
JF	= J.Fowler	YE	= Yorkshire Engine Co.
K	= Kitson		

ROD = Railway Operating Division (locomotives purchased from Government 1919-25).

* indicates that the date or data given is uncertain or not known.

The following points should be noted when consulting the allocation lists:-

1. Engines in works in May/June 1922 are shown at the next working shed (unless condemned at the time).
2. Working sheds are given, but in a few instances engines were on loan from another depot (these are not indicated).
3. Some engines were placed in store for a while before condemnation which in some instances was later than the time that the 'last shed' specified was closed.
4. Nos. 5275-94 (new in 1930) never worked as 2-8-0T, entering service as 2-8-2T Nos. 7200-19. Many of Nos. 6700-49 (new in 1930) spent a couple of years in store; see pp. 161 at the end of the second part of the allocation lists. Many other engines when new were held in store for one or two months, even in British Railways days, such that Nos. 7828/9, new in 12/1950, were not put into traffic until 2/1952!

The author is grateful to Mr. R.C.Riley for help with the 'last sheds' of the GWR diesel cars.

Shed	Pre-Group Railway	Closed to steam
Aberavon	RSB	1922
Aberayron	GWR	1962
Aberbeeg	GWR	1964
Abercynon	TVR	1964
Aberdare	GWR	1965
Aberdare	TVR	1927
Abergavenny	GWR	1932
Aberystwyth	Cam	1965
Aberystwyth (n/g)	Cam	1968
Abingdon	GWR	1954
Alcester	GWR	1939
Andover Junction	MSWJR	1962
Ashburton	GWR	1958
Avonmouth	GWR	1924
Aylesbury	GWR	1962
Bala	GWR	1965
Banbury	GWR	1966
Barnstaple	GWR	1951
Barry	Barry	1964
Basingstoke	GWR	1950
Bassaleg	B&MR	1929
Bath	GWR	1961
Birkenhead	GWR	1967
Blaenrhondda	RSB	1922
Blagdon	GWR	1924
Bodmin	GWR	1962
Branches Fork	GWR	1952
Brecon	B&MR	1962
Bridgend	GWR	1950
Bridgwater	GWR	1960
Bridport	GWR	1959
Brimscombe	GWR	1963
Bristol Bath Road	GWR	1960
Bristol St. Philips Marsh	GWR	1964
Brixham	GWR	1929
Builth Wells	Cam	1957
Bullo Pill	GWR	1931
Burry Port	BPGVR	1962
Cae Harris	GWR/RR	1964
Cardiff	GWR	1962
Cardiff Docks	RR	1931
Cardiff East Dock	-	1965
Cardiff East Moors	Car	1926
Cardigan	GWR	1962
Carmarthen	GWR	1964
Cathays	TVR	1957
Chalford	GWR	1951
Chard	GWR	1924
Cheltenham	GWR	1963
Cheltenham High St	MSWJR	1935
Chester	GWR	1960
Chippenham	GWR	1964
Chipping Norton	GWR	1922
Cirencester	GWR	1959
Cirencester	MSWJR	1924
Clee Hill	GWR	1960
Cleobury Town	CMDPR	1938
Coke Ovens	TVR	1933
Corris	Corris	1948
Corwen	GWR	1927
Cowbridge	TVR	1924
Crewe	GWR	1963
Croes Newydd	GWR	1967
Crwys (Cardiff)	RR	1925
Danygraig	RSB	1960
Didcot	GWR	1965
Dinas Mawddwy	Cam	1931
Dowlais Central	B&MR	1960
Duffryn Yard	GWR	1964
Easton	GWR	1925
Evesham	GWR	1961
Exeter	GWR	1963
Fairford	GWR	1962
Falmouth	GWR	1925
Faringdon	GWR	1951
Ferndale	TVR	1964
Frome	GWR	1963
Garnant	GWR	1931
Gloucester	GWR	1965
Glyncorrog	SWMR	1965
Glyn Neath	GWR	1964
Glyntaff	ADR	1922
Gobowen	GWR	1922
Goodwick	GWR	1963
Helston	GWR	1963
Hemyock	GWR	1929
Henley-on-Thames	GWR	1958
Hereford	GWR	1964
Honeybourne	GWR	1965
Hooton	GWR	1925*
Kerry	Cam	1931
Kidderminster	GWR	1964
Kidwelly	GVR	1923
Kingham	GWR	1962
Kingsbridge	GWR	1961
Kingswear	GWR	1924
Kington	GWR	1951
Laira	GWR	1964
Lambourn	GWR	1937
Landore	GWR	1961
Launceston	GWR	1962
Leamington	GWR	1965
Ledbury	GWR	1964
Leominster	GWR	1962
Llandovery	GWR	1935
Llanelly	GWR	1965
Llanelly Dock	GWR	1925
Llanfyllin	Cam	1952
Llanidloes	Cam	1962
Llantrisant	GWR	1964
Ludgershall	MSWJR	1925
Ludlow	GWR	1951
Lydney	GWR	1964
Machynlleth	Cam	1966
Malmesbury	GWR	1951
Malvern Wells	GWR	1922
Marlborough	GWR	1933
Marlow	GWR	1962
Merthyr	GWR	1964
Merthyr	TVR	1923
Milford Haven	GWR	1962
Minehead	GWR	1956
Moat Lane	Cam	1962
Moretonhampstead	GWR	1947
Moorswater	GWR	1960
Much Wenlock	GWR	1951
Neath	GWR	1965
Neath	N&BR	1964
Newbury	GWR	*
Newcastle Emlyn	GWR	1952
Newport Dock St.	GWR	1929
Newport Ebbw Jct	GWR	1965
Newport Pill	ADR	1963
Newton Abbot	GWR	1962
Newquay	GWR	1930
Neyland	GWR	1963
Oswestry	Cam	1965
Oxford	GWR	1965
Oxley	GWR	1967
Paddington	GWR	1965
Pantyffynon	-	1964
Pembroke Dock	GWR	1963
Penarth Dock	TVR	1929
Penmaenpool	Cam	1965
Penzance	GWR	1962
Plymouth Dock	GWR	1955*
Plymouth Millbay	GWR	1924
Pontrilas	GWR	1953
Pontlottyn	B&MR	1923
Pontypool	GWR	1965
Portmadoc	Cam	1963
Princetown	GWR	1956
Pwllheli	Cam	1966
Pwllrhebog	TVR	1951
Radstock	GWR	1923
Radyr	TVR	1965
Reading	GWR	1965
Rhymney	RR	1965
Roath Branch Jct	TVR	1923
Ross-on-Wye	GWR	1963
St. Blazey	GWR	1962
St. Ives	GWR	1961
Salisbury	GWR	1950
Sandy (Llanelly)	LMMR	1946
Senghenydd	RR	1931
Severn Tunnel Jct	GWR	1965
Shrewsbury	GWR	1967
Slough	GWR	1964
Southall	GWR	1965
Staines	GWR	1952
Stourbridge	GWR	1966
Stratford-on-Avon	GWR	1962
Swansea Burrows Lodge	P&M	1938*
Swansea East Dock	GWR	1964
Swansea Harbour	SHT	1930
Swansea Riverside	SHT	1924
Swindon	GWR	1964
Swindon Town	MSWJR	1924
Taffs Well	RR	1922
Tallyllyn Junction	B&MR	1922
Taunton	GWR	1964
Tenby	GWR	1932
Tetbury	GWR	1959
Tiverton Junction	GWR	1964
Tondu	GWR	1964
Trawsfynydd	GWR	1961
Trehafod	Barry	1925
Treherbert	TVR	1965
Trowbridge	GWR	1923
Truro	GWR	1962
Tyseley	GWR	1966
Wallingford	GWR	1956
Watlington	GWR	1957
Wellington	GWR	1964
Wells	GWR	1963
Welshpool	Cam	1954
Welshpool (n/g)	Cam	1956
Westbury	GWR	1964
Weston-super-Mare	GWR	1960
Weymouth	GWR	1967
Whitland	GWR	1963
Winchester	GWR	1953
Wolverhampton	GWR	1963
Woodstock	GWR	1927
Worcester	GWR	1965
Wrexham	Cam	1925
Yatton	GWR	1960
Yeovil	GWR	1959

Aberwystwyth - the narrow gauge engines are nowadays housed in the former standard gauge shed, the change having made in 1968.

* date of closure uncertain or not known.

Numbering of GWR Locomotives 1922

The renumbering of locomotives absorbed by the GWR in 1922 initially used blank number up to 1390, existing GWR engines not being renumbered. Nos. 1366-75 were deliberately left vacant and numbers 12, 14, 237, 239 and 1277 were not used, no reasons being apparent for their omission. Certain engines of the Alexandra Docks, Cambrian and Cardiff Railways reverted to their former GWR numbers, having been sold to those railways before the Grouping. The result was an untidy mixture and in distinct contrast to the system used for GWR standard classes which tidily sorted out the many classes.

The first stage of renumbering allotted new GWR numbers to locomotives absorbed from the following railways:-

Alexandra Docks (Newport & South Wales) Railway,	ADR
Barry Railway,	Barry
Cambrian Railways,	Cam
Cardiff Railway,	Car
Cleobury, Mortimer & Ditton Prios Light Railway,	CMDPR
Midland & South Western Junction Railway,	MSWJR
Rhymney Railway,	RR
Taff Vale Railway.	TVR

This stage also included the stock of three railways which had been worked as part of the GWR since 1908, their locomotives retaining their old identities until the Grouping:-

Port Talbot Railway,	PTR
Rhondda & Swansea Bay Railway,	RSB
South Wales Mineral Railway.	SWMR

The next stage of absorbtion used further blank numbers up to 2199 and reused a few numbers as well where engines from the above listed companies had already been condemned (but omitting 12, 14, 237, 239, 1366-70, 1480, 1489, 1490, 2057 and 2128), to include the locomotives of the following railways:-

Brecon & Merthyr Tydfil Junction Railway,	B&MR
Burry Port & Gwendraeth Valley Railway,	BPGVR
Neath & Brecon Railway.	N&BR

As above certain engines reverted to their former GWR numbers, having been sold before the Grouping.

Finally the locomotives of four smaller concerns were absorbed in 1923-4, reusing scattered numbers between 26 and 1098:-

Gwendraeth Valley Railway,	GVR
Llanelly & Mynydd Mawr Railway,	LMMR
Swansea Harbour Trust,	SHT
Powlesland & Mason.	P&M

The first stage of renumbering divided locomotives, first by wheel arrangement, in descending order of power, in the following order:-

0-4-4T; 2-6-0; 4-4-4T; 0-6-2T; 0-6-0T; 0-6-0; 4-4-0; 4-4-0T; 2-4-0T; 2-6-2T; 4-4-2T; 2-4-2T; 2-4-0; 0-4-0T; 0-6-4T; 0-8-2T; 0-8-0.

The order of companies was not the same in every wheel arrangement!

In the following renumbering lists the numbers of GWR engines are not repeated in column 3 (former numbers). Column 5 states the former railway class (where used) or the existing GWR class. Column 8 gives either the Swindon or Wolverhampton construction 'Lot Number' or the maker's order number.

GWR Nos.	Former Railway	Former Nos.	Type	Class	Date	Builder	Order Lot No.	See Note
1	GWR	-	2-4-0T	-	1880	GWRS 733	46	
2, 3	Barry	66/7	0-4-4T	G	1892	VF 1348/9	292	
4	Barry	68	0-4-4T	G	1895	SS 4053	E1045	
5, 6	GWR	-	2-4-0T	Metro	1869	GWRS 173/4	18	
7, 8	GWR	-	4-4-0	-	1894	GWRS	-	
9	Barry	69	0-4-4T	G	1895	SS 4054	E1045	
10/1	Cam	3, 5	0-4-4T	3	1895	NW 460/1	-	A
12	-	-	-	-	-	-	-	
13	GWR	-	4-4-0T	-	1886	GWRS 1094	72	
14	-	-	-	-	-	-	-	
15	Cam	7	0-4-4T	3	1895	NW 462	-	
16	GWR	-	0-6-4CT	-	1921	GWRS	219	
17/8	GWR	-	0-6-4CT	-	1901	GWRS 1774/5	121	
19-21	Cam	8, 9, 23	0-4-4T	3	1899	NW 558-60	-	A
22	GWR	-	0-6-0	388	1876	GWRS 624	41	
23	MSWJR	15	0-4-4T	-	1895	BP 3682	7969	
24	MSWJR	16	2-6-0	-	1897	BP 3884	8144	
25	MSWJR	17	4-4-4T	-	1897	SS 4300	E1097	
26	GWR	-	0-6-0	388	1867	GWRS 111	10th Goods	A
27	MSWJR	18	4-4-4T	-	1897	SS 4301	E1097	
28/9	CMDPR	-	0-6-0ST	-	1908	MW 1734/5	61800	28
30-2	RR	1-3	0-6-2T	R	1907	RS 3288-90	E40	
33	RR	16	0-6-2T	M	1904	RS 3130	E12	
34	RR	17	0-6-2T	R	1909	RS 3370	E54	
35/6	RR	39, 40	0-6-2T	R	1921	HC 1431/2	-	
37/8	RR	41/2	0-6-2T	R	1921	HC 1438/9	-	
39-44	RR	43-7, 62	0-6-2T	R	1921	BP 6099-6104	02050	
45	GWR	-	0-4-0ST	-	1880	GWRW 409	H2	
46	RR	97	0-6-2T	R	1909	RS 3371	E54	
47-51	RR	106-10	0-6-2T	M	1904	RS 3125-9	E12	
52-4	RR	10-2	0-6-2T	A1	1910	RS 3387-9	E61	
55	RR	13	0-6-2T	A	1910	RS 3390	E61	
56	RR	14	0-6-2T	A1	1910	RS 3391	E61	
57-61	RR	15, 18-21	0-6-2T	A	1911	HC 946-50	-	
62	RR	22	0-6-2T	A1	1911	HC 951	-	
63-5	RR	23-5	0-6-2T	A1	1914	HC 1063-5	-	
66/7	RR	26/7	0-6-2T	A1	1916	HC 1119/20	-	
68-70	RR	28-30	0-6-2T	A1	1918	RS 3660-2	E100	
71/2	RR	115/6	0-6-2T	A	1910	RS 3392/3	E61	
73	RR	117	0-6-2T	A1	1910	RS 3394	E61	
74/5	RR	118/9	0-6-2T	A	1910	RS 3395/6	E61	
76	RR	5	0-6-2T	P1	1909	RS 3373	E55	
77	RR	31	0-6-2T	P1	1917	HC 1121	-	
78/9	RR	35/6	0-6-2T	AP	1921	HC 1443/4	-	
80/1	RR	37/8	0-6-2T	AP	1921	HC 1456/7	-	
82/3	RR	4, 6	0-6-2T	P	1909	RS 3372/4	E55	
84-6	RR	7-9	0-6-2ST	K	1894	SS 4037-9	E1042	
87-91	RR	57-61	0-6-2ST	K	1890	VF 1284-8	700	
92	GWR	-	0-4-0ST	-	1857	BP 51	169	
93	GWR	-	0-6-0T	850	1875	GWRW	-	
94	GWR	-	0-6-0T	850	1877	GWRW	-	
95	GWR	-	0-4-0ST	-	1857	SS 1026	E334	
96	GWR	-	0-4-0ST	-	1856	SS 964	E315	
97-101	RR	67-71	0-6-2ST	K	1891	VF 1329-33	186	
102	GWR	-	4-4-2	-	1903	SAB 5409	146	102
103/4	GWR	-	4-4-2	-	1905	SAB 5601/2	157	103
105	RR	72	0-6-2ST	K	1891	VF 1334	186	
106-9	RR	73-6	0-6-2ST	K	1894	SS 4041-3/40	E1042	
110	RR	77	0-6-2ST	K	1897	SS 4257	E1098	

GWR Nos.	Former Railway	Former Nos.	Type	Class	Date	Builder	Order Lot No.	See Note
111	GWR	-	4-6-2	111	1908	GWRS 2279	171	
112-5	RR	78-81	0-6-2ST	K	1897	SS 4258-61	E1098	
116	GWR	-	0-6-0	388	1872	GWRS 346	26	
117-9	RR	82-4	0-6-2ST	K	1897	SS 4262-4	E1098	
120/1	GWR	-	0-6-0T	119	1861	GWRS 28/9	4th Goods	
122	RR	85	0-6-2ST	K	1897	SS 4265	E1098	
123/4	GWR	-	0-6-0T	119	1861	GWRS 31/2	4th Goods	
125/6	GWR	-	0-6-0T	119	1862	GWRS 33/4	5th Goods	
127	RR	86	0-6-2ST	K	1897	SS 4266	E1098	
128	GWR	-	0-6-0T	119	1862	GWRS 36	5th Goods	
129-31	RR	87-9	0-6-2ST	K	1899	HC 510-2	-	
132	GWR	-	0-6-0T	131	1862	GWRS 40	6th Goods	
133-5	RR	90-2	0-6-2ST	K	1899	HC 513-5	-	
136-8	RR	93-5	0-6-2ST	K	1899	HC 547-9	-	
139-45	RR	96, 98-103	0-6-2ST	K	1900	NR 5711/3-8	E840	
146	GWR	-	0-6-0T	131	1862	SG 484	-	
147/8	RR	104/5	0-6-2ST	K	1900	NR 5719/20	E840	
149/50	RR	63/4	0-6-2ST	L1	1890	VF 1290/1	701	A
151/2	Car	20/2	0-6-2T	-	1919	K 5180/1	475	
153-5	Car	33-5	0-6-2T	-	1908	K 4595-7	358	
156	Car	1	0-6-2T	-	1894	K 3580	149	
157/8	Car	9, 10	0-6-2T	-	1898	K 3869/70	203	
159	Car	28	0-6-2T	-	1887	K 3069	78	
160/1	Car	11, 21	0-6-2T	-	1905	K 4333/4	305	
162	Car	27	0-6-2T	-	1887	K 3068	78	
163	Car	26	0-6-2T	-	1885	K 2899	-	
164/5	RSB	25/6	0-6-2T	-	1902	K 4106/7	249	
166/7	RSB	27/8	0-6-2T	-	1904	K 4255/6	281	
168/9	RSB	8, 9	0-6-2T	-	1889	K 3211/2	110	
170	RSB	10	0-6-2T	-	1890	K 3213	110	
171-3	RSB	11-3	0-6-2T	-	1894	K 3566-8	147	
174-6	RSB	14-6	0-6-2T	-	1895	K 3569-71	147	
177-9	RSB	20-2	0-6-2T	-	1899	K 3883-5	208	
180	RSB	23	0-6-2T	-	1897	RS 2864	-	
181	RSB	4	0-6-2T	-	1886	K 2896	57	
182	RSB	24	0-6-2T	-	1898	RS 2867	-	
183	PTR	8	0-6-2T	-	1898	RS 2868	-	
184-8	PTR	9-13	0-6-2T	-	1898	RS 2895-9	-	
189	PTR	14	0-6-2T	-	1899	RS 2900	-	
190-2	ADR	29-31	0-6-2ST	-	1908	AB 1154-6	-	
193-7	Barry	117-21	0-6-2T	K	1899	C 2484/2/3/5/6	-	
198-200	Barry	6-8	0-6-2T	B	1888	SS 3454-6	E910	
201	Barry	9	0-6-2T	B	1889	SS 3457	E910	
202	GWR	-	0-4-2T	517	1876	GWRW 301	Z	
203/4	Barry	10/1	0-6-2T	B	1889	SS 3458/9	E910	
205	GWR	-	0-4-2T	517	1876	GWRW 304	Z	
206-14	Barry	12-20	0-6-2T	B	1889	SS 3460-8	E910	
215-22	GWR	-	0-4-2T	517	1876	GWRW 305-12	Z	
223	Barry	23	0-6-2T	B	1889	SS 3571	E948	
224-32	Barry	24-32	0-6-2T	B	1890	SS 3572-80	E948	
233-5	Barry	38-40	0-6-2T	B1	1890	SS 3598-3600	E962	
236	TVR	105	0-6-2T	O4	1908	BP 5132	9779	236
237	-	-	-	-	-	-	-	
238	Barry	41	0-6-2T	B1	1890	SS 3601	E962	
239	-	-	-	-	-	-	-	

GWR Nos.	Former Railway	Former Nos.	Type	Class	Date	Builder	Order Lot No.	See Note
240-4	Barry	42-6	0-6-2T	B1	1890	SS 3602-6	E962	
245-54	Barry	54-63	0-6-2T	B1	1892	VF 1336-45	290	
255-60	Barry	73-8	0-6-2T	B1	1894	SS 4044-9	E1043	
261-72	Barry	105-16	0-6-2T	B1	1900	SS 4607-18	E1138	
273-6	Barry	122-5	0-6-2T	B1	1900	FB 1272-5	-	
277	Barry	126	0-6-2T	B1	1901	FB 1276	-	
278/9	TVR	1, 2	0-6-2T	04	1910	VF 2529/30	16	
280	TVR	6	0-6-2T	04	1908	BP 5127	9779	
281	TVR	8	0-6-2T	04	1910	VF 2531	16	
282	TVR	9	0-6-2T	04	1908	BP 5128	9779	
283/4	TVR	17, 35	0-6-2T	04	1910	VF 2532/3	16	
285	TVR	38	0-6-2T	04	1907	MW 1698	59200	
286	TVR	39	0-6-2T	04	1908	BP 5095	9756	
287	TVR	43	0-6-2T	04	1907	MW 1699	59200	
288	TVR	46	0-6-2T	04	1910	VF 2535	16	
289	TVR	48	0-6-2T	04	1908	BP 5129	9779	
290	TVR	49	0-6-2T	04	1910	BP 5385	0107	
291/2	TVR	56/8	0-6-2T	04	1908	BP 5130/1	9779	
293-6	TVR	59, 67/8, 94	0-6-2T	04	1907	MW 1700-2/4	59200	
297-302	TVR	95/7/8, 101/2/4	0-6-2T	04	1908	BP 5096-5101	9756	
303-7	GWR	-	0-6-0T	302	1864	GWRW 15-9	-	
308/9	GWR	-	0-6-0T	302	1865	GWRW 20/1	-	
310/1	TVR	108/9	0-6-2T	04	1910	VF 2536/7	16	
312	GWR	-	0-6-0	131	1865	GWRS 47	7th Goods	A
313	TVR	110	0-6-2T	04	1910	VF 2538	16	
314	TVR	111	0-6-2T	04	1908	BP 5133	9779	
315	TVR	112	0-6-2T	04	1910	BP 5386	0107	
316	GWR	-	0-6-0	57	1890	GWRW 524	W2	
317-21	TVR	113-6/8	0-6-2T	04	1910	BP 5387-91	0107	
322	GWR	-	0-6-0T	322	1864	BP 463	822	
323	GWR	-	0-6-0T	322	1866	BP 590	937	
324	TVR	119	0-6-2T	04	1910	BP 5392	0107	
325-7	GWR	-	0-6-0T	322	1864	BP 478/67/77	822	
328-32	GWR	-	0-6-0	322	1864	BP 469-73	822	A
333	TVR	121	0-6-2T	04	1910	BP 5393	0107	
334	GWR	-	0-6-0	322	1864	BP 475	822	
335	TVR	7	0-6-2T	A	1914	HL 3058	-	
336	GWR	-	0-6-0	322	1864	BP 468	822	
337	TVR	10	0-6-2T	A	1914	HL 3059	-	
338-41	GWR	-	0-6-0	322	1864	BP 479-82	822	A
342	GWR	-	0-4-0ST	-	1856	BP 27	99	
343/4	TVR	11/2	0-6-2T	A	1914	HL 3060/1	-	
345	TVR	20	0-6-2T	A	1920	HL 3394	-	
346	TVR	45	0-6-2T	A	1915	NBL 21157	L647	
347-9	TVR	75, 80, 90	0-6-2T	A	1921	VF 3492-4	1465	
350	GWR	-	0-6-0	322	1866	BP 581	937	
351	TVR	91	0-6-2T	A	1921	VF 3495	1465	
352	TVR	122	0-6-2T	A	1915	NBL 21159	L647	
353-5	GWR	-	0-6-0	322	1866	BP 584-6	937	
356/7	TVR	123/4	0-6-2T	A	1915	NBL 21160/1	L647	
358	GWR	-	0-6-0	322	1866	BP 589	937	
359	GWR	-	0-6-0	322	1864	BP 464	822	A
360-2	TVR	125/7/8	0-6-2T	A	1916	VF 3178-80	437	
363	GWR	-	0-6-0	360	1866	GWRS 60	8th Goods	
364-6	TVR	129/30/2	0-6-2T	A	1916	VF 3181-3	437	
367	TVR	133	0-6-2T	A	1919	NW 1269	-	

GWR Nos.	Former Railway	Former Nos.	Type	Class	Date	Builder	Order Lot No.	See Note
368	TVR	134	0-6-2T	A	1920	HL 3395	-	
369	GWR	-	0-6-0	360	1866	GWRS 66	8th Goods	
370-4	TVR	135/6/8-40	0-6-2T	A	1919	NW 1270-4	-	
375/6	TVR	144/9	0-6-2T	A	1920	HL 3396/7	-	
377-82	TVR	154/6-60	0-6-2T	A	1919	NW 1275-80	-	
383-5	TVR	162/4/5	0-6-2T	A	1920	HL 3398-3400	-	
386-91	TVR	400-5	0-6-2T	A	1920	HL 3401-6	-	
392	GWR	-	0-6-0	388	1866	GWRS 83	9th Goods	
393/4	TVR	406/7	0-6-2T	A	1920	HL 3407/8	-	
395/6	GWR	-	0-6-0	388	1866	GWRS 86/7	9th Goods	
397	TVR	408	0-6-2T	A	1920	HL 3409	-	
398/9	TVR	409/10	0-6-2T	A	1921	HL 3410/1	-	
400	GWR		0-6-0	388	1867	GWRS 91	9th Goods	
401-3	TVR	411-3	0-6-2T	A	1921	HL 3412-4	-	
404	TVR	414	0-6-2T	A	1921	VF 3496	-	
405	GWR	-	0-6-0	388	1867	GWRS 96	9th Goods	
406	TVR	415	0-6-2T	A	1921	VF 3497	1465	
407	GWR	-	0-6-0	388	1867	GWRS 98	10th Goods	
408	TVR	416	0-6-2T	A	1921	VF 3498	1465	
409	TVR	13	0-6-2T	O4	1910	BP 5384	0107	
410/1	TVR	18/9	0-6-2T	O3	1904	K 4235/6	274	
412/3	TVR	31/2	0-6-2T	O2	1899	NR 5414/11	E811	
414	TVR	36	0-6-2T	O4	1910	VF 2534	16	
415	TVR	44	0-6-2T	O2	1899	NR 5416	E811	
416	TVR	47	0-6-2T	O3	1902	HC 582	-	
417	TVR	55	0-6-2T	O3	1905	VF 1986	155	
418	TVR	57	0-6-2T	O3	1902	HC 583	-	
419	TVR	66	0-6-2T	O2	1899	NR 5415	E811	
420	TVR	69	0-6-2T	O4	1907	MW 1703	59200	
421	TVR	81	0-6-2T	O2	1899	NR 5413	E811	
422	GWR	-	0-6-0	388	1867	GWRS 107	10th Goods	
423-6	TVR	82-5	0-6-2T	O2	1899	NR 5408-10/2	E811	423
427-9	TVR	92/3/6	0-6-2T	O3	1902	HC 584/6/7	-	
430	TVR	103	0-6-2T	O3	1905	VF 1987	155	
431	TVR	117	0-6-2T	O3	1902	HC 585	-	
432	TVR	126	0-6-2T	O3	1904	K 4237	274	
433-5	TVR	131/7/55	0-6-2T	O3	1905	VF 1988-90	155	
436	GWR	-	0-6-0	388	1868	GWRS 126	11th Goods	
437	TVR	161	0-6-2T	O3	1905	VF 1991	155	
438	TVR	3	0-6-2T	A	1914	HL 3057	-	
439/40	TVR	42, 52	0-6-2T	A	1915	NBL 21156/8	L647	
441	TVR	120	0-6-2T	A	1915	HL 3062	-	
442	TVR	4	0-6-2T	M1	1891	TVR	-	
443	TVR	5	0-6-2T	M1	1892	TVR	-	
444/5	TVR	14/5	0-6-2T	M1	1888	K 2987/8	64	
446-8	TVR	21/5/6	0-6-2T	M1	1894	TVR	-	
449	TVR	27	0-6-2T	O1	1894	K 3572	148	
450	TVR	28	0-6-2T	O1	1897	TVR	-	
451	TVR	29	0-6-2T	O1	1894	K 3573	148	
452	TVR	33	0-6-2T	O	1894	TVR	-	
453	TVR	34	0-6-2T	O	1895	TVR	-	
454/5	TVR	37, 41	0-6-2T	O	1894	K 3574/5	148	

GWR Nos.	Former Railway	Former Nos.	Type	Class	Date		Order	See Note
456-61	GWR	-	2-4-0T	Metro	1869	GWRS 156-61	18	
462	TVR	51	0-6-2T	M1	1888	K 2985	64	
463-5	GWR	-	2-4-0T	Metro	1869	GWRS 163-5	18	
466	TVR	54	0-6-2T	M1	1892	TVR	-	
467-70	GWR	-	2-4-0T	Metro	1869	GWRS 167-70	18	
471-5	TVR	60-4	0-6-2T	O1	1897	TVR	-	
476/7	TVR	65, 70	0-6-2T	O1	1894	K 3576/7	148	
478	TVR	71	0-6-2T	M1	1887	K 2978	64	
479/80	TVR	73/8	0-6-2T	O1	1894	K 3578/9	148	
481-4	TVR	86-9	0-6-2T	M1	1891	K 3205-8	107	
485/6	TVR	106/7	0-6-2T	N	1891	K 3391/2	126	
487	TVR	150	0-6-2T	M1	1885	K 2707	48	
488-93	TVR	176-81	0-6-2T	M1	1891	K 3373-8	107	
494-6	TVR	182-4	0-6-2T	M1	1891	K 3383-5	126	
497	GWR	-	0-6-0	388	1870	GWRS 228	21	
498-502	TVR	185-9	0-6-2T	N	1891	K 3386-90	126	
503	TVR	16	0-6-2T	M1	1888	K 2989	64	
504	GWR	-	0-6-0	388	1870	GWRS 232	21	A
505/6	TVR	22/4	0-6-2T	M1	1886	K 2962/3	64	
507/8	TVR	50/3	0-6-2T	M1	1888	K 2984/6	64	
509/10	GWR	-	0-6-0	388	1870	GWRS 237/8	21	
511	TVR	74	0-6-2T	M1	1887	K 2979	64	
512	GWR	-	0-6-0	388	1870	GWRS 240	21	
513	TVR	145	0-6-2T	M1	1885	K2702	48	
514	GWR	-	0-6-0	388	1870	GWRS 242	21	
515/6	TVR	146/7	0-6-2T	M1	1885	K 2703/4	48	
517-9	GWR	-	0-4-2T	517	1868	GWRW 71-3	D	
520	TVR	148	0-6-2T	M1	1885	K 2705	48	
521-8	GWR	-	0-4-2T	517	1868	GWRW 75-82	D	
529-37	GWR	-	0-4-2T	517	1868	GWRW 83-91	E	
538-40	GWR	-	0-4-2T	517	1869	GWRW 92-4	E	
541-51	GWR	-	0-4-2T	517	1869	GWRW 95-105	F	
552	TVR	151	0-6-2T	M1	1895	K 2708	48	
553-9	GWR	-	0-4-2T	517	1869	GWRW 107-13	G	
560	TVR	152	0-6-2T	M1	1885	K 2709	48	
561-4	GWR	-	0-4-2T	517	1869	GWRW 115-8	G	
565/6	GWR	-	0-4-2T	517	1869	GWRW 119/20	H	
567	TVR	153	0-6-2T	M1	1885	K 2710	48	
568	GWR	-	0-4-2T	517	1869	GWRW 122	H	
569-72	GWR	-	0-4-2T	517	1870	GWRW 123-6	H	
573	TVR	163	0-6-2T	M1	1886	K 2959	64	
574-6	GWR	-	0-4-2T	517	1870	GWRW 128-30	H	
577	TVR	166	0-6-2T	M1	1886	K 2974	64	
578-80	TVR	167-9	0-6-2T	M1	1887	K 2976/77/75	64	
581	TVR	190	0-6-2T	O	1895	TVR	-	
582	TVR	344/9	0-6-2T	M1	1885	K 2701/6	48	
584-6	TVR	362/4/5	0-6-2T	M1	1886	K 2961/60/64	64	
587	TVR	23	0-6-2T	U	1895	VF 1441	402	
588	TVR	40	0-6-2T	U1	1896	VF 1459	820	
589-91	TVR	72/6/7	0-6-2T	U	1895	VF 1442-4	402	
592	TVR	79	0-6-2T	U1	1896	VF 1460	820	
593	TVR	191	0-6-2T	U	1895	VF 1437	402	
594	GWR	-	0-6-0	388	1871	GWRS 272	23	
595-7	TVR	192-4	0-6-2T	U	1895	VF 1438-40	402	
598-600	TVR	195/6/8	0-6-2T	U1	1896	VF 1461/2/4	820	
601	GWR	-	0-6-0	388	1871	GWRS 279	23	
602/3	TVR	30, 197	0-6-2T	U1	1896	VF 1458/63	820	
604-6	RR	32-4	0-6-0T	S1	1920	HC 1406/9/10	-	
607	GWR	-	0-6-0	388	1871	GWRS 295	24	
608-11	RR	111-4	0-6-0T	S	1908	HC 846-9	-	

GWR Nos.	Former Railway	Former Nos.	Type	Class	Date	Builder	Order Lot No.	See Note
612	RR	48	0-6-0ST	J	1884	SS 3177	E849	
613	GWR	-	2-4-0T	Metro	1871	GWRS 301	25	
614	RR	49	0-6-0ST	J	1884	SS 3178	E849	
615-7	GWR	-	2-4-0T	Metro	1871	GWRS 303-5	25	
618/9	RR	50/1	0-6-0ST	J	1884	SS 3179/80	E849	
620/1	GWR	-	2-4-0T	Metro	1871	GWRS 308/9	25	
622	RR	52	0-6-0ST	J	1884	SS 3181	E849	
623/4	GWR	-	2-4-0T	Metro	1871	GWRS 311/2	25	
625	RR	53	0-6-0ST	J	1884	SS 3182	E849	
626-8	GWR	-	2-4-0T	Metro	1871	GWRS 314-6	25	
629	RR	54	0-6-0ST	J	1884	SS 3183	E849	
630	GWR	-	2-4-0T	Metro	1871	GWRS 318	25	
631	RR	55	0-6-0ST	J	1884	SS 3184	E849	
632	GWR	-	2-4-0T	Metro	1871	GWRS 320	25	
633-6	GWR	-	0-6-0T	633	1871	GWRW 167-70	M	
637	RSB	29	0-6-0T	633	1871	GWRW 171	M	GWR
638-44	GWR	-	0-6-0T	633	1872	GWRW 172-8	M	
645-54	GWR	-	0-6-0T	645	1872	GWRW 179-88	O	
655	GWR	-	0-6-0T	655	1892	GWRW 563	A3	
656	GWR	-	0-6-0T	645	1872	GWRW 191	P	
657	RR	56	0-6-0ST	J	1884	SS 3185	E849	
658	GWR	-	0-6-0	388	1871	GWRS 322	26	
659/60	RR	033/6	0-6-0ST	I	1874	NW 168/71	-	
661/2	RR	120/1	0-6-0T	B	1907	HC 806/05	-	
663	ADR	26	0-6-2T	-	1868	WE 37	-	663
664/5	ADR	12/3	0-6-0ST	-	1884	RWH 1977/76	-	
666/7	ADR	34/5	0-6-0T	-	1917	KS 3069/73	8459	
668	ADR	15	0-6-0ST	-	1885	RS 2555	-	
669/70	ADR	20/1	0-6-0ST	-	1894	RS 2814/5	-	
671/2	ADR	16/7	0-6-0ST	-	1889	HL 2146/7	-	
673	GWR	-	0-6-0	388	1872	GWRS 337	26	
674/5	ADR	1, 2	0-6-0ST	-	1898	RS 2893/4	-	
676/7	ADR	3, 4	0-6-0ST	-	1900	RS 2978/9	-	
678	ADR	5	0-6-0ST	-	1900	RS 3003	-	
679	ADR	18	0-6-0ST	-	1890	P 465	-	
680	ADR	19	0-6-0ST	-	1886	P 450	-	
681-4	Car	14/6/7, 32	0-6-0ST	-	1920	HC 1404/5/7/8	-	
685/6	Car	7	0-6-0T	-	1919	K 5182	476	
686	Car	3	0-6-0T	-	1895	K 3602	156	
687/8	Car	4, 8	0-6-0T	-	1899	K 3871/2	204	
689	GWR	-	0-6-0	388	1872	GWRS 364	28	
690	Car	13	0-6-0T	-	1895	K 3603	156	
691/2	Car	30, 29	0-6-0T	-	1889	K 3133/32	95	
693	Car	2	0-6-0PT	-	1882	K 2458	27	
694	Car	12	0-6-0ST	-	1872	P&J	-	
695	Car	15	0-6-0ST	-	1873	P&J	-	
696/7	Car	18/9	0-6-0ST	-	1875	P&J	-	A
698	Car	24	0-6-0T	-			-	698
699, 700	Barry	1, 2	0-6-0T	A	1888	SS 3449/50	E928	
701	GWR	-	0-6-0	388	1872	GWRS 376	28	A
702/3	Barry	3, 4	0-6-0T	A	1888	SS 3451/2	E928	
704/5	GWR	-	0-6-0	388	1872	GWRS 379/80	28	A
706	Barry	5	0-6-0T	A	1888	SS 3453	E928	
707	GWR	-	0-6-0	388	1872	GWRS 382	28	
708	Barry	37	0-6-0ST	F	1900	SS 4593	E1137	
709	GWR	-	0-6-0	388	1872	GWRS 384	28	
710-2	Barry	47-9	0-6-0ST	F	1890	SS 3607-9	E963	
713	Barry	52	0-6-0ST	F	1900	SS 4594	E1137	
714/5	Barry	64/5	0-6-0ST	F	1892	VF 1346/7	291	
716/7	Barry	70/1	0-6-0ST	F	1894	SS 4050/1	E1044	

GWR Nos.	Former Railway	Former Nos.	Type	Class	Date	Builder	Order Lot No.	See Note
718	Barry	72	0-6-0ST	F	1895	SS 4052	E1044	
719-24	Barry	99-104	0-6-0ST	F	1900	SS 4595-4600	E1137	
725/6	Barry	127/8	0-6-0ST	F	1905	NBL 16628/9	L91	
727	GWR	-	0-6-0T	1076	1872	GWRS 402	30	
728	RSB	32	0-6-0T	1076	1872	GWRS 403	30	GWR
729	Barry	130	0-6-0ST	F	1905	NBL 16631	L91	
730-4	GWR	-	0-6-0T	1076	1872	GWRS 405-9	30	
735-41	GWR	-	0-6-0T	1076	1873	GWRS 410-6	30	
742	Barry	131	0-6-0ST	F	1905	NBL 16632	L91	
743-6	GWR	-	0-6-0T	1076	1873	GWRS 418-21	30	
747	Barry	132	0-6-0ST	F	1905	NBL 16633	L91	
748-53	GWR	-	0-6-0T	1076	1873	GWRS 423-8	30	
754	Barry	133	0-6-0ST	F	1905	HC 712	-	
755/6	GWR	-	0-6-0T	1076	1873	GWRS 430/1	30	
757-61	GWR	-	0-6-0T	645	1872	GWRW 192/7 -9, 200	P	
762/3	GWR	-	0-6-0T	645	1873	GWRW 201/2	P	
764-6	GWR	-	0-6-0T	645	1873	GWRW 205-7	Q	
767	GWR	-	0-6-0T	655	1892	GWRW 564	A3	
768-75	GWR	-	0-6-0T	645	1873	GWRW 209-16	Q	
776-80	Barry	134-8	0-6-0ST	F	1905	HC 713-7	-	A
781/2	Barry	33/4	0-6-0T	E	1889	HC 331/2	-	
783/4	Barry	50/1	0-6-0T	E	1890	HC 343/4	-	
785	Barry	53	0-6-0T	E	1891	HC 352	-	
786/7	TVR	99, 100	0-6-0ST	V	1899	K 3820/1	197	
788-91	TVR	275/80/90/ 1	0-6-0ST	V	1899	K 3816-9	197	
792-4	TVR	141-3	0-6-0T	H	1884	K 2697-9	46	
795	TVR	264	0-6-0ST	E	1873	SS 2282	E622	A
796	TVR	265	0-6-0ST	E	1873	SS 2312	E634	
797	TVR	250	0-6-0ST	D	1867	SS 1788	E525	
798	TVR	270	0-6-0ST	D	1865	SS 1578	E475	
799	RSB	1	0-6-0T	-	1884	BP 2435	6400	
800	GWR	-	0-6-0	388	1873	GWRS 456	31	
801	RSB	3	0-6-0T	-	1884	BP 2677	6673	
802	RSB	5	0-6-0T	-	1889	BP 3102	7201	
803/4	GWR	-	0-6-0	388	1873	GWRS 459/60	31	
805/6	RSB	6, 7	0-6-0T	-	1889	BP3103/4	7201	
807	Barry	129	0-6-0ST	F	1905	NBL 16630	L91	807
808/9	PTR	22/3	0-6-0ST	-	1900	HC 551/2	-	
810	GWR	-	2-4-0	806	1873	GWRS 466	32	
811-4	PTR	24-7	0-6-0ST	-	1900	HC 553-6	-	
815/6	PTR	3, 15	0-6-0ST	-	1898	RS 2863, 2901	-	
817/8	SWMR	6, 7	0-6-0ST	-	1873	AE 991/88	SDN	
819	Cam	24	0-6-0T	-	1903	HE 811	25700	
820/1	Cam	26, 35	0-6-0T	-	1898	CF 1161/2	-	
822/3	Cam	1, 2	0-6-0T	-	1902	BP 3496/7	8868	NG
824	Cam	30	0-6-0ST	-	1865	MW 140	1680	
825	MSWJR	13	0-6-0T	-	1894	D 3168	3168E	
826-35	GWR	-	0-4-2T	517	1873	GWRW 217-26	R	
836/7	GWR	-	0-4-2T	517	1874	GWRW 227/8	R	
838-42	GWR	-	0-4-2T	517	1874	GWRW 229-33	S	
843	MSWJR	14	0-6-0T	-	1894	D 3164	3164E	
844	Cam	15	0-6-0	15	1918	BP 5944	0981	
845	GWR	-	0-4-2T	517	1874	GWRW 236	S	
846-8	GWR	-	0-4-2T	517	1875	GWRW 237-9	S	
849	Cam	29	0-6-0	15	1918	BP 5945	0981	
850-4	GWR	-	0-6-0T	850	1874	GWRW 241-5	T	
855	Cam	31	0-6-0	15	1919	BP 5946	0981	

GWR Nos.	Former Railway	Former Nos.	Type	Class	Date	Builder	Order Lot No.	See Note
856-61	GWR	-	0-6-0T	850	1874	GWRW 247-52	T	
862/3	GWR	-	0-6-0T	850	1874	GWRW 253/4	V	
864	Cam	38	0-6-0	15	1908	BP 5031	9683	
865/6	GWR	-	0-6-0T	850	1874	GWRW 256/7	V	
867-72	GWR	-	0-6-0T	850	1875	GWRW 258-63	V	
873/4	Cam	42, 54	0-6-0	15	1919	BP 5947/8	0981	
875/6	Cam	73/4	0-6-0	73	1894	N 4691/2	E724	
877	GWR	-	0-6-0	388	1874	GWRS 486	35	
878-80	Cam	75-7	0-6-0	73	1894	N 4693-5	E724	
881-3	Cam	78-80	0-6-0	73	1895	VF 1445-7	490	
884/5	Cam	87/8	0-6-0	73	1899	NR 5401/2	E809	
886	GWR	-	0-6-0	388	1874	GWRS 495	35	A
887-9	Cam	89-91	0-6-0	15	1903	RS 3089-91	E3	A
890	GWR	-	0-6-0	388	1874	GWRS 499	35	
891/2	Cam	92/3	0-6-0	15	1903	RS 3092/3	E3	
893-6	Cam	99-102	0-6-0	15	1908	BP 5029/30/ 2/3	9683	
897	Cam	4	0-6-0	1	1872	SS 2232	E611	
898	Cam	14	0-6-0	1	1875	SS 2511	E697	
899	Cam	40	0-6-0	1	1863	SS 1446	E442	
900/1	Cam	45/6	0-6-0	1	1864	SS 1530/1	E442	
902/3	GWR	-	0-6-0T	645	1872	GWRW 196/94	P	
904	GWR	-	0-6-0T	645	1872	GWRW 189	O	
905-7	GWR	-	0-6-0T	1854	1895	GWRS 1433-5	98	
908/9	Cam	48/9	0-6-0	1	1873	SS 2339/47	E649	
910/1	Cam	51/2	0-6-0	1	1865	SS 1590/7	E480	
912	TVR	219	0-6-0	L	1884	TVR	-	
913	TVR	253	0-6-0	L	1882	TVR	-	
914	TVR	259	0-6-0	L	1884	K 2658	42	
915	TVR	261	0-6-0	K	1874	K 1976	-	
916	TVR	281	0-6-0	L	1885	K 2842	53	
917	TVR	284	0-6-0	K	1874	K 1974	-	
918	TVR	288	0-6-0	L	1875	K 1978	-	
919	TVR	298	0-6-0	L	1875	K 2027	-	
920	TVR	337	0-6-0	K	1884	K 2664	42	
921	TVR	210	0-6-0	K	1881	K 2444	24	A
922	TVR	217	0-6-0	K	1880	K 2354	16	
923	TVR	220	0-6-0	L	1884	TVR	-	
924/5	TVR	235/6	0-6-0	K	1880	K 2352/3	16	A
926	TVR	239	0-6-0	K	1875	TVR	-	A
927-9	TVR	242/5/52	0-6-0	K	1883	K 2581-3	36	A
930	TVR	283	0-6-0	K	1874	TVR	-	A
931-3	TVR	297, 301/2	0-6-0	K	1875	K 2026/30/1	-	A
934	GWR	-	0-6-0	927	1874	GWRS 510	36	
935	TVR	304	0-6-0	K	1875	K 2033	-	A
936	TVR	313	0-6-0	K	1880	K 2346	16	
937	GWR	-	0-6-0	927	1874	GWRS 513	36	A
938/9	TVR	314/6	0-6-0	K	1880	K 2347/51	16	
940	GWR	-	0-6-0	927	1874	GWRS 516	36	
941	TVR	320	0-6-0	K	1881	K 2443	24	
942	TVR	322	0-6-0	K	1883	K 2584	36	A
943/4	TVR	325/7	0-6-0	K	1884	K 2587/9	36	A
945	GWR	-	0-6-0	927	1874	GWRS 521	36	
946	TVR	328	0-6-0	K	1884	K 2590	36	
947	GWR	-	0-6-0T	1076	1874	GWRS 523	37	
948	TVR	335	0-6-0	K	1884	K 2662	42	
949-66	GWR	-	0-6-0T	1076	1874	GWRS 525-42	37	
967	GWR	-	2-4-0T	Metro	1874	GWRS 543	38	
968-70	TVR	336/9/40	0-6-0	K	1884	K 2663/6/7	42	A
971-3	GWR	-	2-4-0T	Metro	1874	GWRS 547-9	38	

GWR Nos.	Former Railway	Former Nos.	Type	Class	Date	Builder	Order Lot No.	See Note
974	TVR	354	0-6-0	K	1885	K 2841	53	A
975-7	GWR	-	2-4-0T	Metro	1874	GWRS 551-3	38	
978	TVR	356	0-6-0	K	1885	K 2843	53	
979-83	GWR	-	2-4-0T	Metro	1874	GWRS 555-9	38	
984	TVR	357	0-6-0	K	1885	K 2844	53	
985/6	GWR	-	2-4-0T	Metro	1874	GWRS 561/2	38	
987-92	GWR	-	0-6-0T	850	1875	GWRW 265-70	X	
993	ADR	33	0-6-0T	850	1875	GWRW 271	X	GWR
994-6	GWR	-	0-6-0T	850	1875	GWRW 272-4	X	
997/8	GWR	-	0-6-0T	850	1876	GWRW 275/6	X	
999	TVR	287	4-4-0T	I	1885	TVR	-	999
1000-2	TVR	358-60	0-6-0	K	1885	K 2845-7	53	
1003-8	MSWJR	19-24	0-6-0	-	1899	BP 4097-4102	8467	
1009-11	MSWJR	25-7	0-6-0	-	1902	BP 4440-2	8902	
1012	GWR	-	0-6-0	388	1867	GWRS 114	1st Renewals	
1013	MSWJR	28	0-6-0	-	1902	BP 4443	8902	
1014	Cam	94	4-4-0	94	1904	RS 3131	E13	
1015	GWR	-	0-6-0	360	1866	GWRS 67	8th Goods	
1016-27	GWR	-	0-6-0T	1016	1867	GWRW 47-58	B	
1028	GWR	-	0-6-0T	1016	1867	GWRW 59	C	
1029	Cam	96	4-4-0	94	1904	RS 3133	E13	
1030-4	GWR	-	0-6-0T	1016	1867	GWRW 61-5	C	
1035	Cam	97	4-4-0	94	1904	RS 3134	E13	
1036-9	GWR	-	0-6-0T	1016	1868	GWRW 67-70	C	
1040-2	GWR	-	0-6-0T	1016	1870	GWRW 131-3	J	
1043	Cam	98	4-4-0	94	1904	RS 3135	E13	
1044-51	GWR	-	0-6-0T	1016	1870	GWRW 135-42	J	
1052-5	GWR	-	0-6-0T	1016	1870	GWRW 143-6	K	
1056-63	GWR	-	0-6-0T	1016	1871	GWRW 147-54	K	
1064-7	GWR	-	0-6-0T	1016	1871	GWRW 155-8	L	
1068	Cam	11	4-4-0	61	1904	Cam	-	
1069-75	GWR	-	0-6-0T	1016	1871	GWRW 160-6	L	
1076-81	GWR	-	0-6-0T	1076	1870	GWRS 245-50	22	
1082	Cam	19	4-4-0	61	1901	Cam	-	
1083	GWR	-	0-6-0	388	1870	GWRS 225	21	
1084	Cam	81	4-4-0	61	1895	SS 4070	E1051	A
1085/6	Cam	32, 47	4-4-0	61	1897	RS 2871/2	-	A
1087	GWR	-	0-6-0	388	1870	GWRS 252	23	
1088	Cam	61	4-4-0	61	1893	SS 3901	E1018	
1089	GWR	-	0-6-0	388	1870	GWRS 254	23	
1090/1	Cam	62/3	4-4-0	61	1893	SS 3902/3	E1018	
1092	GWR	-	0-6-0	388	1870	GWRS 257	23	
1093	Cam	64	4-4-0	61	1893	SS 3904	E1018	
1094/5	GWR	-	0-6-0	388	1870	GWRS 259/60	23	
1096/7	Cam	65/6	4-4-0	61	1893	SS 3905/6	E1018	
1098/9	GWR	-	0-6-0	388	1870	GWRS 263/4	23	A
1100/1	Cam	67/8	4-4-0	61	1893	SS 3907/8	E1018	
1102-5	Cam	69-72	4-4-0	61	1894	SS 3976-9	E1029	
1106/7	Cam	83/4	4-4-0	61	1895	SS 4072/3	E1051	
1108/9	Cam	85/6	4-4-0	61	1898	RS 2873/4	-	
1110	Cam	50	4-4-0	16	1891	SS 3696	E989	
1111	GWR	-	0-6-0	388	1871	GWRS 286	24	
1112	Cam	60	4-4-0	16	1891	SS 3697	E989	
1113/4	Cam	34/6	4-4-0	-	1864	BP 424/6	773	A
1115/6	Cam	16/7	4-4-0	16	1878	SS 2789/90	E750	
1117/8	Cam	20/1	4-4-0	16	1886	SS 3356/7	E879	A
1119	MSWJR	1	4-4-0	-	1905	NBL 16817	L114	
1120/1	MSWJR	2, 3	4-4-0	-	1909	NBL 18791/2	L333	

GWR Nos.	Former Railway	Former Nos.	Type	Class	Date	Builder	Order Lot No.	See Note
1122	MSWJR	4	4-4-0	-	1914	NBL 20539	L576	
1123	MSWJR	5	4-4-0	-	1912	NBL 19755	L479	
1124	MSWJR	6	4-4-0	-	1910	NBL 19133	L388	
1125	MSWJR	7	4-4-0	-	1911	NBL 19314	L412	
1126	MSWJR	8	4-4-0	-	1912	NBL 19756	L479	
1127	MSWJR	9	4-4-0	-	1893	D 3076	3076E	
1128	MSWJR	31	4-4-0	-	1914	NBL 20540	L576	
1129-31	Cam	2, 12, 33	4-4-0T	2	1864	BP 421-3	773	
1132	Cam	37	4-4-0T	2	1885	BP 2675	6648	
1133	TVR	285	4-4-0T	I	1884	TVR	-	
1134-43	GWR	-	0-6-0T	1076	1874	GWRS 563-72	39	
1144-53	GWR	-	0-6-0T	1076	1875	GWRS 573-82	39	
1154-8	GWR	-	0-4-2T	517	1875	GWRW 277-81	W	
1159-65	GWR	-	0-4-2T	517	1876	GWRW 282-8	W	
1166	GWR	-	0-6-0T	1076	1875	GWRS 603	40	
1167	RSB	34	0-6-0T	1076	1875	GWRS 604	40	GWR
1168-83	GWR	-	0-6-0T	1076	1875	GWRS 605-20	40	
1184	TVR	286	4-4-0T	I	1885	TVR	-	
1185	GWR	-	0-6-0T	1076	1875	GWRS 622	40	
1186-8	GWR	-	0-6-0	388	1876	GWRS 633-5	41	
1189	PTR	37	2-4-0T	-	1898	SS 3610	E960	
1190	Cam	44	2-4-0T	-	1864	SS 1488	E452	
1191	Cam	56	2-4-0T	-	1865	SS 1656	E508	
1192	Cam	57	2-4-0T	-	1866	SS 1681	E510	
1193-5	GWR	-	0-6-0	388	1876	GWRS 640-2	41	
1196/7	Cam	58/9	2-4-0T	-	1866	SS 1682/3	E510	
1198	Cam	3	2-4-0T	-	1896	WGB 1497	-	NG
1199	ADR	25	2-6-2T	-	1892	K 3393	127	
1200	GWR	-	0-6-0	388	1876	GWRS 647	42	
1201	ADR	10	2-6-2T	-	1888	BP 2868	6925	
1202/3	GWR	-	0-6-0	388	1876	GWRS 649/50	42	
1204	ADR	11	2-6-2T	-	1888	BP 2870	6925	
1205/6	ADR	36/7	2-6-2T	-	1921	HL 3427/8	-	
1207/8	ADR	6, 7	2-6-2T	-	1887	BP 2866/65	6925	
1209	ADR	8	2-6-2T	-	1888	BP 2869	6925	
1210	GWR	-	0-6-0	388	1876	GWRS 657	42	
1211	ADR	9	2-6-2T	-	1887	BP 2867	6925	
1212/3	Cam	1, 2	2-6-2T	-	1902	D&M 1, 2	-	NG
1214/5	GWR	-	0-6-0	388	1876	GWRS 661/2	42	
1216-9	GWR	-	0-6-0T	850	1876	GWRW 289-92	Y	
1220-7	GWR	-	0-6-0T	850	1877	GWRW 293-300	Y	
1228-39	GWR	-	0-6-0T	1076	1876	GWRS 663-74	43	
1240-7	GWR	-	0-6-0T	1076	1877	GWRS 675-82	43	
1248-67	GWR	-	0-6-0T	1076	1877	GWRS 683-702	44	
1268-76	GWR	-	0-6-0T	1076	1877	GWRS 703-11	45	
1277	-	-	-	-	-	-	-	A
1278-83	GWR	-	0-6-0T	1076	1877	GWRS 713-8	45	
1284-97	GWR	-	0-6-0T	1076	1878	GWRS 719-32	45	
1298	GWR	-	2-4-0T	1298	1878	GWRS	-	
1299	GWR	-	2-4-0CT	1298	1878	GWRS	-	
1300	GWR	-	2-4-0T	1298	1878	GWRS	-	
1301-3	TVR	170-2	4-4-2T	C	1888	VF 1229-31	215	
1304-6	TVR	175/73/74	4-4-2T	C	1891	VF 1314/12/13	970	
1307	RSB	17	2-4-2T	-	1895	K 3581	150	
1308	GWR	-	2-4-0T	1308	1902	AB 956	-	
1309/10	RSB	18/9	2-4-2T	-	1895	K 3582/3	150	
1311-3	Barry	86-8	2-4-2T	J	1897	HC 473-5	-	
1314-6	Barry	89-91	2-4-2T	J	1898	SS 4367-9	E1110	
1317-21	Barry	94-8	2-4-2T	J	1899	SS 4497-4501	E1136	
1322/3	Barry	21/2	2-4-2T	C	1889	SS 3528/9	E937	

GWR Nos.	Former Railway	Former Nos.	Type	Class	Date	Builder	Order Lot No.	See Note
1324/5	RR	65/6	2-4-2ST	L	1890	VF 1292/3	701	
1326	PTR	36	2-4-2T	-	1890	SS 3626	E979	
1327	Car	36	2-4-2T	-	1898	LNWR 3869	E119	A
1328/9	Cam	10, 1	2-4-0	-	1883	GWRW	-	
1330/1	Cam	41/3	2-4-0	28	1864	SS 1485/6	E452	
1332	Cam	53	2-4-0	28	1865	SS 1633	E484	
1333	Cam	55	2-4-0	28	1865	SS 1655	E508	
1334-6	MSWJR	10-2	2-4-0	-	1894	D 3165-7	3165E	
1337	GWR	-	0-6-0ST	-	1889	MW 1127	27900	
1338	Car	5	0-4-0ST	-	1898	K 3799	194	
1339	Car	6	0-4-0ST	-	1899	K 3969	222	
1340	ADR	-	0-4-0ST	-	1897	AE 1386	-	1340
1341	ADR	-	0-4-0ST	-			-	1341
1342	TVR	267	0-4-0ST	S	1876	HC 166	-	
1343	TVR	266	0-4-0ST	T	1876	HC 159	-	
1344	ADR	22	0-6-4T	-	1886	BP 2606	6532	
1345/6	ADR	23/4	0-6-4T	-	1885	BP 2603/02	6532	
1347-55	Barry	139-47	0-6-4T	L	1914	HL 3038-46	-	
1356	ADR	32	0-6-0T	-	1873	FJ 122	-	GWR
1357	Barry	148	0-6-4T	L	1914	HL 3047	-	
1358-60	PTR	17-9	0-8-2T	-	1901	SS 4794-6	E1179	
1361-5	GWR	-	0-6-0ST	1361	1910	GWRS 2375-9	179	
1366-75	-	-	-	-	-	-	-	A
1376	GWR	-	0-6-0T	-	1874	BER	-	
1377	GWR	-	0-6-0T	-	1875	BER	-	
1378/9	PTR	20/1	0-8-2T	-	1899	C 2492/3	-	
1380-6	Barry	79-85	0-8-2T	H	1896	SS 4182-8	E1071	
1387/8	Barry	35/6	0-8-0	D	1888	SS 3446/7	E956	1387
1389	Barry	92	0-8-0	D	1886	SS 3365	E1103	1389
1390	Barry	93	0-8-0	D	1887	SS 3394	E1103	1390
1391/2	-	-	-	-	-	-	-	A
1393-9	GWR	-	0-6-0T	1393	1873	SS 2350-5/7/9	E643	1393
1400	-	-	-	-	-	-	-	A
1401	GWR	-	2-4-0T	Metro	1878	GWRS 734	47	
1402	-	-	-	-	-	-	-	A
1403-11	GWR	-	2-4-0T	Metro	1878	GWRS 736-44	47	
1412	-	-	-	-	-	-	-	A
1413-20	GWR	-	2-4-0T	Metro	1878	GWRS 746-53	47	
1421-5	GWR	-	0-4-2T	517	1877	GWRW 313-7	I	
1426	ADR	14	0-4-2T	517	1877	GWRW 318	I	
1427-32	GWR	-	0-4-2T	517	1877	GWRW 319-24	I	
1433-40	GWR	-	0-4-2T	517	1877	GWRW 349-56	C2	
1441-4	GWR	-	0-4-2T	517	1878	GWRW 357-60	C2	
1445-51	GWR	-	2-4-0T	Metro	1881	GWRS 886-92	57	
1452	-	-	-	-	-	-	-	A
1453-7	GWR	-	2-4-0T	Metro	1882	GWRS 894-8	57	
1458	-	-	-	-	-	-	-	A
1459	GWR	-	2-4-0T	Metro	1882	GWRS 900	57	
1460	-	-	-	-	-	-	-	A
1461-4	GWR	-	2-4-0T	Metro	1882	GWRS902-5	57	
1465-76	GWR	-	0-4-2T	517	1883	GWRW 434-45	M2	
1477-9	GWR	-	0-4-2T	517	1884	GWRW 458-60	P2	
1480	-	-	-	-	-	-	-	
1481	GWR	-	0-4-2T	517	1884	GWRW 462	P2	
1482-8	GWR	-	0-4-2T	517	1885	GWRW 463-9	P2	
1489/90	-	-	-	-	-	-	-	
1491-1500	GWR	-	2-4-0T	Metro	1892	GWRS 1351-60	91	
1501-12	GWR	-	0-6-0T	1501	1878	GWRW 325-36	A2	
1513-8	GWR	-	0-6-0T	1501	1878	GWRW 337-42	B2	

GWR Nos.	Former Railway	Former Nos.	Type	Class	Date	Builder	Order Lot No.	See Note
1519-24	GWR	-	0-6-0T	1501	1879	GWRW 343-8	B2	
1525-36	GWR	-	0-6-0T	1501	1879	GWRW 361-72	D2	
1537-9	GWR	-	0-6-0T	1501	1879	GWRW 373-5	E2	
1540-5	GWR	-	0-6-0T	1501	1880	GWRW 376-81	E2	
1546	SWMR	5	0-6-0T	1501	1880	GWRW 382	E2	
1547/8	GWR	-	0-6-0T	1501	1880	GWRW 383/4	E2	
1549-60	GWR	-	0-6-0T	1501	1880	GWRW 385-96	F2	
1561/2	GWR	-	0-6-0T	1076	1878	GWRS 754/5	48	
1563	-	-	-	-	-	-	-	A
1564-8	GWR	-	0-6-0T	1076	1878	GWRS 757-61	48	
1569-80	GWR	-	0-6-0T	1076	1879	GWRS 762-73	48	
1581-90	GWR	-	0-6-0T	1076	1879	GWRS 774-83	49	
1591	-	-	-	-	-	-	-	A
1592-1600	GWR	-	0-6-0T	1076	1879	GWRS 785-93	49	
1601-6	GWR	-	0-6-0T	1076	1879	GWRS 794-9	50	
1607-20	GWR	-	0-6-0T	1076	1880	GWRS 800-13	50	
1621-40	GWR	-	0-6-0T	1076	1880	GWRS 824-43	52	
1641-6	GWR	-	0-6-0T	1076	1881	GWRS 846-51	55	
1647-51	GWR	-	0-6-0T	1076	1881	GWRS 852-6	55	
1652	RSB	31	0-6-0T	1076	1881	GWRS 857	55	GWR
1653-9	GWR	-	0-6-0T	1076	1881	GWRS 858-64	55	
1660	RSB	2	0-6-0T	1076	1881	GWRS 865	55	GWR
1661-6	GWR	-	0-6-0T	1661	1886	GWRS 1053-8	69	
1667	Car	31	0-6-0T	1661	1886	GWRS 1059	69	GWR
1668	-	-	-	-	-	-	-	A
1669	GWR	-	0-6-0T	1661	1886	GWRS 1061	69	
1670	-	-	-	-	-	-	-	A
1671-3	GWR	-	0-6-0T	1661	1886	GWRS 1063-5	69	
1674	-	-	-	-	-	-	-	A
1675	GWR	-	0-6-0T	1661	1886	GWRS 1067	69	
1676	Car	23	0-6-0T	1661	1886	GWRS 1068	69	GWR
1677	-	-	-	-	-	-	-	A
1678	GWR	-	0-6-0T	1661	1886	GWRS 1070	69	
1679	ADR	27	0-6-0T	1661	1886	GWRS 1071	69	
1680	GWR	-	0-6-0T	1661	1886	GWRS 1072	69	
1681/2	GWR	-	0-6-0T	1661	1886	GWRS 1074/5	71	
1683	ADR	28	0-6-0T	1661	1886	GWRS 1076	71	GWR
1684	GWR	-	0-6-0T	1661	1886	GWRS 1077	71	
1685	-	-	-	-	-	-	-	A
1686	GWR	-	0-6-0T	1661	1886	GWRS 1079	71	
1687/8	GWR	-	0-6-0T	1661	1887	GWRS 1080/1	71	
1689	Car	25	0-6-0T	1661	1887	GWRS 1082	71	GWR
1690/1	GWR	-	0-6-0T	1661	1887	GWRS 1083/4	71	
1692-4	-	-	-	-	-	-	-	A
1695-1700	GWR	-	0-6-0T	1661	1887	GWRS 1088-93	71	
1701-9	GWR	-	0-6-0T	1854	1891	GWRS 1241-9	85	
1710	RSB	36	0-6-0T	1854	1891	GWRS 1250	85	GWR
1711-4	GWR	-	0-6-0T	1854	1891	GWRS 1251-4	85	
1715	-	-	-	-	-	-	-	A
1716-20	GWR	-	0-6-0T	1854	1891	GWRS 1256-60	85	
1721-40	GWR	-	0-6-0T	1854	1892	GWRS 1301-20	88	
1741-50	GWR	-	0-6-0T	655	1892	GWRW 565-74	A3	
1751-5	GWR	-	0-6-0T	1854	1892	GWRS 1321-5	89	
1756	RSB	35	0-6-0T	1854	1892	GWRS 1326	89	GWR
1757-64	GWR	-	0-6-0T	1854	1892	GWRS 1327-34	89	
1765-70	GWR	-	0-6-0T	1854	1893	GWRS 1335-40	89	
1771/2	GWR	-	0-6-0T	655	1892	GWRW 575/6	B3	
1773-87	GWR	-	0-6-0T	655	1893	GWRW 577-91	B3	
1788-90	GWR	-	0-6-0T	655	1894	GWRW 592-4	B3	
1791-1800	GWR	-	0-6-0T	1854	1895	GWRS 1436-45	98	

GWR Nos.	Former Railway	Former Nos.	Type	Class	Date	Builder	Order Lot No.	See Note
1801–5	GWR	–	0–6–0T	1501	1881	GWRW 397–401	G2	
1806	SWMR	3	0–6–0T	1501	1881	GWRW 402	G2	GWR
1807–10	GWR	–	0–6–0T	1501	1881	GWRW 403–6	G2	
1811	SWMR	1	0–6–0T	1501	1881	GWRW 407	G2	GWR
1812	GWR	–	0–6–0T	1501	1881	GWRW 408	G2	
1813–24	GWR	–	0–6–0T	1813	1882	GWRS 906–17	59	
1825	RSB	37	0–6–0T	1813	1882	GWRS 918	59	GWR
1826/7	GWR	–	0–6–0T	1813	1882	GWRS 919/20	59	
1828–32	GWR	–	0–6–0T	1813	1883	GWRS 921–5	59	
1833	–	–	–	–	–	–	–	A
1834	RSB	30	0–6–0T	1813	1883	GWRS 926	60	GWR
1835–52	GWR	–	0–6–0T	1813	1883	GWRS 927–44	60	
1853	GWR	–	0–6–0T	1813	1884	GWRS 945	60	
1854–73	GWR	–	0–6–0T	1854	1890	GWRS 1159–78	79	
1874–81	GWR	–	0–6–0T	1854	1890	GWRS 1201–8	83	
1882	–	–	–	–	–	–	–	A
1883–7	GWR	–	0–6–0T	1854	1890	GWRS 1210–4	83	
1888–93	GWR	–	0–6–0T	1854	1891	GWRS 1215–20	83	
1894–1900	GWR	–	0–6–0T	1854	1895	GWRS 1446–52	98	
1901–8	GWR	–	0–6–0T	850	1881	GWRW 410–7	J2	
1909–12	GWR	–	0–6–0T	850	1882	GWRW 418–21	J2	
1913–24	GWR	–	0–6–0T	850	1882	GWRW 422–33	L2	
1925	GWR	–	0–6–0T	850	1883	GWRW 446	O2	
1926–36	GWR	–	0–6–0T	850	1884	GWRW 447–57	O2	
1937–42	GWR	–	0–6–0T	850	1886	GWRW 470–5	Q2	
1943–8	GWR	–	0–6–0T	850	1887	GWRW 476–81	Q2	
1949–60	GWR	–	0–6–0T	850	1888	GWRW 482–93	R2	
1961–6	GWR	–	0–6–0T	850	1889	GWRW 500–5	T2	
1967–72	GWR	–	0–6–0T	850	1890	GWRW 506–11	T2	
1973–81	GWR	–	0–6–0T	850	1890	GWRW 512–20	V2	
1982–4	GWR	–	0–6–0T	850	1891	GWRW 521–3	V2	
1985–96	GWR	–	0–6–0T	850	1891	GWRW 527–38	X2	
1997–2001	GWR	–	0–6–0T	850	1891	GWRW 539–43	Y2	
2002–8	GWR	–	0–6–0T	850	1892	GWRW 544–50	Y2	
2009–14	GWR	–	0–6–0T	850	1894	GWRW 551–6	Z2	
2015–20	GWR	–	0–6–0T	850	1895	GWRW 557–62	Z2	
2021–30	GWR	–	0–6–0T	2021	1897	GWRW 625–34	D3	
2031–4	GWR	–	0–6–0T	2021	1897	GWRW 635–8	F3	
2035–40	GWR	–	0–6–0T	2021	1898	GWRW 639–44	F3	
2041–54	GWR	–	0–6–0T	2021	1898	GWRW 645–58	G3	
2055/6	GWR	–	0–6–0T	2021	1899	GWRW 659/60	G3	
2057	–	–	–	–	–	–	–	
2058–60	GWR	–	0–6–0T	2021	1899	GWRW 622–4	G3	
2061–74	GWR	–	0–6–0T	2021	1899	GWRW 665–78	H3	
2075–80	GWR	–	0–6–0T	2021	1900	GWRW 679–84	H3	
2081–5	GWR	–	0–6–0T	2021	1900	GWRW 685–9	J3	
2086–2100	GWR	–	0–6–0T	2021	1901	GWRW 690–704	J3	
2101–19	GWR	–	0–6–0T	2021	1902	GWRW 705–23	K3	
2120	GWR	–	0–6–0T	2021	1903	GWRW 724	K3	
2121–7	GWR	–	0–6–0T	2021	1903	GWRW 725–31	L3	
2128	–	–	–	–	–	–	–	
2129–38	GWR	–	0–6–0T	2021	1903	GWRW 733–42	L3	
2139/40	GWR	–	0–6–0T	2021	1904	GWRW 743/4	L3	
2141–56	GWR	–	0–6–0T	2021	1904	GWRW 745–60	M3	
2157–60	GWR	–	0–6–0T	2021	1905	GWRW 761–4	M3	
2161–99	–	–	–	–	–	–	–	A
2221	GWR	–	4–4–2T	2221	1905	GWRS 2076	151	
2222–30	GWR	–	4–4–2T	2221	1906	GWRS 2077–85	151	
2231–8	GWR	–	4–4–2T	2221	1908	GWRS 2320–7	175	
2239/40	GWR	–	4–4–2T	2221	1909	GWRS 2328/9	175	

GWR Nos.	Former Railway	Former Nos.	Type	Class	Date	Builder	Order Lot No.	See Note
2241-50	GWR	-	4-4-2T	2221	1912	GWRS 2466-75	188	
2301-7	GWR	-	0-6-0	2301	1883	GWRS 946-52	61	
2309-20	GWR	-	0-6-0	2301	1883	GWRS 954-65	61	
2321-33	GWR	-	0-6-0	2301	1884	GWRS 966-78	62	
2335-40	GWR	-	0-6-0	2301	1884	GWRS 980-5	62	
2341-60	GWR	-	0-6-0	2301	1884	GWRS 986-1005	63	
2361-70	GWR	-	0-6-0	2361	1885	GWRS 1032-41	67	
2371-80	GWR	-	0-6-0	2361	1886	GWRS 1042-51	67	
2381-6	GWR	-	0-6-0	2301	1890	GWRS 1181-6	82	
2388-2400	GWR	-	0-6-0	2301	1890	GWRS 1188-1200	82	
2401-19	GWR	-	0-6-0	2301	1891	GWRS 1271-89	87	
2421	GWR	-	0-6-0	2301	1891	GWRS 1291	87	
2422-30	GWR	-	0-6-0	2301	1892	GWRS 1292-1300	87	
2431-47	GWR	-	0-6-0	2301	1893	GWRS 1361-77	92	
2449/50	GWR	-	0-6-0	2301	1893	GWRS 1379/80	92	
2451/2	GWR	-	0-6-0	2301	1895	GWRS 1453/4	99	
2454-63	GWR	-	0-6-0	2301	1895	GWRS 1456-65	99	
2464-70	GWR	-	0-6-0	2301	1896	GWRS 1466-72	99	
2471-90	GWR	-	0-6-0	2301	1896	GWRS 1473-92	100	
2511-30	GWR	-	0-6-0	2301	1897	GWRS 1552-71	107	
2531-41	GWR	-	0-6-0	2301	1897	GWRS 1572-82	108	
2543-50	GWR	-	0-6-0	2301	1897	GWRS 1584-91	108	
2551-60	GWR	-	0-6-0	2301	1897	GWRS 1632-41	111	
2561/2	GWR	-	0-6-0	2301	1898	GWRS 1642/3	111	
2564-76	GWR	-	0-6-0	2301	1898	GWRS 1645-57	111	
2577-80	GWR	-	0-6-0	2301	1899	GWRS 1658-61	111	
2600	GWR	-	2-6-0	2600	1900	GWRS 1886	128	
2601	GWR	-	2-6-0	2600	1906	GWRS 2125	156	
2602	GWR	-	2-6-0	2600	1907	GWRS	166	
2603-10	GWR	-	2-6-0	2600	1906	GWRS	166	
2611-20	GWR	-	2-6-0	2600	1903	GWRS 1796-1805	123	
2621-40	GWR	-	2-6-0	2600	1901	GWRS 1908-27	131	
2641-53	GWR	-	2-6-0	2600	1901	GWRS 1929-41	133	
2654-60	GWR	-	2-6-0	2600	1902	GWRS 1942-8	133	
2661-80	GWR	-	2-6-0	2600	1902	GWRS 1949-68	135	
2700	GWR	-	0-6-0T	2721	1901	GWRS 1906	129	
2701-19	GWR	-	0-6-0T	655	1896	GWRW 605-23	E3	
2720	GWR	-	0-6-0T	655	1897	GWRW 624	E3	
2721	GWR	-	0-6-0T	2721	1897	GWRS 1662	112	
2722-40	GWR	-	0-6-0T	2721	1898	GWRS 1663-81	112	
2741-55	GWR	-	0-6-0T	2721	1899	GWRS 1703-17	115	
2756	RSB	33	0-6-0T	2721	1899	GWRS 1718	115	GWR
2757-60	GWR	-	0-6-0T	2721	1899	GWRS 1719-22	115	
2761-80	GWR	-	0-6-0T	2721	1900	GWRS 1776-95	122	
2781-99	GWR	-	0-6-0T	2721	1901	GWRS 1887-1905	129	
2800	GWR	-	2-8-0	2800	1903	GWRS 1991	139	
2801-10	GWR	-	2-8-0	2800	1905	GWRS 2096-2105	153	
2811-20	GWR	-	2-8-0	2800	1905	GWRS 2115-24	155	
2821-30	GWR	-	2-8-0	2800	1907	GWRS 2158-67	160	
2831-5	GWR	-	2-8-0	2800	1911	GWRS 2390-4	181	
2836-45	GWR	-	2-8-0	2800	1912	GWRS 2436-45	186	
2846-9	GWR	-	2-8-0	2800	1912	GWRS 2486-9	190	
2850-5	GWR	-	2-8-0	2800	1913	GWRS 2490-5	190	
2856-75	GWR	-	2-8-0	2800	1918	GWRS 2762-81	210	
2876-83	GWR	-	2-8-0	2800	1919	GWRS 2782-9	210	

GWR Nos.	Former Railway	Former Nos.	Type	Class	Date	Builder	Order Lot No.	See Note
2900	GWR	-	4-6-0	2900	1902	GWRS 1928	132	
2901-10	GWR	-	4-6-0	2900	1906	GWRS 2199-2208	164	
2911-30	GWR	-	4-6-0	2900	1907	GWRS 2259-78	170	
2931-40	GWR	-	4-6-0	2900	1911	GWRS 2426-35	185	
2941-50	GWR	-	4-6-0	2900	1912	GWRS 2476-85	189	
2951-5	GWR	-	4-6-0	2900	1913	GWRS 2506-10	192	
2971	GWR	-	4-6-0	2900	1903	GWRS 2024	145	
2972-80	GWR	-	4-6-0	2900	1905	GWRS 2106-14	154	
2981-90	GWR	-	4-6-0	2900	1905	GWRS 2128-37	158	
2998	GWR	-	4-6-0	2900	1903	GWRS 1990	138	
3000-11	GWR	-	2-8-0	ROD	1919	NBL 22104-15	L710	ROD
3012-9	GWR	-	2-8-0	ROD	1919	NBL 22119-26	L711	ROD
3100	GWR	-	2-6-2T	3100	1903	GWRS 1992	140	
3111-20	GWR	-	2-6-2T	3100	1905	GWRS 2066-75	150	
3121-30	GWR	-	2-6-2T	3100	1905	GWRS 2086-95	152	
3131-49	GWR	-	2-6-2T	3100	1906	GWRS 2138-56	159	
3150	GWR	-	2-6-2T	3150	1906	GWRS 2157	159	
3151-70	GWR	-	2-6-2T	3150	1907	GWRS 2239-58	169	
3171-84	GWR	-	2-6-2T	3150	1907	GWRS 2280-93	172	
3185-90	GWR	-	2-6-2T	3150	1908	GWRS 2294-9	172	
3201	GWR	-	2-4-0	3201	1884	GWRS 1026	65	
3202-4	GWR	-	2-4-0	3201	1885	GWRS 1027-9	65	
3206-25	GWR	-	2-4-0	3206	1889	GWRS 1137-56	75	
3231	GWR	-	2-4-0	3226	1889	GWRW 499	S2	
3232-41	GWR	-	2-4-0	3232	1892	GWRS 1341-50	90	
3242	GWR	-	2-4-0	3232	1893	GWRS 1381	93	
3244-51	GWR	-	2-4-0	3232	1893	GWRS 1383-90	93	
3252	GWR	-	4-4-0	3252	1895	GWRS 1431	97	
3253-60	GWR	-	4-4-0	3252	1895	GWRS 1493-1500	101	
3261-5	GWR	-	4-4-0	3252	1896	GWRS 1504-6/9/10	102	
3266-9	GWR	-	4-4-0	3252	1896	GWRS 1531/3-5	105	
3270-80	GWR	-	4-4-0	3252	1897	GWRS 1536/7/40/2-4/6-50	105	
3281-91	GWR	-	4-4-0	3252	1899	GWRS 1683-5/7/9-91/3/6/8/9	113	
3300	GWR	-	4-4-0	3300	1895	GWRS 1432	97	
3301-5	GWR	-	4-4-0	3300	1896	GWRS 1501-3/7/8	102	
3306	GWR	-	4-4-0	3300	1896	GWRS 1532	105	
3307-10	GWR	-	4-4-0	3300	1897	GWRS 1538/9/41/5	105	
3311	GWR	-	4-4-0	3300	1898	GWRS 1682	113	
3312-9	GWR	-	4-4-0	3300	1899	GWRS 1686/8/92/4/5/7, 1700/1	113	
3320-5	GWR	-	4-4-0	3300	1899	GWRS 1743-8	118	
3326-39	GWR	-	4-4-0	3300	1900	GWRS 1749-62	118	
3340	GWR	-	4-4-0	3300	1899	GWRS 1773	120	
3341-60	GWR	-	4-4-0	3300	1900	GWRS 1806-25	124	
3361/2	GWR	-	4-4-0	3300	1902	GWRS 1970/1	137	
3363-80	GWR	-	4-4-0	3300	1903	GWRS 1972-89	137	
3381-90	GWR	-	4-4-0	3300	1903	GWRS 2003-12	142	
3391-3410	GWR	-	4-4-0	3300	1904	GWRS 2036-55	148	
3411-25	GWR	-	4-4-0	3300	1906	GWRS 2169-83	162	
3426-40	GWR	-	4-4-0	3300	1906	GWRS 2184-98	163	

GWR Nos.	Former Railway	Former Nos.	Type	Class	Date	Builder	Order Lot No.	See Note
3441-50	GWR	-	4-4-0	3300	1909	GWRS 2350-9	177	
3451-5	GWR	-	4-4-0	3300	1910	GWRS 2360-4	177	
3500	GWR	-	2-4-0T	Metro	1899	GWRS 1772	119	
3501-8	GWR	-	2-4-0	3201	1885	GWRS 1006-13	64	
3510-6	GWR	-	2-4-0	3201	1885	GWRS 1015-21	64	
3518/9	GWR	-	2-4-0	3201	1885	GWRS 1023/4	64	
3521	Cam	82	4-4-0	3521	1887	GWRS 1095	73	GWR
3522-34	GWR	-	4-4-0	3521	1887	GWRS 1096-1108	73	
3535-40	GWR	-	4-4-0	3521	1888	GWRS 1109-14	73	
3542-5	GWR	-	4-4-0	3521	1888	GWRS 1118-21	76	
3546	Cam	95	4-4-0	3521	1888	GWRS 1122	76	GWR
3547-53	GWR	-	4-4-0	3521	1888	GWRS 1123-9	76	
3554-60	GWR	-	4-4-0	3521	1889	GWRS 1130-6	76	
3561-70	GWR	-	2-4-0T	Metro	1894	GWRS 1421-30	96	
3571-9	GWR	-	0-4-2T	3571	1895	GWRW 595-603	C3	
3580	GWR	-	0-4-2T	3571	1897	GWRW 604	C3	
3581-90	GWR	-	2-4-0T	Metro	1899	GWRS 1733-42	117	
3591/2	GWR	-	2-4-0T	Metro	1899	GWRS 1763/4	119	
3593	GWR	-	2-4-2T	Metro	1899	GWRS 1765	119	
3594-9	GWR	-	2-4-0T	Metro	1899	GWRS 1766-71	119	
3600	GWR	-	2-4-2T	3600	1900	GWRS 1907	130	
3601-20	GWR	-	2-4-2T	3600	1902	GWRS 1866-85	134	
3621-30	GWR	-	2-4-2T	3600	1903	GWRS 2013-22	143	
3700-9	GWR	-	4-4-0	3700	1901	GWRS 1853-62	126	
3710-9	GWR	-	4-4-0	3700	1903	GWRS 1993-2002	141	
3800	GWR	-	4-4-0	3800	1904	GWRS 2056	149	
3801-20	GWR	-	4-4-0	3800	1906	GWRS 2209-28	165	
3821-5	GWR	-	4-4-0	3800	1911	GWRS 2416-20	184	
3826-30	GWR	-	4-4-0	3800	1912	GWRS 2421-5	184	
3831-9	GWR	-	4-4-0	3800	1904	GWRS 2057-65	149	
3901-20	GWR	-	2-6-2T	3901	1896	GWRS 1511-30	107	
4000	GWR	-	4-6-0	4000	1906	GWRS 2168	161	
4001-10	GWR	-	4-6-0	4000	1907	GWRS 2229-38	168	
4011-20	GWR	-	4-6-0	4000	1908	GWRS 2300-9	173	
4021-30	GWR	-	4-6-0	4000	1909	GWRS 2365-74	178	
4031-7	GWR	-	4-6-0	4000	1910	GWRS 2380-6	180	
4038-40	GWR	-	4-6-0	4000	1911	GWRS 2387-9	180	
4041-5	GWR	-	4-6-0	4000	1913	GWRS 2536-40	195	
4046-60	GWR	-	4-6-0	4000	1914	GWRS 2572-86	199	
4061/2	GWR	-	4-6-0	4000	1922	GWRS 2915/6	217	
4100	GWR	-	4-4-0	4100	1897	GWRS 1592	109	
4101-17	GWR	-	4-4-0	4100	1898	GWRS 1593-1609	109	
4118/9	GWR	-	4-4-0	4100	1899	GWRS 1610/1	109	
4120-38	GWR	-	4-4-0	4100	1900	GWRS 1826-34/6-45	125	
4139-48	GWR	-	4-4-0	4100	1901	GWRS 1846-52/63-5	126	
4149-68	GWR	-	4-4-0	4100	1908	GWRS 2330-49	176	
4169/70	GWR	-	4-4-0	4100	1894	GWRS	-	
4201	GWR	-	2-8-0T	4200	1910	GWRS 2395	182	
4202-21	GWR	-	2-8-0T	4200	1912	GWRS 2446-55	187	
4222-31	GWR	-	2-8-0T	4200	1913	GWRS 2541-50	196	
4232-41	GWR	-	2-8-0T	4200	1914	GWRS 2587-96	200	
4242-51	GWR	-	2-8-0T	4200	1916	GWRS 2632-41	203	
4252-61	GWR	-	2-8-0T	4200	1917	GWRS 2642-51	203	
4262-70	GWR	-	2-8-0T	4200	1919	GWRS 2842-50	213	
4271-85	GWR	-	2-8-0T	4200	1920	GWRS 2851-65	213	

GWR Nos.	Former Railway	Former Nos.	Type	Class	Date	Builder	Order Lot No.	See Note
4286-95	GWR	-	2-8-0T	4200	1921	GWRS	220	
4300	GWR	-	2-6-0	4300	1916	GWRS 2671	204	
4301-20	GWR	-	2-6-0	4300	1911	GWRS 2396-2415	183	
4321-30	GWR	-	2-6-0	4300	1913	GWRS 2516-25	193	
4331-40	GWR	-	2-6-0	4300	1913	GWRS 2526-35	194	
4341/2	GWR	-	2-6-0	4300	1913	GWRS 2552/3	198	
4343-60	GWR	-	2-6-0	4300	1914	GWRS 2554-71	198	
4361-80	GWR	-	2-6-0	4300	1915	GWRS 2612-31	202	
4381-99	GWR	-	2-6-0	4300	1916	GWRS 2652-70	204	
4400	GWR	-	2-6-2T	4400	1904	GWRS 2023	144	
4401-4	GWR	-	2-6-2T	4400	1905	GWRW 765-8	147	
4405-10	GWR	-	2-6-2T	4400	1906	GWRW 769-74	147	
4500-2	GWR	-	2-6-2T	4500	1906	GWRW 775-7	N3	
4503-15	GWR	-	2-6-2T	4500	1907	GWRW 778-90	N3	
4516-9	GWR	-	2-6-2T	4500	1908	GWRW 791-4	N3	
4520-4	GWR	-	2-6-2T	4500	1909	GWRS 2310-4	174	
4525-9	GWR	-	2-6-2T	4500	1910	GWRS 2315-9	174	
4530-9	GWR	-	2-6-2T	4500	1913	GWRS 2496-2505	191	
4540-7	GWR	-	2-6-2T	4500	1914	GWRS 2597-2604	201	
4548-54	GWR	-	2-6-2T	4500	1915	GWRS 2605-11	201	
4600	GWR	-	4-4-2T	4600	1913	GWRS 2551	197	
4700	GWR	-	2-8-0	4700	1919	GWRS 2866	214	
4701-5	GWR	-	2-8-0	4700	1922	GWRS	221	
5300-4	GWR	-	2-6-0	4300	1916	GWRS 2672-6	205	
5305-9	GWR	-	2-6-0	4300	1917	GWRS 2677-81	205	
5310-29	GWR	-	2-6-0	4300	1917	GWRS 2682-2701	206	
5330-7	GWR	-	2-6-0	4300	1917	GWRS 2702-9	207	
5338-49	GWR	-	2-6-0	4300	1918	GWRS 2710-21	207	
5350-9	GWR	-	2-6-0	4300	1918	GWRS 2722-31	208	
5360-9	GWR	-	2-6-0	4300	1919	GWRS 2732-41	208	
5370-9	GWR	-	2-6-0	4300	1919	GWRS 2742-51	209	
5380-9	GWR	-	2-6-0	4300	1920	GWRS 2752-61	209	
5390-9	GWR	-	2-6-0	4300	1920	GWRS 2790-9	211	
6300-6	GWR	-	2-6-0	4300	1920	GWRS 2800-6	211	
6307-17	GWR	-	2-6-0	4300	1921	GWRS 2807-17	211	
6318-41	GWR	-	2-6-0	4300	1921	GWRS 2818-41	212	
6370-99	GWR	-	2-6-0	4300	1921	RS 3802-31	E114	6370
7300-2	GWR	-	2-6-0	4300	1921	RS 3832-4	E114	6370
7303/4	GWR	-	2-6-0	4300	1922	RS 3835/6	E114	6370
7305-18	GWR	-	2-6-0	4300	1921	GWRS	222	
7319	GWR	-	2-6-0	4300	1922	GWRS	222	

NOTES:

A Number, or some numbers, used again in 1922 for locomotives taken into stock from B&MR and N&BR and in 1923-4 for locomotives of the GVR, LMMR, SHT and P&M.

GWR Former GWR engine which resumed its old GWR number.

NG Narrow gauge engines - Nos. 822/3 2' 6" (Welshpool & Llanfair) and Nos. 1198, 1212/3 1' 11½" (Vale of Rheidol).

ROD Taken into stock under Lot 215 when purchased in May 1919 (the former Railway Operating Division numbers were 2008-13/5-20/4-31).

28 CMDPR engines had only names, being CLEOBURY and BURWARTON respectively.

102 Lot number given was the Swindon Lot number.

103 Lot number given was the Swindon Lot number.

236 Originally left vacant; TVR 0-6-2T No. 105 was allotted number 303 but as the GWR engine of that number was still in traffic number 236 was used instead.

423 Actual order of NR progressive numbers 5409/12/10/08.

663 Built as 0-6-0T and still recorded as such by the ADR, hence inclusion in the 0-6-0T series.

698 Building details unknown (this number was used again in 1922).

807 Barry 0-6-0ST No. 129 was initially allotted number 728 but as RSB 0-6-0T No. 32 was formerly GWR No. 728 it took that number. Therefore the Barry engine took the vacated number in the batch allotted to RSB engines.

999 Originally left vacant; TVR 4-4-0T No. 287 was allotted number 1186 but as the GWR engine of that number was still in traffic number 999 was used instead.

1340 Not numbered on ADR, but named TROJAN.

1341 Not numbered on ADR, but named ALEXANDRA; building details not known.

1387 Built by SS to their order E919; purchased by the Barry in 1889 (as order E956).

1389 Built by SS to their order E890; purchased by the Barry in 1897 (as order E1103).

1390 Built by SS to their order E901; purchased by the Barry in 1897 (as order E1103).

1393 Actual order of SS progressive numbers 2351/54*/52*/53*/55/59/57* (* exact order uncertain).

6370 Swindon Lot 218 was issued for RS order E114; fifteen engines ordered from RS (Progressive numbers 3837-51) were built at Swindon, Nos. 7305-19.

Swindon Works progressive numbers - the last engines to carry such numbers were those built to Lot 183, but numbers were issued up to and including Lot 217.

Locomotives renumbered - the locomotive stock is given as numbered, or to be renumbered, in May 1922. Many GWR engines (and some of those of other railways) had other numbers when built (for details see the appropriate part of 'The Locomotives of the Great Western Railway' by the Railway Correspondence and Travel Society). A number of engines were renumbered by the GWR after the completion of the 1922 renumbering scheme and details are given later in this volume.

Locomotives taken into stock from the B&MR, BPGVR and N&BR in August 1922 (using vacant, or vacated numbers in the range 11-2199).

GWR Nos.	Former Railway	Former Nos.	Type	Class	Date	Builder	Order Lot No.	See Note
11	B&MR	36	0-6-2T	-	1909	RS 3379	E59	
21	B&MR	37	0-6-2T	-	1909	RS 3380	E59	
332	B&MR	38	0-6-2T	-	1909	RS 3381	E59	
504	B&MR	39	0-6-2T	-	1909	RS 3382	E59	
698	B&MR	40	0-6-2T	-	1914	RS 3577	E91	
888	B&MR	41	0-6-2T	-	1914	RS 3578	E91	
1084	B&MR	42	0-6-2T	-	1914	RS 3579	E91	
1113	B&MR	43	0-6-2T	-	1914	RS 3580	E91	
1114	N&BR	11	0-6-2T	-	1906	RS 3212	E26	
1117	N&BR	12	0-6-2T	-	1908	RS 3351	E49	
1277	N&BR	13	0-6-2T	-	1910	RS 3397	E62	
1327/71	N&BR	9, 10	0-6-2T	-	1897	RS 2866/65	-	
1372-4	B&MR	45-7	0-6-2T	-	1915	RS 3645-7	E97	
1375	B&MR	48	0-6-2T	-	1921	RS 3798	E112	
1391	B&MR	44	4-4-2T	-	1879	BP 1840	3739	
1392	N&BR	5	4-4-0T	-	1871	YE 161	E23	
1400	N&BR	6	2-4-0T	-	1893	SS 3884	E1013	

GWR Nos.	Former Railway	Former Nos.	Type	Class	Date	Builder	Order Lot No.	See Note
1402/12	B&MR	9, 10	2-4-0T	-	1888	RS 2656/7	-	
1452	B&MR	12	2-4-0T	-	1889	RS 2659	-	
1458	B&MR	25	2-4-0T	-	1898	RS 2878	-	
1460	B&MR	11	2-4-0T	-	1889	RS 2658	-	
1563	N&BR	14	0-6-0ST	1076	1878	GWRS 756	48	GWR
1591	N&BR	15	0-6-0ST	1076	1879	GWRS 784	49	GWR
1668/70	B&MR	49, 50	0-6-2T	-	1921	RS 3799, 3800	E112	
1674/7	B&MR	19, 20	0-6-2ST	-	1905	VF 2040/1	734	
1685	B&MR	32	0-6-0ST	1661	1886	GWRS 1078	71	GWR
1692	B&MR	23	0-6-2ST	-	1894	VF 1405	66	
1693/4	B&MR	33/4	0-6-0ST	1661	1887	GWRS 1086/7	71	GWR
1715	N&BR	16	0-6-0ST	1854	1891	GWRS 1255	85	GWR
1833	B&MR	26	0-6-2ST	-	1894	VF 1406	66	
1882	N&BR	3	0-6-0ST	1854	1890	GWRS 1209	83	GWR
2161	B&MR	35	0-6-0T	-	1917	KS 3070	8459	
2162	BPGVR	2	0-6-0T	-	1914	HC 1066	-	
2163	BPGVR	9	0-6-0T	-	1909	HC 893	-	
2164	BPGVR	11	0-6-0T	-	1912	HC 969	-	
2165	BPGVR	12	0-6-0T	-	1913	HC 1024	-	
2166	BPGVR	13	0-6-0T	-	1916	HC 1222	-	
2167	BPGVR	14	0-6-0T	-	1919	HC 1385	-	
2168	BPGVR	15	0-6-0T	-	1916	HC 1164	-	
2169/70	B&MR	22/4	0-6-0ST	-	1896	K 3667/8	168	
2171-3	B&MR	27-9	0-6-0ST	-	1900	NW 584-6	-	
2174/5	N&BR	7, 8	0-6-0ST	-	1899	NW 561/2	-	
2176	BPGVR	7	0-6-0ST	-	1907	AE 1535	1680	
2177/8	B&MR	1, 2	0-6-0ST	-	1885	JF 4995/6	-	
2179/80	B&MR	3, 4	0-6-0ST	-	1886	JF 5054/5	-	
2181	B&MR	5	0-6-0ST	-	1884	RS 2498	-	
2182/3	B&MR	6, 7	0-6-0ST	-	1884	RS 2447/8	-	
2184	B&MR	8	0-6-0ST	-	1884	RS 2497	-	
2185/6	B&MR	13/4	0-6-0ST	-	1886	JF 5056/7	-	
2187/8	B&MR	15/6	0-6-0ST	-	1884	RS 2499, 2500	-	
2189	N&BR	2	0-6-0ST	-	1872	AE 943	NH	
2190/1	B&MR	17/8	0-6-0ST	-	1881	SS 2975/74	E799	
2192	BPGVR	1	0-6-0ST	-	1900	CF 1197	-	
2193	BPGVR	3	0-6-0ST	-	1901	CF 1209	-	
2194	BPGVR	4	0-6-0ST	-	1903	AE 1463	1312	
2195	BPGVR	5	0-6-0ST	-	1905	AE 1491	1475	
2196	BPGVR	6	0-6-0ST	-	1906	AE 1519	1622	
2197	BPGVR	8	0-6-0T	-	1909	HC 871	-	
2198	BPGVR	10	0-6-0T	-	1910	HC 924	-	
2199	N&BR	1	0-6-0ST	-	1872	AE 942	NH	

NOTE:

GWR Former GWR engine which resumed its old GWR number.

Locomotives taken into stock from the GVR and LMMR in January 1923 (using vacated numbers in the range 26-969).

GWR Nos.	Former Railway	Former Nos.	Type	Class	Date	Builder	Order Lot No.	See Note
26	GVR	1	0-6-0ST	-	1905	HC 721	-	
312	LMMR	-	0-6-0T	-	1907	AB 1111	-	LMMR
339	LMMR	-	0-6-0T	-	1913	HC 1032	-	LMMR
359	LMMR	-	0-6-0ST	-	1917	HC 1214	-	LMMR

GWR Nos.	Former Railway	Former Nos.	Type	Class	Date	Builder	Order Lot No.	See Note
704	LMMR	-	0-6-0T	-	1920	MW 1982	78400	LMMR
803	LMMR	-	0-6-0T	-	1911	HC 930	-	LMMR
937	LMMR	-	0-6-0T	-	1912	HC 977	-	LMMR
944	LMMR	-	0-6-0T	-	1902	AE 1448	1222	LMMR
969	LMMR	-	0-6-0T	-	1875	FW 279	-	LMMR

NOTES:

LMMR Named only on the LMMR:-

312	GEORGE WADDELL	704	VICTORY	944	GREAT MOUNTAIN
339	TARNDUNE	803	RAVELSTON	969	SEYMOUR CLARKE
359	HILDA	937	MERKLAND		

In addition GVR No. 2 MARGRET 0-6-0ST 1878 FW 410 was in the hands of the GWR until its sale, arranged before the GWR takeover, was completed in 1923 - no GWR number was allotted.

Locomotives taken into stock from the SHT and P&M in July 1923 and January 1924 respectively (using vacated numbers in the range 150-1098).

GWR Nos.	Former Railway	Former Nos.	Type	Class	Date	Builder	Order Lot No.	See Note
150	SHT	3	0-4-0ST	-	1905	HC 725	-	
696	P&M	3	0-4-0ST	-	1913	P 1328	-	
701	SHT	5	0-4-0ST	-	1905	AB 1045	-	
779	P&M	4	0-4-0ST	-	1916	P 1449	-	
795	P&M	5	0-4-0ST	-	1903	BE 301	-	
886	SHT	7	0-4-0ST	-	1899	P 739	-	
921	P&M	6	0-4-0ST	-	1906	BE 314	-	
925	P&M	7	0-4-0ST	-	1874	AE 1053	CR	
926	SHT	8	0-4-0ST	-	1899	P 761	-	
927	P&M	11	0-4-0ST	-	1907	P 1054	-	
928	P&M	14	0-4-0ST	-	1912	AB 1273	-	
929	SHT	11	0-4-0ST	-	1906	P 1053	-	
930	SHT	9	0-4-0ST	-	1902	P 908	-	
933	SHT	10	0-4-0ST	-	1904	P 974	-	
935	P&M	12	0-4-0ST	-	1912	P 1179	-	
937	SHT	17	0-6-0ST	-	1915	P 1397	-	
942	P&M	-	0-4-0ST	-	1905	HL 2558	-	942
943	SHT	14	0-4-0ST	-	1911	HC 939	-	
968	SHT	12	0-4-0ST	-	1908	P 1105	-	
974	SHT	13	0-4-0ST	-	1909	HL 2781	-	
1085	SHT	15	0-6-0ST	-	1912	P 1282	-	
1086	SHT	16	0-6-0ST	-	1913	P 1302	-	
1098	SHT	18	0-4-0ST	-	1918	P 1522	-	

NOTE:

942 Named only on P&M - DOROTHY.

GWR Locomotive Stock May 1922

The following list of engines and their allocation uses the GWR numbers applicable in 1922 but those marked (+) after the number were renumbered subsequently. The engines concerned and their new numbers were (some reverted to their earlier numbers later):-

1st No.	New No.	Date	Old No.	1st No.	New No.	Date	Old No.	1st No.	New No.	Date	Old No.
7	4171	2/23	-	2157	2190	1/40	-	3148	5148	4/28	-
8	4172	2/23	-	2832	4806	11/46	4/49	3149	5149	10/28	-
300	200	12/46	-	2834	4808	7/47	1/50	3254	9054	10/46	-
301	201	-	-	2839	4804	11/46	10/48	3264	9064	10/46	-
302	202	-	-	2845	4809	8/47	12/49	3265	9065	10/46	-
310	203	12/48	-	2847	4811	9/47	6/49	3272	9072	10/46	-
311	204	8/47	-	2848	4807	6/47	7/49	3273	9073	10/46	-
313	205	7/47	-	2849	4803	11/46	4/49	3276	9076	10/46	-
314	206	-	-	2853	4810	8/47	6/49	3283	9083	11/46	-
315	207	10/47	-	2854	4801	11/46	2/49	3284	9084	10/46	-
317	208	8/48	-	2862	4802	12/46	9/48	3287	9087	11/46	-
318	209	8/46	-	2863	4805	11/46	5/49	3289	9089	10/46	-
319	210	12/47	-	2872	4800	12/46	9/48	3291	9091	10/46	-
320	211	5/48	-	3100	5100	4/29	-	5300	8300	2/28	7/44
321	215	4/48	-	3111	5111	12/28	-	5301	8301	1/28	-
324	216	3/48	-	3112	5112	6/28	-	5302	8302	1/28	2/45
333	217	4/49	-	3113	5113	5/28	-	5304	8304	2/28	-
401	303	1/47	-	3114	5114	5/28	-	5305	8305	2/28	6/44
402	304	4/48	-	3115	5115	5/28	-	5307	8307	1/28	8/44
403	305	5/47	-	3116	5116	5/28	-	5308	8308	1/28	-
404	306	7/49	-	3117	5117	5/28	-	5309	8309	1/28	7/44
406	307	7/49	-	3118	5118	5/28	-	5313	8313	1/28	7/44
408	308	12/48	-	3119	5119	6/28	-	5314	8314	1/28	6/44
409	218	1/50	-	3120	5120	5/28	-	5315	8315	2/28	7/44
414	219	5/48	-	3121	5121	5/29	-	5318	8318	1/28	8/44
420	220	11/48	-	3122	5122	9/28	-	5320	8320	2/28	6/44
438	309	6/49	-	3123	5123	11/28	-	5322	8322	1/28	6/44
439	312	9/49	-	3124	5124	3/29	-	5325	8325	3/28	7/44
440	316	6/50	-	3125	5125	11/28	-	5326	8326	1/28	6/44
441	322	3/47	-	3126	5126	8/29	-	5327	8327	2/28	8/44
604	90	12/47	-	3127	5127	10/30	-	5328	8328	1/28	6/44
605	91	3/48	-	3128	5128	12/28	-	5329	8329	2/28	-
606	92	5/47	-	3129	5129	6/29	-	5331	8331	1/28	9/44
608	93	9/47	-	3130	5130	6/29	-	5332	8332	1/28	6/44
609	94	9/46	-	3131	5131	5/29	-	5333	8333	3/28	7/44
610	95	9/48	-	3132	5132	7/29	-	5334	8334	1/28	6/44
611	96	12/49	-	3133	5133	11/28	-	5335	8335	3/28	8/44
792	193	6/48	-	3134	5134	9/28	-	5337	8337	3/28	10/44
793	194	9/48	-	3135	5135	5/29	-	5338	8338	3/28	6/44
794	195	2/49	-	3136	5136	9/28	-	5340	8340	1/28	6/44
1106	1110	2/26	-	3137	5137	10/29	-	5341	8341	1/28	11/44
1213	9	3/49	-	3138	5138	8/28	-	5342	8342	1/28	-
2074	2183	3/39	-	3139	5139	6/29	-	5343	8343	1/28	8/44
2087	2188	10/39	-	3140	5140	8/29	-	5344	8344	1/28	8/44
2105	2189	12/39	-	3141	5141	1/29	-	5350	8350	1/28	8/44
2118	2186	7/39	-	3142	5142	7/29	-	5351	8351	1/28	6/44
2125	2182	7/39	-	3143	5143	6/29	-	5352	8352	2/28	-
2133	2181	2/39	-	3144	5144	7/29	-	5353	8353	1/28	6/44
2143	2187	8/39	-	3145	5145	3/28	-	5354	8354	1/28	-
2145	2184	6/39	-	3146	5146	6/29	-	5357	8357	2/28	10/44
2149	2185	7/39	-	3147	5147	11/28	-	5358	8358	1/28	6/44

1st No.	New No.	Date	Old No.	1st No.	New No.	Date	Old No.	1st No.	New No.	Date	Old No.
5359	8359	1/28	2/46	5369	8369	3/28	6/44	5383	8383	2/28	-
5360	8360	1/28	6/44	5372	8372	3/28	8/44	5384	8484	1/28	6/44
5361	8361	1/28	8/44	5373	8373	1/28	6/44	5386	8386	2/28	9/44
5362	8362	1/28	6/44	5374	8374	1/28	7/44	5387	8387	3/28	-
5363	8363	1/28	-	5376	8376	2/28	8/44	5388	8388	1/28	9/44
5364	8364	2/28	8/44	5378	8378	3/28	6/44	5389	8389	2/28	-
5365	8365	2/28	8/44	5379	8379	1/28	7/44	5390	8390	3/28	9/44
5366	8366	1/28	-	5381	8381	1/28	8/44	5391	8391	1/28	9/44
5368	8368	2/28	6/44	5382	8382	1/28	2/45	5393	8393	1/28	9/48

(Nos. 301/2/14 were not actually renumbered).

The 'Old No.' column is the date of reversion to the first number.

In late 1927 2-6-0 Nos. 4351/86/95 ran briefly as Nos. 8300/35/44.

Engines marked (*) in the allocation list were condemned without being renumbered. Where no shed is shown in the 1922 Shed column the locomotive concerned was either already condemned (ex-Cambrian engines) or was in works and soon afterwards condemned, but they are included as GWR numbers were allotted.

No.	Date	1922 Shed	Last GWR Shed	Wdn GWR	Last BR shed	Wdn BR
1	5/80	Chester	Wolverhampton	7/24	-	-
2	8/92	Barry	Barry	5/25	-	-
3	8/92	Barry	Barry	4/29	-	-
4	3/95	Barry	Barry	10/26	-	-
5	5/69	Barnstaple	Ludlow	10/32	-	-
6	5/69	Stourbridge	Hereford	10/28	-	-
7+	3/94	Wolverhampton	Chester	9/28	-	-
8+	5/94	Leamington	Bristol SPM	4/29	-	-
9	3/95	Barry	Barry	12/27	-	-
10	6/95	Llanfyllin	Aberystwyth	10/32	-	-
11*	7/95	CONDEMNED	-	5/22	-	-
13	10/86	Moorswater	Swindon	5/26	-	-
15*	8/95	Oswestry	Oswestry	10/22	-	-
16	4/21	CME (Swindon)	CME (Swindon)	9/36	-	-
17	4/01	CME (W'ton)	CME (Swindon)	9/36	-	-
18	4/01	CME (Swindon)	CME (Swindon)	9/36	-	-
19	6/99	Machynlleth	Aberystwyth	10/32	-	-
20	6/99	Wrexham	Aberystwyth	6/28	-	-
21*	7/99	CONDEMNED	-	5/22	-	-
22	1/76	Chester	Oxley	5/32	-	-
23	5/95	Swindon Town	Swindon	2/30	-	-
24	5/97	Swindon Town	Swindon	7/30	-	-
25	-/97	Andover Jct	Swindon	10/27	-	-
26	12/67	Kidderminster	Worcester	4/23	-	-
27	-/97	Swindon Town	Kidderminster	9/29	-	-
28	8/08	Cleobury M.	Kidderminster	-	Ebbw Junction	11/53
29	8/08	Cleobury M.	Kidderminster	-	Kidderminster	2/54
30	12/07	Cardiff Docks	Radyr	-	Radyr	5/49
31	12/07	Cardiff Docks	Cardiff E Dock	-	Radyr	3/51
32	12/07	Cardiff Docks	Radyr	-	Radyr	2/50
33	9/04	Cardiff Docks	Cardiff E Dock	-	Cardiff E Dock	2/51
34	9/09	Cardiff Docks	Radyr	-	Radyr	11/49
35	7/21	Cardiff Docks	Radyr	-	Cardiff E Dock	11/56
36	7/21	Cardiff Docks	Cardiff E Dock	-	Cardiff E Dock	10/57
37	9/21	Cardiff Docks	Cardiff E Dock	-	Cardiff E Dock	9/56
38	9/21	Cardiff Docks	Radyr	-	Cardiff E Dock	10/57

No.	Date	1922 Shed	Last GWR Shed	Wdn GWR	Last BR shed	Wdn BR
39	12/21	Cardiff Docks	Rhymney	-	Cardiff E Dock	8/55
40	12/21	Cardiff Docks	Radyr	-	Cardiff E Dock	10/53
41	12/21	Cardiff Docks	Radyr	-	Cardiff E Dock	5/56
42	12/21	Cardiff Docks	Radyr	-	Cardiff E Dock	9/57
43	12/21	Cardiff Docks	Radyr	-	Cardiff E Dock	2/57
44	12/21	Cardiff Docks	Radyr	-	Cardiff E Dock	7/56
45	6/80	Croes Newydd	Croes Newydd	4/38	-	-
46	10/09	Cardiff Docks	Radyr	-	Radyr	7/50
47	6/04	Cardiff E M	Cardiff E Dock	-	Cardiff E Dock	4/49
48	6/04	Cardiff E M	Cardiff E Dock	5/47	-	-
49	7/04	Cardiff Docks	Cardiff E Dock	4/38	-	-
50	8/04	Cardiff Docks	Radyr	1/35	-	-
51	8/04	Cardiff Docks	Cardiff E Dock	-	Cardiff E Dock	10/48
52	5/10	Senghenydd	Cardiff E Dock	-	Cardiff E Dock	11/49
53	5/10	Senghenydd	Cardiff E Dock	-	Cardiff E Dock	6/49
54	5/10	Merthyr	Cardiff E Dock	-	Cardiff E Dock	4/48
55	5/10	Cardiff Docks	Cardiff E Dock	-	Cardiff E Dock	2/53
56	5/10	Merthyr	Radyr	-	Cardiff E Dock	9/53
57	4/11	Cardiff Docks	Barry	-	Radyr	4/52
58	5/11	Merthyr	Barry	-	Cardiff E Dock	9/54
59	6/11	Cardiff Docks	Barry	-	Cardiff E Dock	7/55
60	7/11	Cardiff Docks	Danygraig	-	Danygraig	3/51
61	8/11	Cardiff Docks	Cardiff E Dock	-	Cardiff E Dock	2/50
62	9/11	Merthyr	Pill	-	Pill	12/48
63	6/14	Cardiff Docks	Radyr	-	Cardiff E Dock	9/52
64	7/14	Cardiff E M	Radyr	-	Radyr	5/50
65	8/14	Cardiff Docks	Aberdare	-	Cardiff E Dock	2/54
66	3/16	Senghenydd	Cardiff E Dock	-	Cardiff E Dock	7/55
67	7/16	Merthyr	Rhymney	-	Cardiff E Dock	9/52
68	9/18	Cardiff Docks	Cardiff E Dock	-	Cardiff E Dock	6/54
69	9/18	Merthyr	Duffryn Yard	-	Duffryn Yard	7/55
70	9/18	Cardiff Docks	Duffryn Yard	-	Duffryn Yard	7/55
71	6/10	Cardiff Docks	Danygraig	-	Danygraig	12/48
72	7/10	Merthyr	Cardiff E Dock	-	Cardiff E Dock	2/52
73	7/10	Senghenydd	Cardiff E Dock	-	Radyr	7/52
74	7/10	Cardiff Docks	Cardiff E Dock	-	Cardiff E Dock	1/51
75	7/10	Cardiff Docks	Neath	-	Neath	10/53
76	7/09	Cardiff Docks	Rhymney	-	Rhymney	11/50
77	4/17	Cardiff Docks	Rhymney	-	Rhymney	11/53
78	8/21	Rhymney	Barry	-	Cardiff E Dock	8/55
79	8/21	Cardiff E M	Rhymney	-	Radyr	7/55
80	11/21	Cardiff Docks	Rhymney	-	Radyr	2/54
81	11/21	Cardiff E M	Rhymney	-	Rhymney	5/54
82	7/09	Cardiff Docks	Rhymney	-	Cardiff E Dock	5/54
83	8/09	Cardiff Docks	Cae Harris	-	Cardiff E Dock	5/55
84	-/94	Cardiff Docks	Cae Harris	4/28	-	-
85	-/94	Cae Harris	Cardiff Docks	12/27	-	-
86	-/94	Merthyr	Cardiff Docks	10/26	-	-
87	-/90	Radyr	Senghenydd	3/27	-	-
88	-/90	Rhymney	Cardiff Docks	10/26	-	-
89	-/90	Cardiff Docks	Rhymney	8/26	-	-
90	-/90	Rhymney	Cae Harris	6/28	-	-
91	-/90	Cardiff Docks	Cardiff Docks	1/27	-	-
92	4/57	Croes Newydd	Wellington	7/42	-	-
93	10/60	Hereford	Hereford	2/31	-	-
94	10/60	Bristol	Swindon	7/32	-	-
95	11/57	Bristol	Yatton	11/24	-	-
96	9/56	Wolverhampton	Croes Newydd	11/35	-	-
97	-/91	Rhymney	Dowlais	5/32	-	-
98	-/91	Radyr	Cardiff Docks	10/26	-	-

No.	Date	1922 Shed	Last GWR Shed	Wdn GWR	Last BR shed	Wdn BR
99	-/91	Rhymney	Cardiff Docks	6/28	-	-
100	-/91	Rhymney	Senghenydd	10/26	-	-
101	-/91	Rhymney	Cardiff Docks	10/26	-	-
102	10/03	Oxford	Oxford	10/26	-	-
103	6/05	Oxford	Oxford	3/27	-	-
104	6/05	Oxford	Oxford	9/28	-	-
105	-/91	Cardiff Docks	Cardiff Docks	5/25	-	-
106	-/94	Cardiff Docks	Cardiff Docks	10/26	-	-
107	-/94	Rhymney	Senghenydd	12/27	-	-
108	-/94	Senghenydd	Cardiff Docks	10/26	-	-
109	-/94	Cardiff Docks	Cardiff Docks	10/26	-	-
110	-/97	Merthyr	Cardiff Docks	5/25	-	-
111	2/08	Paddington	Paddington	1/24	-	-
112	-/97	Cardiff Docks	Cardiff Docks	7/27	-	-
113	-/97	Merthyr	Cardiff Docks	10/26	-	-
114	-/97	Senghenydd	Senghenydd	5/28	-	-
115	-/97	Rhymney	Senghenydd	6/28	-	-
116	2/72	Hereford	Hereford	11/27	-	-
117	-/97	Cardiff E M	Rhymney	9/28	-	-
118	-/97	Senghenydd	Cardiff Docks	12/27	-	-
119	-/97	Cardiff E M	Cae Harris	6/28	-	-
120	12/61	Bridgend	Oswestry	9/33	-	-
121	12/61	Tondu	Tondu	8/28	-	-
122	-/97	Rhymney	Cae Harris	5/32	-	-
123	12/61	Newport	Dock Street	11/27	-	-
124	12/61	Paddington	Paddington	5/27	-	-
125	5/62	Bridgend	Tondu	1/26	-	-
126	5/62	Bridgend	Tondu	7/28	-	-
127	-/97	Merthyr	Rhymney	4/28	-	-
128	6/62	Newport	Ebbw Junction	1/27	-	-
129	-/99	Rhymney	Cardiff Docks	10/26	-	-
130	-/99	Merthyr	Cardiff Docks	10/26	-	-
131	-/99	Rhymney	Rhymney	12/27	-	-
132	9/62	Stourbridge	Stourbridge	8/24	-	-
133	-/99	Cardiff Docks	Rhymney	10/26	-	-
134	-/99	Rhymney	Cardiff Docks	5/26	-	-
135	-/99	Merthyr	Cardiff Docks	12/27	-	-
136	-/00	Senghenydd	Cathays	6/34	-	-
137	-/00	Cardiff Docks	Cardiff Docks	12/27	-	-
138	-/00	Senghenydd	Cae Harris	7/34	-	-
139	-/00	Cardiff Docks	Cae Harris	6/34	-	-
140	-/00	Cardiff E M	Rhymney	8/26	-	-
141	-/00	Senghenydd	Abercynnon	7/34	-	-
142	-/00	Rhymney	Rhymney	10/26	-	-
143	-/00	Senghenydd	Rhymney	6/31	-	-
144	-/00	Cardiff Docks	Radyr	9/28	-	-
145	-/00	Senghenydd	Cae Harris	4/31	-	-
146	7/62	Stourbridge	Shrewsbury	4/25	-	-
147	-/00	Senghenydd	Rhymney	12/27	-	-
148	-/00	Cardiff Docks	Rhymney	12/27	-	-
149*	-/90	WORKS	-	10/22	-	-
150*	-/90	Cardiff Docks	Barry	3/23	-	-
151	3/19	Cardiff E M	Barry	12/30	-	-
152	3/19	Cardiff E M	Barry	12/36	-	-
153	12/08	Cardiff E M	Barry	12/30	-	-
154	12/08	Cardiff E M	Barry	6/34	-	-
155	12/08	Cardiff E M	Cardiff E Dock	-	Cardiff E Dock	9/53
156	11/94	Cardiff E M	Cathays	2/31	-	-
157	12/98	Cardiff E M	Senghenydd	10/28	-	-
158	12/98	Cardiff E M	Radyr	5/32	-	-

No.	Date	1922 Shed	Last GWR Shed	Wdn GWR	Last BR shed	Wdn BR
159	12/87	Cardiff E M	Cathays	2/31	-	-
160	5/05	Cardiff E M	Pill	4/30	-	-
161	5/05	Cardiff E M	Cardiff Docks	10/29	-	-
162	11/87	Cardiff E M	Danygraig	7/32	-	-
163	12/85	Cardiff E M	Ferndale	7/32	-	-
164	2/02	Danygraig	Duffryn Yard	5/36	-	-
165	2/02	Danygraig	Danygraig	12/27	-	-
166	4/04	Danygraig	Danygraig	11/32	-	-
167	4/04	Danygraig	Danygraig	9/36	-	-
168	11/89	Danygraig	Danygraig	4/32	-	-
169	11/89	Danygraig	Danygraig	12/34	-	-
170	6/90	Danygraig	Danygraig	6/28	-	-
171	12/94	Danygraig	Duffryn Yard	2/36	-	-
172	12/94	Danygraig	Duffryn Yard	1/35	-	-
173	12/94	Danygraig	Duffryn Yard	9/36	-	-
174	1/95	Danygraig	Danygraig	1/29	-	-
175	2/95	Danygraig	Danygraig	9/28	-	-
176	2/95	Danygraig	Danygraig	10/26	-	-
177	10/99	Landore	Danygraig	6/35	-	-
178	10/99	Landore	Danygraig	6/36	-	-
179	10/99	Danygraig	Duffryn Yard	6/36	-	-
180	6/97	Danygraig	Duffryn Yard	10/28	-	-
181	3/86	Danygraig	Danygraig	1/29	-	-
182	2/98	Danygraig	Duffryn Yard	10/29	-	-
183	5/98	Duffryn Yard	Duffryn Yard	3/29	-	-
184	9/98	Duffryn Yard	Duffryn Yard	-	Duffryn Yard	10/48
185	11/98	Duffryn Yard	Duffryn Yard	8/29	-	-
186	11/98	Duffryn Yard	Duffryn Yard	12/34	-	-
187	12/98	Duffryn Yard	Duffryn Yard	2/35	-	-
188	12/98	Duffryn Yard	Duffryn Yard	9/47	-	-
189	2/99	Duffryn Yard	Duffryn Yard	1/30	-	-
190	9/08	Pill	Pill	-	Pill	4/48
191	10/08	Pill	Pill	11/34	-	-
192	11/08	Pill	Pill	10/46	-	-
193	11/99	Barry	Barry	2/29	-	-
194	11/99	Barry	Barry	7/30	-	-
195	12/99	Barry	Barry	4/27	-	-
196	12/99	Barry	Barry	5/32	-	-
197	12/99	Barry	Barry	6/28	-	-
198	12/88	Llantrisant	Radyr	-	Radyr	1/48
199	12/88	Barry	Barry	8/32	-	-
200	12/88	Llantrisant	Barry	11/45	-	-
201*	1/89	WORKS	-	10/22	-	-
202	4/76	Chipping N	Worcester	9/28	-	-
203	1/89	Barry	Radyr	7/32	-	-
204	2/98	Barry	Cardiff	5/32	-	-
205	5/76	Newquay	Abercynon	1/34	-	-
206	3/89	Trehafod	Radyr	7/34	-	-
207	3/89	Llantrisant	Radyr	12/45	-	-
208	3/89	Barry	Cathays	7/32	-	-
209	3/89	Llantrisant	Senghenydd	5/31	-	-
210	4/89	Llantrisant	Barry	7/38	-	-
211	4/89	Trehafod	Radyr	8/32	-	-
212	5/89	Barry	Cardiff E Dock	-	Cardiff E Dock	7/48
213	5/89	Llantrisant	Barry	-	Barry	1/49
214	5/89	Llantrisant	Cathays	10/47	-	-
215	6/76	Frome	Goodwick	1/34	-	-
216	6/76	Neath	Reading	12/36	-	-
217	6/76	Newton Abbot	Swindon	2/34	-	-
218	7/76	Cirencester	Westbury	9/30	-	-

No.	Date	1922 Shed	Last GWR Shed	Wdn GWR	Last BR shed	Wdn BR
219	7/76	Aberayron	Weymouth	8/34	-	-
220	7/76	Bristol	Wallingford	12/29	-	-
221	8/76	Carmarthen	Machynlleth	9/29	-	-
222	8/76	Bullo Pill	Staines	11/28	-	-
223	12/89	Barry	Cardiff	8/32	-	-
224	1/90	Trehafod	Barry	7/47	-	-
225	1/90	Barry	Radyr	7/32	-	-
226	1/90	Trehafod	Radyr	11/37	-	-
227	1/90	Trehafod	Radyr	7/34	-	-
228	1/90	Barry	Barry	5/32	-	-
229	1/90	Trehafod	Barry	11/34	-	-
230	2/90	Trehafod	Radyr	11/36	-	-
231	2/90	Barry	Barry	-	Barry	10/49
232	2/90	Barry	Cardiff	8/32	-	-
233	6/90	Barry	Barry	8/32	-	-
234	6/90	Barry	Barry	6/36	-	-
235	6/90	Barry	Barry	6/37	-	-
236	6/08	Aberdare TV	Abercynon	-	Cardiff E Dock	1/53
238	6/90	Barry	Cardiff	-	Cardiff	6/48
240	7/90	Barry	Radyr	-	Barry	3/51
241	7/90	Barry	Cardiff	8/32	-	-
242	7/90	Barry	Swindon	11/47	-	-
243	7/90	Trehafod	Barry	12/37	-	-
244	7/90	Llantrisant	Barry	6/39	-	-
245	3/92	Barry	Cardiff	5/32	-	-
246	3/92	Trehafod	Radyr	-	Radyr	1/49
247	3/92	Barry	Llantrisant	8/47	-	-
248	3/92	Barry	Barry	-	Barry	7/48
249	4/92	Barry	Radyr	7/32	-	-
250	4/92	Barry	Radyr	12/47	-	-
251	4/92	Barry	Barry	8/32	-	-
252	4/92	Barry	Barry	12/38	-	-
253	5/92	Barry	Barry	12/36	-	-
254	6/92	Barry	Barry	1/39	-	-
255	12/94	Barry	Cardiff	5/32	-	-
256	12/94	Barry	Barry	12/38	-	-
257	12/94	Barry	Barry	8/45	-	-
258	12/94	Barry	Radyr	-	Barry	11/49
259	12/94	Barry	Cardiff E Dock	-	Cardiff E Dock	8/48
260	12/94	Barry	Barry	11/34	-	-
261	5/00	Barry	Barry	-	Barry	8/48
262	5/00	Barry	Barry	-	Barry	3/50
263	6/00	Barry	Barry	-	Barry	3/51
264	6/00	Barry	Barry	12/34	-	-
265	6/00	Barry	Barry	-	Barry	11/49
266	6/00	Barry	Barry	1/47	-	-
267	6/00	Barry	Barry	-	Barry	3/51
268	6/00	Barry	Barry	-	Barry	4/48
269	6/00	Barry	Cardiff E Dock	-	Cardiff E Dock	10/49
270	6/00	Barry	Barry	-	Barry	4/51
271	6/00	Barry	Barry	-	Barry	4/51
272	6/00	Barry	Barry	-	Barry	3/50
273	3/00	Barry	Barry	11/45	-	-
274	3/00	Barry	Barry	-	Barry	4/51
275	4/00	Barry	Barry	-	Barry	4/48
276	4/00	Trehafod	Barry	-	Barry	4/51
277	4/00	Barry	Barry	-	Barry	4/49
278	7/10	Cathays	Ferndale	-	Ferndale	5/51
279	7/10	Treherbert	Treherbert	-	Cardiff E Dock	5/54
280	7/08	Cathays	Cardiff	-	Cardiff	5/49

No.	Date	1922 Shed	Last GWR Shed	Wdn GWR	Last BR shed	Wdn BR
281	8/10	Cardiff E M	Abercynon	-	Abercynon	8/50
282	7/08	Aberdare TV	Aberdare	-	Cardiff E Dock	6/54
283	8/10	Penarth Dock	Ferndale	-	Barry	3/49
284	8/10	Cathays	Aberdare	-	Aberdare	11/52
285	4/07	Cathays	Cardiff E Dock	-	Barry	8/53
286	6/08	Cathays	Barry	-	Barry	10/50
287	4/07	Coke Ovens	Abercynon	-	Abercynon	12/49
288	9/10	Coke Ovens	Abercynon	-	Abercynon	3/50
289	7/08	Cardiff E M	Swansea E Dock	-	Swansea E Dock	8/49
290	10/10	Cathays	Ferndale	-	Cardiff E Dock	7/55
291	7/08	Cardiff E M	Duffryn Yard	-	Duffryn Yard	1/50
292	7/08	Coke Ovens	Dowlais	-	Dowlais	4/52
293	6/07	Cathays	Radyr	-	Radyr	7/51
294	7/07	Cardiff E M	Duffryn Yard	-	Duffryn Yard	7/50
295	7/07	Cardiff E M	Abercynon	-	Abercynon	6/51
296	9/07	Penarth Dock	Duffryn Yard	-	Duffryn Yard	9/49
297	7/08	Cathays	Cardiff E Dock	-	Treherbert	12/49
298	7/08	Aberdare TV	Ferndale	-	Ferndale	7/48
299	7/08	Cathays	Ferndale	-	Treherbert	6/51
300+	7/08	Cathays	Cardiff	-	Cardiff	7/48
301+	6/08	Cathays	Cathays	-	Cathays	12/48
302+	7/08	Abercynon	Treherbert	-	Treherbert	4/48
303	10/64	Pontypool	Aberayron	7/32	-	-
304	10/64	Pontypool	Bridgend	5/28	-	-
305	11/64	Severn Tunnel	Ebbw Junction	4/31	-	-
306	11/64	Newport	Oswestry	7/32	-	-
307	11/64	Milford Haven	Didcot	5/31	-	-
308	12/64	Llanelly	Llanelly	11/29	-	-
309	2/65	Bristol	Yeovil	4/31	-	-
310+	9/10	Aberdare TV	Cardiff	-	Cardiff	1/52
311+	9/10	Cardiff E M	Aberdare	-	Cardiff E Dock	7/55
312	1/65	Crewe	Crewe	2/23	-	-
313+	9/10	Cathays	Cardiff	-	Cardiff	7/54
314+	8/08	Cardiff E M	Aberdare	-	Aberdare	12/49
315+	10/10	Cardiff E M	Ferndale	-	Treherbert	11/52
316	8/90	Stourbridge	Stourbridge	11/27	-	-
317+	10/10	Aberdare TV	Cardiff	-	Cardiff E Dock	7/55
318+	10/10	Treherbert	Cardiff	-	Cardiff	8/52
319+	10/10	Cardiff E M	Cardiff E Dock	-	Cardiff E Dock	3/55
320+	10/10	Cathays	Dowlais	-	Cardiff E Dock	5/55
321+	10/10	Cathays	Ferndale	-	Cardiff E Dock	7/55
322	10/64	Tyseley	Tyseley	10/30	-	-
323	4/66	Tyseley	Stourbridge	7/32	-	-
324+	10/10	Penarth Dock	Ferndale	-	Cardiff E Dock	1/55
325	11/64	Stourbridge	Stourbridge	3/30	-	-
326	10/64	Wolverhampton	Stourbridge	10/28	-	-
327	11/64	Stourbridge	Stourbridge	3/30	-	-
328	10/64	Shrewsbury	Stourbridge	4/29	-	-
329	10/64	Pontypool	Pontypool	9/26	-	-
330	10/64	Shrewsbury	Stourbridge	9/29	-	-
331	10/64	Stourbridge	Stourbridge	9/23	-	-
332	11/64	Shrewsbury	Shrewsbury	8/22	-	-
333+	10/10	Cathays	Merthyr	-	Treherbert	9/52
334	11/64	Stourbridge	Oxley	1/28	-	-
335	11/14	Treherbert	Cardiff	-	Cardiff E Dock	3/54
336	10/64	Tyseley	Stourbridge	11/29	-	-
337	12/14	Cathays	Abercynon	-	Abercynon	2/53
338	12/64	Stourbridge	Shrewsbury	8/24	-	-
339	12/64	Wolverhampton	Wolverhampton	2/23	-	-
340	12/64	Stourbridge	Shrewsbury	10/25	-	-

No.	Date	1922 Shed	Last GWR Shed	Wdn GWR	Last BR shed	Wdn BR
341	12/64	Wolverhampton	Stourbridge	10/24	-	-
342	10/56	Croes Newydd	Chester	8/31	-	-
343	12/14	Penarth Dock	Cathays	-	Cathays	10/55
344	12/14	Penarth Dock	Cathays	-	Cathays	11/52
345	5/20	Cathays	Barry	-	Cathays	9/55
346	8/15	Coke Ovens	Cathays	-	Cathays	10/53
347	12/21	Cathays	Cathays	-	Cathays	12/56
348	12/21	Ferndale	Cathays	-	Cathays	5/56
349	12/21	Coke Ovens	Pontypool	-	Abercynon	3/57
350	2/66	Shrewsbury	Shrewsbury	8/26	-	-
351	12/21	Treherbert	Abercynon	-	Abercynon	7/56
352	8/15	Abercynon	Treherbert	-	Cardiff E Dock	3/55
353	3/66	Leamington	Shrewsbury	3/28	-	-
354	3/66	Birkenhead	Leamington	8/34	-	-
355	3/66	Merthyr	Shrewsbury	4/31	-	-
356	8/15	Merthyr TV	Abercynon	-	Abercynon	1/55
357	8/15	Treherbert	Cardiff	-	Barry	1/56
358	4/66	Stourbridge	Shrewsbury	5/30	-	-
359	10/64	Shrewsbury	Shrewsbury	3/23	-	-
360	7/16	Penarth Dock	Cathays	-	Cathays	2/55
361	7/16	Abercynon	Barry	-	Cathays	1/57
362	7/16	Coke Ovens	Aberdare	-	Cathays	5/56
363	5/66	Crewe	Swindon	3/33	-	-
364	7/16	Merthyr TV	Cathays	-	Cathays	3/57
365	7/16	Cathays	Treherbert	-	Abercynon	10/55
366	8/16	Cathays	Treherbert	-	Treherbert	10/55
367	2/19	Coke Ovens	Cathays	-	Cathays	3/56
368	5/20	Treherbert	Treherbert	-	Abercynon	10/56
369	6/66	Didcot	Didcot	8/25	-	-
370	2/19	Cathays	Rhymney	-	Abercynon	8/57
371	3/19	Penarth Dock	Cathays	-	Cardiff E Dock	1/55
372	3/19	Cathays	Barry	-	Barry	2/55
373	4/19	Treherbert	Treherbert	-	Abercynon	8/57
374	4/19	Aberdare TV	Aberdare	-	Cardiff E Dock	9/55
375	5/20	Cathays	Rhymney	-	Barry	11/54
376	5/20	Cathays	Cathays	-	Cathays	1/57
377	5/19	Cathays	Barry	-	Barry	5/56
378	5/19	Coke Ovens	Treherbert	-	Cathays	12/56
379	5/19	Ferndale	Barry	-	Abercynon	2/56
380	7/19	Cathays	Abercynon	-	Abercynon	10/56
381	7/19	Cathays	Cardiff	-	Abercynon	8/57
382	7/19	Cathays	Barry	-	Barry	4/56
383	5/20	Cathays	Cathays	-	Abercynon	8/57
384	6/20	Treherbert	Cathays	-	Cathays	2/56
385	6/20	Merthry TV	Pontypool	-	Cathays	4/57
386	6/20	Penarth Dock	Barry	-	Abercynon	4/56
387	7/20	Penarth Dock	Barry	-	Cathays	11/56
388	8/20	Penarth Dock	Barry	-	Cathays	10/56
389	9/20	Cathays	Barry	-	Cardiff E Dock	1/56
390	9/20	Merthyr TV	Cathays	-	Abercynon	8/57
391	10/20	Cathays	Cathays	-	Cathays	9/56
392	12/66	Hereford	Gloucester	10/28	-	-
393	11/20	Cathays	Cathays	-	Abercynon	5/57
394	12/20	Cathays	Barry	-	Barry	2/56
395	12/66	Pontypool	Pontypool	11/23	-	-
396	12/66	Stourbridge	Shrewsbury	8/28	-	-
397	12/20	Aberdare TV	Cardiff E Dock	-	Abercynon	4/57
398	1/21	Cathays	Abercynon	-	Abercynon	8/57
399	1/21	Cathays	Treherbert	-	Abercynon	9/56
400	1/67	Birkenhead	Leamington	4/29	-	-

No.	Date	1922 Shed	Last GWR Shed	Wdn GWR	Last BR shed	Wdn BR
401+	2/21	Cathays	Treherbert	-	Barry	5/56
402+	2/21	Aberdare TV	Abercynon	-	Abercynon	8/57
403+	3/21	Cathays	Cathays	-	Cathays	4/57
404+	12/21	Penarth Dock	Barry	-	Barry	4/56
405	2/67	Leamington	Stafford Road	1/26	-	-
406+	12/21	Cathays	Cathays	-	Cathays	3/56
407	2/67	Whitland	Didcot	7/27	-	-
408+	12/21	Cathays	Swansea E Dock	-	Cardiff E Dock	12/55
409+	10/10	Cathays	Ferndale	-	Treherbert	12/52
410	1/04	Cathays	Cardiff	-	Cardiff	3/48
411	2/04	Cardiff E M	Cardiff	-	Cardiff	2/48
412	6/99	Ferndale	Coke Ovens	11/27	-	-
413	5/99	Coke Ovens	Coke Ovens	11/28	-	-
414+	8/10	Aberdare TV	Abercynon	-	Abercynon	6/51
415	6/99	Coke Ovens	Coke Ovens	8/28	-	-
416	2/02	Ferndale	Ferndale	7/32	-	-
417	1/05	Cathays	Aberdare	5/28	-	-
418	2/02	Ferndale	Coke Ovens	6/31	-	-
419	6/99	Cathays	Coke Ovens	8/28	-	-
420+	8/07	Cathays	Cardiff	-	Cardiff	4/51
421	5/99	Coke Ovens	Coke Ovens	4/28	-	-
422	7/67	Hereford	Worcester	1/29	-	-
423	5/99	Ferndale	Ferndale	8/26	-	-
424	5/99	Coke Ovens	Radyr	10/26	-	-
425	5/99	Coke Ovens	Radyr	8/26	-	-
426	5/99	Coke Ovens	Coke Ovens	1/27	-	-
427	3/02	Ferndale	Coke Ovens	6/31	-	-
428	6/02	Cathays	Ferndale	5/31	-	-
429	6/02	Cathays	Radyr	6/31	-	-
430	1/05	Cardiff E M	Ferndale	7/34	-	-
431	5/02	Cathays	Radyr	5/31	-	-
432	2/04	Cathays	Treherbert	11/28	-	-
433	1/05	Ferndale	Treherbert	9/29	-	-
434	2/05	Radyr	Ferndale	6/31	-	-
435	2/05	Cathays	Aberdare	5/30	-	-
436	3/68	Barnstaple	Exeter	2/30	-	-
437	2/05	Ferndale	Coke Ovens	4/31	-	-
438+	10/14	Merthyr TV	Swansea E Dock	-	Swansea E Dock	2/53
439+	8/15	Abercynon	Barry	-	Cathays	6/56
440+	8/15	Cathays	Barry	-	Abercynon	7/56
441+	1/15	Ferndale	Barry	-	Abercynon	10/54
442	12/91	Cathays	Cathays	6/25	-	-
443	1/92	Cathays	Cathays	8/26	-	-
444	9/88	Cathays	Coke Ovens	10/25	-	-
445	9/88	Penarth Dock	Radyr	10/26	-	-
446	9/94	Cathays	Cathays	8/29	-	-
447	9/94	Abercynon	Cathays	6/25	-	-
448	11/94	Cowbridge	Cathays	2/30	-	-
449	10/94	Treherbert	Cathays	9/26	-	-
450	5/97	Abercynon	Cathays	10/26	-	-
451	10/94	Cowbridge	Aberdare TV	6/25	-	-
452	11/94	Cathays	Ferndale	6/25	-	-
453	1/95	Llantrisant	Coke Ovens	8/26	-	-
454	10/94	Treherbert	Treherbert	10/26	-	-
455	11/94	Cathays	Cathays	10/26	-	-
456	1/69	Taunton	Pontypool	1/31	-	-
457	1/69	Ross-on-Wye	Oxford	11/34	-	-
458	2/69	Gloucester	Southall	4/33	-	-
459	2/69	Aberdare	Merthyr	10/33	-	-
460	2/69	Bristol	Bath Road	5/32	-	-

No.	Date	1922 Shed	Last GWR Shed	Wdn GWR	Last BR shed	Wdn BR
461	2/69	Taunton	Taunton	12/30	-	-
462	3/88	Cathays	Cathays	6/25	-	-
463	3/69	Tondu	Worcester	6/29	-	-
464	3/69	Cheltenham	Worcester	12/34	-	-
465	3/69	Wolverhampton	Bala	2/31	-	-
466	3/92	Cathays	Radyr	6/28	-	-
467	4/69	Bala	Bala	12/30	-	-
468	4/69	Kidderminster	Truro	2/32	-	-
469	4/69	Tondu	Henley	9/28	-	-
470	4/69	Bristol	Newton Abbot	2/34	-	-
471	7/97	Cowbridge	Cathays	4/27	-	-
472	7/97	Abercynon	Radyr	6/25	-	-
473	9/97	Abercynon	Ferndale	3/28	-	-
474	10/97	Treherbert	Cathays	4/27	-	-
475	10/97	Ferndale	Cathays	2/30	-	-
476	11/94	Treherbert	Treherbert	12/25	-	-
477	11/94	Cathays	Ferndale	6/31	-	-
478	12/87	Cathays	Radyr	6/25	-	-
479	11/94	Llantrisant	Radyr	12/25	-	-
480	12/94	Cathays	Ferndale	10/28	-	-
481	4/91	Merther TV	Radyr	4/28	-	-
482	4/91	Abercynon	Radyr	11/27	-	-
483	4/91	Aberdare TV	Radyr	11/27	-	-
484	4/91	Radyr	Coke Ovens	12/32	-	-
485	7/91	Ferndale	Radyr	11/28	-	-
486	7/91	Treherbert	Treherbert	11/34	-	-
487	1/85	Coke Ovens	Cardiff	12/34	-	-
488*	4/91	Treherbert	Cathays	7/25	-	-
489	4/91	Radyr	Radyr	1/25	-	-
490	4/91	Cathays	Radyr	12/25	-	-
491	4/91	Penarth Dock	Cathays	11/28	-	-
492	4/91	Coke Ovens	Ferndale	7/27	-	-
493	4/91	Coke Ovens	Cathays	11/28	-	-
494	6/91	Aberdare TV	Radyr	10/28	-	-
495	6/91	Coke Ovens	Radyr	9/28	-	-
496	6/91	Aberdare TV	Radyr	9/28	-	-
497	7/70	Chester	Oxley	11/25	-	-
498	7/91	Treherbert	Cathays	12/29	-	-
499	7/91	Merthyr TV	Radyr	6/30	-	-
500	7/91	Abercynon	Treherbert	10/28	-	-
501	7/91	Radyr	Radyr	6/28	-	-
502	7/91	Ferndale	Cathays	8/30	-	-
503	9/88	Aberdare TV	Abercynon	10/25	-	-
504	8/70	Taunton	Taunton	7/22	-	-
505	4/86	Cathays	Aberdare TV	6/25	-	-
506	5/86	Abercynon	Cathays	12/34	-	-
507	3/88	Treherbert	Coke Ovens	12/26	-	-
508	3/88	Treherbert	Radyr	6/25	-	-
509	9/70	Worcester	Hereford	5/30	-	-
510	9/70	Wolverhampton	Oxley	10/29	-	-
511	12/87	Abercynon	Coke Ovens	5/32	-	-
512	9/70	Chester	Oxley	3/31	-	-
513	1/85	Ferndale	Coke Ovens	11/27	-	-
514	9/70	Didcot	Didcot	8/24	-	-
515	1/85	Coke Ovens	Radyr	7/25	-	-
516	1/85	Abercynon	Cathays	5/25	-	-
517	4/68	Shrewsbury	Much Wenlock	2/34	-	-
518	5/68	Southall	Frome	2/28	-	-
519	5/68	Cirencester	Malmesbury	4/33	-	-
520	1/85	Radyr	Radyr	11/27	-	-

No.	Date	1922 Shed	Last GWR Shed	Wdn GWR	Last BR shed	Wdn BR
521	6/68	Shrewsbury	Malmesbury	4/30	-	-
522	6/68	Swindon	Builth Wells	10/35	-	-
523	7/68	Tyseley	Oswestry	7/34	-	-
524	7/68	Banbury	Oswestry	10/30	-	-
525	7/68	Stourbridge	Chippenham	1/34	-	-
526	8/68	Trowbridge	Reading	9/33	-	-
527	8/68	Bristol	Bath Road	9/28	-	-
528	9/68	Bristol	Weymouth	10/35	-	-
529	9/68	Swindon	Swindon	9/30	-	-
530	10/68	Truro	Newton Abbot	5/32	-	-
531	10/68	Swindon	Staines	10/29	-	-
532	10/68	Shrewsbury	Malmesbury	12/29	-	-
533	11/68	Kidderminster	Kington	1/34	-	-
534	11/68	Worcester	Weymouth	7/32	-	-
535	12/68	Swindon	Yatton	3/29	-	-
536	12/68	Ross-on-Wye	Carmarthen	9/29	-	-
537	12/68	Aberdare	Leamington	10/32	-	-
538	1/69	Llanelly	Carmarthen	12/34	-	-
539	1/69	Bristol	Swindon	5/33	-	-
540	1/69	Llanfyllin	Bristol SPM	5/33	-	-
541	2/69	Oxford	Bath Road	10/29	-	-
542	2/69	Bristol	Wallingford	4/34	-	-
543	3/69	Bristol	Bath Road	3/29	-	-
544	3/69	Tyseley	Croes Newydd	10/28	-	-
545	4/69	Taunton	Swindon	9/28	-	-
546	4/69	Weymouth	Bullo Pill	9/28	-	-
547	4/69	Banbury	Westbury	5/30	-	-
548	5/69	Bristol	Abercynon	10/34	-	-
549	5/69	Weymouth	Swindon	8/29	-	-
550	6/69	Leominster	Worcester	1/29	-	-
551	6/69	Bristol	Swindon	8/29	-	-
552	1/85	Cathays	Treherbert	8/30	-	-
553	7/69	Didcot	Gloucester	6/32	-	-
554	7/69	Blagdon	Southall	5/33	-	-
555	8/69	Ludlow	Wellington	7/30	-	-
556	8/69	Goodwick	Weymouth	9/33	-	-
557	8/69	Much Wenlock	Ebbw Junction	5/33	-	-
558	9/69	Southall	Neath	9/34	-	-
559	9/69	Exeter	Carmarthen	2/34	-	-
560	1/85	Cowbridge	Cathays	6/25	-	-
561	10/69	Weymouth	Westbury	2/29	-	-
562	10/69	Gloucester	Kidderminster	4/29	-	-
563	11/69	Trowbridge	Merthyr	6/29	-	-
564	11/69	Tyseley	Leamington	11/33	-	-
565	11/69	Chard	Leamington	12/30	-	-
566	12/69	Pontrilas	Ebbw Junction	5/34	-	-
567	2/85	Aberdare TV	Cathays	7/32	-	-
568	12/69	Bristol	Severn Tunnel	3/32	-	-
569	1/70	Banbury	Alcester	5/33	-	-
570	1/70	Wolverhampton	Reading	10/34	-	-
571	3/70	Brixham	Swindon	2/36	-	-
572	3/70	Trowbridge	Worcester	9/28	-	-
573	3/86	Treherbert	Cathays	12/34	-	-
574	4/70	Aberdare	Reading	2/34	-	-
575	4/70	Newport	Southall	12/31	-	-
576	5/70	Shrewsbury	Bath	7/31	-	-
577	7/86	Abercynon	Cathays	1/34	-	-
578	6/87	Cathays	Aberdare TV	2/27	-	-
579	6/87	Abercynon	Radyr	10/26	-	-
580	6/87	Cowbridge	Cathays	5/25	-	-

No.	Date	1922 Shed	Last GWR Shed	Wdn GWR	Last BR shed	Wdn BR
581	5/95	Treherbert	Radyr	8/30	-	-
582	1/85	Cathays	Radyr	7/30	-	-
583	1/85	Radyr	Coke Ovens	8/33	-	-
584	3/86	Treherbert	Cardiff	12/34	-	-
585	3/86	Cathays	Coke Ovens	12/26	-	-
586	7/86	Cathays	Cathays	6/25	-	-
587	10/95	Cathays	Abercynon	5/31	-	-
588	3/96	Merthyr TV	Coke Ovens	3/28	-	-
589	10/95	Cathays	Coke Ovens	12/30	-	-
590	10/95	Cathays	Cathays	9/28	-	-
591	11/95	Merthyr TV	Cathays	3/28	-	-
592	3/96	Cathays	Cathays	9/28	-	-
593	7/95	Cathays	Treherbert	8/30	-	-
594	2/71	Llanelly	Llanelly	7/27	-	-
595	7/95	Cathays	Abercynon	9/28	-	-
596	9/95	Cathays	Cathays	4/27	-	-
597	9/95	Cathays	Abercynon	9/28	-	-
598	4/96	Cathays	Cathays	9/28	-	-
599	4/96	Cathays	Cathays	11/27	-	-
600	5/96	Cathays	Treherbert	10/28	-	-
601	3/71	Chester	Banbury	9/29	-	-
602	3/96	Cathays	Treherbert	5/31	-	-
603	5/96	Cathays	Cathays	3/28	-	-
604+	6/20	Merthyr	Cardiff E Dock	-	Cardiff E Dock	3/54
605+	7/20	Merthyr	Cardiff E Dock	-	Cardiff E Dock	6/54
606+	7/20	Merthyr	Cardiff E Dock	-	Cardiff E Dock	6/54
607	6/71	Landore	Landore	9/28	-	-
608+	6/08	Cardiff Docks	Cardiff E Dock	-	Cardiff E Dock	5/54
609+	7/08	Cardiff Docks	Cardiff E Dock	-	Cardiff E Dock	5/54
610+	7/08	Cardiff Docks	Cardiff E Dock	-	Cardiff E Dock	9/53
611+	8/08	Cardiff Docks	Cardiff E Dock	-	Cardiff E Dock	4/54
612	-/84	Cardiff Docks	Cardiff Docks	12/27	-	-
613	7/71	Bristol	Westbury	2/31	-	-
614	-/84	Cardiff Docks	Cardiff Docks	12/27	-	-
615	8/71	Henley	Oxford	2/31	-	-
616	8/71	Oxford	Henley	2/31	-	-
617	8/71	Stourbridge	Llantrisant	5/34	-	-
618	-/84	Cardiff Docks	Cardiff Docks	6/25	-	-
619	-/84	Cardiff Docks	Cardiff Docks	5/25	-	-
620	8/71	Kidderminster	Stratford	11/28	-	-
621	9/71	Tondu	Taunton	4/31	-	-
622	-/84	Cardiff Docks	Cardiff Docks	6/25	-	-
623	9/71	Oxford	Aberdare	10/30	-	-
624	9/71	Wells	Pontypool	7/29	-	-
625	-/84	Cardiff Docks	Cardiff Docks	5/25	-	-
626	9/71	Yeovil	Oxford	12/32	-	-
627	10/71	Worcester	Stratford	10/28	-	-
628	10/71	Cheltenham	Worcester	3/30	-	-
629	-/84	Cardiff Docks	Cardiff Docks	6/25	-	-
630	10/71	Yatton	Yeovil	1/29	-	-
631	-/84	Cardiff Docks	Cardiff E M	10/26	-	-
632	10/71	Pontypool	Taunton	11/29	-	-
633	11/71	Paddington	Paddington	10/33	-	-
634	11/71	Paddington	Paddington	3/34	-	-
635	12/71	Paddington	Paddington	8/28	-	-
636	12/71	Neath	Goodwick	3/33	-	-
637	12/71	Danygraig	Goodwick	6/34	-	-
638	1/72	Frome	Frome	9/30	-	-
639	1/72	Pontypool	Goodwick	3/34	-	-
640	2/72	Swansea E Dock	Goodwick	5/34	-	-

No.	Date	1922 Shed	Last GWR Shed	Wdn GWR	Last BR shed	Wdn BR
641	2/72	Paddington	Paddington	11/33	-	-
642	3/72	Paddington	Paddington	3/34	-	-
643	3/72	Paddington	Paddington	2/34	-	-
644	4/72	Carmarthen	Goodwick	2/32	-	-
645	4/72	Didcot	Didcot	10/32	-	-
646	5/72	Stourbridge	Birkenhead	7/34	-	-
647	5/72	Whitland	Pantyffynon	5/32	-	-
648	6/72	Wolverhampton	Stourbridge	7/32	-	-
649	6/72	Llanelly	Neath	7/32	-	-
650	6/72	Worcester	Gloucester	4/30	-	-
651	7/72	Paddington	Neath	5/32	-	-
652	7/72	Goodwick	Milford Haven	3/29	-	-
653	8/72	Whitland	Goodwick	5/33	-	-
654	8/72	Corwen	Wellington	10/28	-	-
655	2/92	Corwen	Wellington	3/29	-	-
656	9/72	Tyseley	Croes Newydd	2/33	-	-
657	-/84	Cardiff Docks	Cardiff Docks	7/27	-	-
658	11/71	Hereford	Worcester	7/27	-	-
659	-/74	Cardiff Docks	Cardiff Docks	6/25	-	-
660	-/74	Cardiff Docks	Cardiff Docks	2/25	-	-
661	9/07	Cardiff Docks	Cardiff Docks	2/25	-	-
662	9/07	Senghenydd	Cathays	6/25	-	-
663	-/68	Pill	Pill	10/26	-	-
664	3/84	Pill	Pill	6/30	-	-
665	2/84	Pill	Ebbw Junction	9/26	-	-
666	8/17	Pill	Pill	-	Pill	4/55
667	8/17	Pill	Pill	-	Pill	11/54
668	12/85	Pill	Ebbw Junction	2/29	-	-
669	11/94	Pill	Pill	7/25	-	-
670	11/94	Pill	Pill	10/26	-	-
671	10/89	Pill	Pill	8/37	-	-
672	11/89	Pill	Pill	6/24	-	-
673	1/72	Chester	Birkenhead	2/24	-	-
674	6/98	Pill	Pill	7/29	-	-
675	6/98	Pill	Pill	6/26	-	-
676	4/00	Pill	Pill	5/29	-	-
677	4/00	Pill	Pill	8/29	-	-
678	12/00	Pill	Ebbw Junction	3/25	-	-
679	1/90	Pill	Weymouth	9/29	-	-
680	-/86	Pill	Oswestry	-	Oswestry	12/48
681	2/20	Cardiff E M	Cardiff E Dock	-	Cardiff E Dock	2/55
682	2/20	Cardiff E M	Cardiff E Dock	-	Cardiff E Dock	10/53
683	3/20	Cardiff E M	Cardiff E Dock	-	Cardiff E Dock	12/54
684	3/20	Cardiff E M	Cardiff E Dock	-	Cardiff E Dock	5/54
685	4/19	Cardiff E M	Cardiff Docks	10/31	-	-
686	6/95	Cardiff E M	Cardiff E M	6/25	-	-
687	7/99	Cardiff E M	Cardiff E M	6/25	-	-
688	7/99	Cardiff E M	Cardiff Docks	12/26	-	-
689	5/72	Stourbridge	Wellington	3/33	-	-
690	7/95	Cardiff E M	Cardiff E M	10/26	-	-
691	3/89	Cardiff E M	Cardiff Docks	3/29	-	-
692	4/89	Cardiff E M	Cardiff Docks	3/29	-	-
693	2/82	Cardiff E M	Cardiff E M	2/25	-	-
694	-/72	Cardiff E M	Cardiff E M	10/26	-	-
695	-/73	Cardiff E M	Cardiff Docks	10/26	-	-
696	-/75	Cardiff E M	Cardiff E M	11/23	-	-
697	-/75	Cardiff E M	Cardiff E M	6/25	-	-
698*	*	Cardiff E M	-	5/22	-	-
699	11/88	Barry	Cathays	4/31	-	-
700	11/88	Barry	Barry	10/26	-	-

No.	Date	1922 Shed	Last GWR Shed	Wdn GWR	Last BR shed	Wdn BR
701	6/72	Croes Newydd	Chester	4/23	-	-
702	11/88	Barry	Cardiff	4/31	-	-
703	11/88	Barry	Barry	5/32	-	-
704	8/72	Taunton	Taunton	3/23	-	-
705	8/72	Wolverhampton	Stafford Road	11/29	-	-
706	11/88	Barry	Cardiff	11/30	-	-
707	9/72	Slough	Slough	9/22	-	-
708	3/00	Barry	Cathays	8/32	-	-
709	9/72	Wolverhampton	Shrewsbury	12/28	-	-
710	7/90	Barry	Cathays	5/32	-	-
711	8/90	Barry	Cathays	6/34	-	-
712	8/90	Barry	Cathays	5/32	-	-
713	3/00	Barry	Pill	7/36	-	-
714	6/92	Barry	Cathays	9/34	-	-
715	7/92	Barry	Barry	5/32	-	-
716	12/94	Barry	Cardiff	5/32	-	-
717	12/94	Barry	Cathays	11/34	-	-
718	1/95	Barry	Cardiff E Dock	12/34	-	-
719	3/00	Barry	Cathays	9/34	-	-
720	3/00	Barry	Pill	11/36	-	-
721	3/00	Barry	Cardiff E Dock	12/34	-	-
722	3/00	Barry	Duffryn Yard	11/35	-	-
723	3/00	Barry	Cathays	8/34	-	-
724	3/00	Barry	Cardiff Docks	5/33	-	-
725	2/05	Barry	Cathays	11/34	-	-
726	2/05	Barry	Cardiff Docks	8/32	-	-
727	10/72	Carmarthen	Carmarthen	12/29	-	-
728	10/72	Danygraig	Duffryn Yard	6/29	-	-
729	3/05	Llantrisant	Cardiff Docks	8/32	-	-
730	11/72	Newport	Tondu	9/30	-	-
731	11/72	Southall	Paddington	9/29	-	-
732	12/72	Pontypool	Tondu	7/32	-	-
733	12/72	Neyland	Duffryn Yard	10/34	-	-
734	12/72	Aberbeeg	Cardiff	3/31	-	-
735	1/73	Chippenham	Frome	5/30	-	-
736	1/73	Aberdare	Cardiff	8/32	-	-
737	1/73	Exeter	Exeter	7/32	-	-
738	1/73	Paddington	Landore	8/36	-	-
739	1/73	Westbury	Weymouth	4/30	-	-
740	1/73	Reading	Paddington	9/29	-	-
741	2/73	Cardiff	Merthyr	4/31	-	-
742	4/05	Trehafod	Barry	5/32	-	-
743	2/73	Aberbeeg	Pontypool	12/34	-	-
744	2/73	Neyland	Pontypool	5/32	-	-
745	2/73	Llanelly	Duffryn Yard	4/33	-	-
746	2/73	Gloucester	Cheltenham	9/27	-	-
747	4/05	Barry	Cardiff Docks	5/33	-	-
748	3/73	Paddington	Aberdare	8/32	-	-
749	3/73	Llantrisant	Neyland	6/28	-	-
750	3/73	Newport	Ebbw Junction	10/30	-	-
751	3/73	Weymouth	Westbury	2/31	-	-
752	4/73	Llanelly	Llanelly	8/34	-	-
753	4/73	Worcester	Ebbw Junction	2/35	-	-
754	2/05	Barry	Duffryn Yard	1/37	-	-
755	4/73	Milford Haven	Llanelly	5/29	-	-
756	4/73	Landore	Llanelly	12/28	-	-
757	10/72	Birkenhead	Birkenhead	8/33	-	-
758	11/72	Wolverhampton	Oxley	1/31	-	-
759	12/72	Wellington	Croes Newydd	9/35	-	-
760	12/72	Wolverhampton	Ross-on-Wye	5/32	-	-

No.	Date	1922 Shed	Last GWR Shed	Wdn GWR	Last BR shed	Wdn BR
761	12/72	Tyseley	Kidderminster	5/30	-	-
762	1/73	Garnant	Duffryn Yard	6/34	-	-
763	1/73	Aberdare	Slough	3/33	-	-
764	2/73	Tyseley	Banbury	3/30	-	-
765	2/73	Stourbridge	Merhtyr	8/32	-	-
766	2/73	Wolverhampton	Chester	9/35	-	-
767	2/92	Tyseley	Croes Newydd	10/34	-	-
768	3/73	Trawsfynydd	Oxley	9/35	-	-
769	4/73	Bristol	Aberdare	4/30	-	-
770	4/73	Carmarthen	Goodwick	10/36	-	-
771	4/73	Wolverhampton	Tyseley	7/26	-	-
772	5/73	Birkenhead	Croes Newydd	1/34	-	-
773	5/73	Croes Newydd	Tyseley	3/29	-	-
774	6/73	Croes Newydd	Chester	7/32	-	-
775	6/73	Tyseley	Chester	3/33	-	-
776	3/05	Barry	Cardiff	8/32	-	-
777	3/05	Barry	Barry	11/26	-	-
778	4/05	Barry	Barry	4/32	-	-
779*	5/05	Barry	Barry	8/23	-	-
780	5/05	Barry	Cardiff E Dock	12/34	-	-
781	8/89	Barry	Cardiff Docks	7/32	-	-
782	9/89	Barry	Barry	6/39	-	-
783	1/90	Barry	Barry	-	Barry	8/48
784	1/90	Barry	Barry	-	Barry	8/49
785	3/91	Barry	Barry	5/32	-	-
786	9/99	Penarth Dock	Penarth Dock	11/27	-	-
787	9/99	Penarth Dock	Radyr	10/26	-	-
788	5/99	Aberdare TV	Penarth Dock	12/27	-	-
789	5/99	Penarth Dock	Cardiff	4/30	-	-
790	5/99	Penarth Dock	Barry	9/28	-	--
791	6/99	Aberdare TV	Penarth Dock	11/27	-	-
792+	12/84	Treherbert	Treherbert	-	Treherbert	2/52
793+	12/84	Treherbert	Treherbert	-	Treherbert	11/53
794+	12/84	Treherbert	Treherbert	-	Treherbert	11/51
795*	-/73	Cathays	Cathays	7/23	-	-
796	-/73	Cathays	Cathays	7/27	-	-
797	4/67	Cathays	Cathays	6/25	-	-
798	3/65	Cathays	Cathays	10/26	-	-
799	2/85	Aberavon	Cathays	12/27	-	-
800	11/73	Worcester	Worcester	12/26	-	-
801	2/86	Aberavon	Neath	8/28	-	-
802	10/89	Danygraig	Duffryn Yard	12/34	-	-
803	12/73	Hereford	Cathays	3/23	-	-
804	12/73	Tyseley	Hereford	12/24	-	-
805	11/89	Aberavon	Danygraig	12/34	-	-
806	11/89	Danygraig	Duffryn Yard	4/40	-	-
807	3/05	Barry	Cathays	7/32	-	-
808	12/00	Duffryn Yard	Duffryn Yard	12/33	-	-
809	12/00	Duffryn Yard	Duffryn Yard	5/28	-	-
810	5/73	Shrewsbury	Shrewsbury	12/26	-	-
811	2/00	Duffryn Yard	Duffryn Yard	10/33	-	-
812	3/00	Duffryn Yard	Duffryn Yard	6/34	-	-
813	6/00	Duffryn Yard	Duffryn Yard	1/34	-	-
814	6/00	Duffryn Yard	Duffryn Yard	8/30	-	-
815	3/98	Duffryn Yard	Duffryn Yard	11/30	-	-
816	11/98	Duffryn Yard	Duffryn Yard	7/29	-	-
817	12/73	Duffryn Yard	Duffryn Yard	10/26	-	-
818	12/73	Duffryn Yard	Swansea E Dock	5/32	-	-
819	6/03	Kerry	Oswestry	3/46	-	-
820	11/98	Kerry	Oswestry	3/30	-	-

No.	Date	1922 Shed	Last GWR Shed	Wdn GWR	Last BR shed	Wdn BR
821	10/98	Oswestry	Cardiff	10/32	-	-
822	9/02	Welshpool W&L	Welshpool W&L	-	Welshpool W&L	8/61
823	9/02	Welshpool W&L	Welshpool W&L	-	Welshpool W&L	7/62
824	11/64	Oswestry	Oswestry	9/40	-	-
825	-/94	Swindon Town	Swindon	6/26	-	-
826	9/73	Croes Newydd	Stafford Road	10/28	-	-
827	9/73	Leominster	Llanfyllin	5/33	-	-
828	10/73	Southall	Ross-on-Wye	11/34	-	-
829	10/73	Tyseley	Ross-on-Wye	3/34	-	-
830	10/73	Bullo Pill	Newton Abbot	1/36	-	-
831	11/73	Neyland	Newton Abbot	12/35	-	-
832	11/73	Exeter	Helston	4/30	-	-
833	11/73	Southall	Weymouth	3/34	-	-
834	12/73	Taunton	Kington	5/33	-	-
835	12/73	Worcester	Swindon	2/35	-	-
836	1/74	Ross-on-Wye	Kington	12/30	-	-
837	1/74	Bristol	Oswestry	2/34	-	-
838	2/74	Taunton	Newton Abbot	12/29	-	-
839	3/74	Farringdon	Penmaenpool	6/32	-	-
840	3/74	Malmesbury	Worcester	8/29	-	-
841	4/74	Tyseley	Gloucester	2/31	-	-
842	4/74	Newport	Bath Road	9/30	-	-
843	-/94	Andover Jct	Swindon	8/26	-	-
844	10/18	Oswestry	Moat Lane	-	Oswestry	8/54
845	12/74	Helston	N'castle Emlyn	12/34	-	-
846	1/75	Dinas Mawddwy	Machynlleth	12/32	-	-
847	2/75	N'castle Emlyn	Newton Abbot	5/33	-	-
848	2/75	Southall	Oswestry	6/45	-	-
849	11/18	Machynlleth	Oswestry	-	Machynlleth	10/54
850	2/74	Newport	Pill	12/34	-	-
851	5/74	Bristol	Swindon	3/31	-	-
852	6/74	Paddington	Paddington	4/31	-	-
853	6/74	Penzance	Penzance	6/30	-	-
854	6/74	Exeter	Taunton	2/35	-	-
855	5/19	Oswestry	Oswestry	-	Oswestry	10/54
856	7/74	Chester	Chester	5/32	-	-
857	8/74	Chester	Shrewsbury	11/33	-	-
858	8/74	Swindon	Worcester	3/31	-	-
859	9/74	Landore	Sandy	12/34	-	-
860	9/74	Whitland	Whitland	8/28	-	-
861	10/74	Whitland	Burry Port	7/32	-	-
862	10/74	Croes Newydd	Birkenhead	4/32	-	-
863	11/74	Exeter	Taunton	5/32	-	-
864	3/08	Oswestry	Machynlleth	-	Machynlleth	11/52
865	12/74	Stourbridge	Shrewsbury	10/33	-	-
866	12/74	Bristol	Weymouth	7/32	-	-
867	1/75	Newton Abbot	Exeter	11/33	-	-
868	2/75	Paddington	Bristol SPM	9/35	-	-
869	3/75	Swindon	Bristol SPM	10/34	-	-
870	4/75	Paddington	Taunton	11/34	-	-
871	4/75	Whitland	Ebbw Junction	7/31	-	-
872	5/75	Whitland	Llanelly	9/36	-	-
873	6/19	Machynlleth	Brecon	-	Oswestry	3/54
874	5/19	Oswestry	Machynlleth	3/33	-	-
875	5/94	Brecon	Portmadoc	12/36	-	-
876	5/94	Moat Lane	Moat Lane	5/46	-	-
877	1/74	Tyseley	Shrewsbury	3/29	-	-
878	5/94	Builth Wells	Oswestry	10/38	-	-
879	6/94	Brecon	Oswestry	5/32	-	-
880	6/94	Llanidloes	Oswestry	11/38	-	-

No.	Date	1922 Shed	Last GWR Shed	Wdn GWR	Last BR shed	Wdn BR
881	8/95	Machynlleth	Machynlleth	7/32	-	-
882	8/95	Portmadoc	Oswestry	1/35	-	-
883	9/95	Brecon	Brecon	11/26	-	-
884	1/99	Oswestry	Oswestry	8/47	-	-
885	1/99	Pwllheli	Oswestry	8/44	-	-
886	1/74	Banbury	Didcot	4/23	-	-
887	4/03	Oswestry	Oswestry	-	Oswestry	11/52
888*	5/03	CONDEMNED	-	5/22	-	-
889*	5/03	CONDEMNED	-	8/22	-	-
890	2/74	Wolverhampton	Banbury	12/25	-	-
891*	5/03	CONDEMNED	-	8/22	-	-
892	6/03	Machynlleth	Brecon	-	Machynlleth	4/53
893	3/08	Portmadoc	Brecon	-	Oswestry	2/53
894	3/08	Oswestry	Portmadoc	-	Machynlleth	4/53
895	3/08	Oswestry	Builth Wells	-	Oswestry	10/54
896	3/08	Oswestry	Oswestry	-	Oswestry	4/53
897*	6/72	CONDEMNED	-	7/22	-	-
898	9/75	Oswestry	Oswestry	11/47	-	-
899*	6/63	CONDEMNED	-	7/22	-	-
900	8/64	Portmadoc	Oswestry	9/45	-	-
901*	8/64	CONDEMNED	-	7/22	-	-
902	11/72	Llantrisant	Pontypool	7/32	-	-
903	10/72	Carmarthen	Danygraig	4/31	-	-
904	8/72	Tyseley	Stafford Road	8/32	-	-
905	5/95	Bridgend	Pill	11/47	-	-
906	5/95	Carmarthen	Neath	-	Neath	4/48
907	5/95	Cardiff	Didcot	-	Didcot	3/51
908	11/73	Oswestry	Oswestry	12/38	-	-
909*	12/73	CONDEMNED	-	7/22	-	-
910	4/65	Oswestry	Oswestry	5/35	-	-
911*	5/65	CONDEMNED	-	7/22	-	-
912	3/84	Ferndale	Cardiff	10/27	-	-
913	12/82	Radyr	Radyr	6/25	-	-
914	8/84	Radyr	Cathays	6/25	-	-
915	6/74	Cathays	Cathays	10/26	-	-
916*	2/85	CONDEMNED	-	7/22	-	-
917	5/74	Radyr	Aberdare TV	8/27	-	-
918*	1/75	CONDEMNED	-	7/22	-	-
919	5/75	Cathays	Aberdare	2/30	-	-
920	9/84	Radyr	Penarth Dock	12/27	-	-
921*	10/81	Radyr	Radyr	6/23	-	-
922	11/80	Penarth Dock	Cardiff	10/26	-	-
923*	3/84	Aberdare TV	Cathays	9/23	-	-
924	12/80	Cathays	Aberdare TV	2/27	-	-
925*	12/80	Cathays	Cathays	10/23	-	-
926*	10/75	Radyr	Cathays	3/23	-	-
927*	10/83	Abercynon	Abercynon	9/23	-	-
928*	11/83	Aberdare TV	Aberdare TV	10/23	-	-
929*	11/83	Radyr	Radyr	4/23	-	-
930*	12/74	Ferndale	Cathays	3/23	-	-
931*	5/75	CONDEMNED	-	7/22	-	-
932	10/75	Radyr	Radyr	12/25	-	-
933*	10/75	Aberdare TV	Aberdare TV	4/23	-	-
934	3/74	Chester	Stourbridge	11/28	-	-
935*	10/75	Radyr	Radyr	6/23	-	-
936	7/80	Radyr	Treherbert	12/25	-	-
937	4/74	Birkenhead	Birkenhead	2/23	-	-
938	7/80	Radyr	Abercynon	5/25	-	-
939	10/80	Treherbert	Cathays	6/24	-	-
940	4/74	Birkenhead	Birkenhead	8/25	-	-

No.	Date	1922 Shed	Last GWR Shed	Wdn GWR	Last BR shed	Wdn BR
941*	9/81	Radyr	Radyr	10/23	-	-
942	11/83	Cathays	Cathays	12/23	-	-
943*	1/84	Ferndale	Ferndale	4/23	-	-
944*	1/84	Abercynon	Cathays	2/23	-	-
945	5/74	Stourbridge	Birkenhead	1/25	-	-
946*	2/84	Radyr	Cathays	9/23	-	-
947	5/74	Duffryn Yard	Neath N&B	7/32	-	-
948	9/84	Radyr	Cathays	6/25	-	-
949	5/74	Landore	Bridgend	10/34	-	-
950	6/74	Didcot	Swindon	5/32	-	-
951	6/74	Landore	Llanelly	10/30	-	-
952	6/74	Swansea E Dock	Neyland	3/31	-	-
953	6/74	Didcot	Paddington	9/29	-	-
954	6/74	Tondu	Aberdare	2/31	-	-
955	6/74	Aberbeeg	Ebbw Junction	4/31	-	-
956	6/74	Southall	Reading	5/30	-	-
957	7/74	Milford Haven	Swansea E Dock	8/32	-	-
958	7/74	Newport	Aberbeeg	11/33	-	-
959	7/74	Taunton	Taunton	5/30	-	-
960	7/74	Aberbeeg	Cardiff	5/32	-	-
961	7/74	Landore	Neath	7/30	-	-
962	7/74	Gloucester	Gloucester	3/31	-	-
963	8/74	Abergavenny	Pontypool	6/36	-	-
964	8/74	Duffryn Yard	Swansea E Dock	4/29	-	-
965	8/74	Cardiff	Severn Tunnel	1/30	-	-
966	8/74	Taunton	Westbury	1/29	-	-
967	8/74	Bristol	Llantrisant	4/34	-	-
968*	9/84	Radyr	Cathays	3/23	-	-
969*	9/84	Cathays	Cathays	2/23	-	-
970*	9/84	Radyr	Radyr	8/23	-	-
971	9/74	Stratford	Ebbw Junction	11/34	-	-
972	9/74	Southall	Slough	2/32	-	-
973	9/74	Tyseley	Bala	5/32	-	-
974*	1/85	Penarth Dock	Cathays	4/23	-	-
975	9/74	Taunton	Ebbw Junction	4/34	-	-
976	10/74	Worcester	Southall	11/30	-	-
977	10/74	Bristol	Bath Road	3/31	-	-
978	2/85	Penarth Dock	Penarth Dock	7/25	-	-
979	10/74	Pontypool	Leominster	9/33	-	-
980	10/74	Kidderminster	Truro	8/33	-	-
981	10/74	Cheltenham	Pontypool	7/30	-	-
982	10/74	Worcester	Oxford	12/33	-	-
983	11/74	Reading	Goodwick	3/34	-	-
984	2/85	Cathays	Cathays	2/24	-	-
985	11/74	Landore	Aberdare	12/30	-	-
986	11/74	Frome	Bristol SPM	3/34	-	-
987	6/75	Croes Newydd	Stourbridge	4/36	-	-
988	7/75	Swindon	Sandy	11/28	-	-
989	8/75	Llanelly	Danygraig	4/36	-	-
990	9/75	Swindon	Hereford	3/31	-	-
991	9/75	Landore	Llanelly	2/31	-	-
992	9/75	Paddington	Swindon	-	Swindon	2/51
993	10/75	Pill	Pill	9/30	-	-
994	11/75	Swindon	Swindon	6/30	-	-
995	11/75	Gloucester	Bristol SPM	3/31	-	-
996	12/75	Newport	Ebbw Junction	2/31	-	-
997	1/76	Neath	Sandy	12/34	-	-
998	2/76	Worcester	Hereford	7/32	-	-
999	9/85	Cathays	Penarth Dock	12/25	-	-
1000*	2/85	Abercynon	Cathays	1/23	-	-

No.	Date	1922 Shed	Last GWR Shed	Wdn GWR	Last BR shed	Wdn BR
1001	2/85	Cathays	Radyr	12/26	-	-
1002	2/85	Cathays	Cathays	6/25	-	-
1003	8/99	Chel High St.	Gloucester	10/36	-	-
1004	8/99	Swindon Town	Gloucester	12/34	-	-
1005	8/99	Chel High St.	Gloucester	3/38	-	-
1006	8/99	Chel High St.	Gloucester	12/34	-	-
1007	8/99	Swindon Town	Swindon	6/37	-	-
1008	9/99	Ciren' MSWJR	Bristol SPM	12/36	-	-
1009	8/02	Chel High St.	Gloucester	1/35	-	-
1010	8/02	Chel High St.	Gloucester	12/34	-	-
1011	9/02	Swindon Town	Bristol SPM	5/37	-	-
1012	10/67	Shrewsbury	Stafford Road	9/29	-	-
1013	9/02	Chel High St.	Bristol SPM	6/37	-	-
1014	7/04	Oswestry	Oswestry	4/28	-	-
1015	6/66	Stourbridge	Oxley	7/29	-	-
1016	2/67	Newport	Ebbw Junction	7/30	-	-
1017	1/67	Neath	Duffryn Yard	1/29	-	-
1018	3/67	Llanelly	Llanelly	8/32	-	-
1019	3/67	Swindon	Ebbw Junction	8/34	-	-
1020	4/67	Aberdare	Merthyr	1/28	-	-
1021	5/67	Aberbeeg	Merthyr	8/32	-	-
1022	6/67	Llantrisant	Cardiff	12/25	-	-
1023	6/67	Wolverhampton	Stafford Road	8/28	-	-
1024	7/67	Duffryn Yard	Neath	2/29	-	-
1025	7/67	Newport	Ebbw Junction	12/27	-	-
1026	8/67	Reading	Oxford	4/31	-	-
1027	9/67	Stourbridge	Shrewsbury	4/30	-	-
1028	9/67	Shrewsbury	Stafford Road	8/28	-	-
1029	11/04	Oswestry	Oswestry	4/28	-	-
1030	10/67	Aberdare	Aberbeeg	4/30	-	-
1031	11/67	Tyseley	Tyseley	8/28	-	-
1032	12/67	Aberdare	Abergavenny	4/28	-	-
1033	12/67	Goodwick	Neath	11/29	-	-
1034	12/67	Llanelly	Gloucester	7/30	-	-
1035	11/04	Oswestry	Aberystwyth	5/28	-	-
1036	1/68	Cheltenham	Gloucester	3/31	-	-
1037	2/68	Leominster	Kidderminster	3/31	-	-
1038	2/68	Croes Newydd	Banbury	10/29	-	-
1039	3/68	Aberbeeg	Pontypool	11/28	-	-
1040	5/70	Gloucester	Kidderminster	1/29	-	-
1041	6/70	Shrewsbury	Shrewsbury	6/34	-	-
1042	6/70	Bristol	Aberbeeg	5/32	-	-
1043	11/04	Oswestry	Machynlleth	1/33	-	-
1044	7/70	Oxford	Oxford	12/29	-	-
1045	8/70	Banbury	Shrewsbury	1/35	-	-
1046	8/70	Wolverhampton	Stafford Road	9/31	-	-
1047	9/70	Tyseley	Stafford Road	7/35	-	-
1048	9/70	Aberdare	Merthyr	6/30	-	-
1049	10/70	Neath	Neath	6/28	-	-
1050	10/70	Newport	Ebbw Junction	12/31	-	-
1051	11/70	Bath	Ebbw Junction	7/30	-	-
1052	11/70	Llantrisant	Paddington	9/29	-	-
1053	11/70	Landore	Worcester	2/31	-	-
1054	12/70	Landore	Neath	8/31	-	-
1055	12/70	Llanelly	Pantyffynon	8/32	-	-
1056	1/71	Merthyr	Aberdare	6/30	-	-
1057	1/71	Newport	Tondu	12/28	-	-
1058	2/71	Whitland	Whitland	5/28	-	-
1059	2/71	Croes Newydd	Chester	5/33	-	-
1060	3/71	Hereford	Southall	4/32	-	-

No.	Date	1922 Shed	Last GWR Shed	Wdn GWR	Last BR shed	Wdn BR
1061	3/71	Wolverhampton	Stafford Road	3/32	-	-
1062	4/71	Hereford	Paddington	4/31	-	-
1063	4/71	Aberbeeg	Merthyr	3/28	-	-
1064	5/71	Newport	Paddington	2/29	-	-
1065	5/71	Bridport	Yeovil	4/31	-	-
1066	6/71	Landore	Reading	4/32	-	-
1067	6/71	Severn Tunnel	Ebbw Junction	3/28	-	-
1068*	7/04	Pwllheli	Portmadoc	1/24	-	-
1069	7/71	Neath	Neath N&B	8/32	-	-
1070	7/71	Shrewsbury	Chester	8/32	-	-
1071	7/71	Llanelly	Whitland	4/31	-	-
1072	8/71	Stourbridge	Oxley	5/34	-	-
1073	9/71	Wolverhampton	Croes Newydd	2/29	-	-
1074	9/71	Cardiff	Cardiff	2/28	-	-
1075	10/71	Tyseley	Ludlow	6/34	-	-
1076	5/70	Tondu	St. Blazey	7/30	-	-
1077	5/70	Neath	Aberdare	7/31	-	-
1078	5/70	Tyseley	Stafford Road	10/28	-	-
1079	6/70	Aberbeeg	Pill	10/30	-	-
1080	6/70	Tyseley	Croes Newydd	5/34	-	-
1081	6/70	Paddington	Oxford	2/31	-	-
1082	7/01	Oswestry	Machynlleth	4/28	-	-
1083	6/70	Pontypool	Pontypool	9/22	-	-
1084*	5/95	CONDEMNED	-	5/22	-	-
1085*	10/97	Llanidloes	Brecon	3/23	-	-
1086*	12/97	Machynlleth	Machynlleth	5/23	-	-
1087	10/70	Shrewsbury	Stourbridge	9/29	-	-
1088	4/93	Llanidloes	Oswestry	3/26	-	-
1089	10/70	Shrewsbury	Oxley	10/29	-	-
1090	4/93	Aberystwyth	Machynlleth	9/28	-	-
1091	4/93	Wrexham	Machynlleth	10/30	-	-
1092	11/70	Tyseley	Gloucester	7/29	-	-
1093	4/93	Brecon	Brecon	4/26	-	-
1094	11/70	Wolverhampton	Stourbridge	3/34	-	-
1095	11/70	Llanelly	Llanelly	12/30	-	-
1096*	4/93	Llanidloes	Llanidloes	11/22	-	-
1097	5/93	Oswestry	Oswestry	11/27	-	-
1098	12/70	Swansea E Dock	Llanelly	6/23	-	-
1099	12/70	Pontypool	Pontypool	3/28	-	-
1100	5/93	Builth Wells	Oswestry	7/30	-	-
1101*	5/93	Welshpool	Welshpool	10/22	-	-
1102	2/94	Llanidloes	Brecon	12/25	-	-
1103*	3/94	CONDEMNED	-	7/22	-	-
1104	3/94	Moat Lane	Llanidloes	9/25	-	-
1105*	3/94	CONDEMNED	-	7/22	-	-
1106+	5/95	Brecon	Oswestry	4/31	-	-
1107	5/95	Whitchurch	Moat Lane	5/25	-	-
1108	4/98	Builth Wells	Machynlleth	11/27	-	-
1109	4/98	Oswestry	Oswestry	1/24	-	-
1110	5/91	Portmadoc	Machynlleth	1/25	-	-
1111	4/71	Shrewsbury	Tyseley	10/28	-	-
1112	5/91	Machynlleth	Oswestry	6/28	-	-
1113*	7/64	CONDEMNED	-	5/22	-	-
1114*	8/64	CONDEMNED	-	5/22	-	-
1115	8/78	Machynlleth	Machynlleth	1/25	-	-
1116	8/78	Oswestry	Wrexham	9/24	-	-
1117*	7/86	CONDEMNED	-	5/22	-	-
1118	7/86	Aberystwyth	Brecon	10/30	-	-
1119	7/05	Chel High St.	Andover Jct	5/35	-	-
1120	4/09	Chel High St.	Swindon	8/31	-	-

No.	Date	1922 Shed	Last GWR Shed	Wdn GWR	Last BR shed	Wdn BR
1121	4/09	Andover Jct	Swindon	3/36	-	-
1122	6/14	Chel High St.	Chel High St.	11/35	-	-
1123	4/12	Chel High St.	Cheltenham	4/38	-	-
1124	4/10	Andover Jct	Swindon	8/35	-	-
1125	2/11	Chel High St.	Gloucester	2/32	-	-
1126	4/12	Chel High St.	Cheltenham	12/38	-	-
1127*	-/93	CONDEMNED	-	1/24	-	-
1128	6/14	Chel High St.	Chel High St.	10/35	-	-
1129*	7/64	CONDEMNED	-	9/22	-	-
1130	7/64	Oswestry	Oswestry	6/23	-	-
1131*	7/64	CONDEMNED	-	7/22	-	-
1132*	10/85	Oswestry	Oswestry	1/23	-	-
1133	7/84	Cathays	Coke Ovens	12/25	-	-
1134	11/74	Tenby	Oxford	11/34	-	-
1135	11/74	Frome	Westbury	4/31	-	-
1136	11/74	Stourbridge	Croes Newydd	2/35	-	-
1137	12/74	Aberbeeg	Ebbw Junction	10/33	-	-
1138	12/74	Newport	Ebbw Junction	5/30	-	-
1139	12/74	Penzance	Kidderminster	6/30	-	-
1140	12/74	Stourbridge	Stourbridge	9/28	-	-
1141	12/74	Neath	Duffryn Yard	8/32	-	-
1142	12/74	Tenby	Duffryn Yard	12/35	-	-
1143	12/74	Cardiff	Cardiff	8/32	-	-
1144	1/75	Tondu	Merthyr	3/31	-	-
1145	1/75	Duffryn Yard	Duffryn Yard	5/34	-	-
1146	1/75	Aberbeeg	Dock Street	4/29	-	-
1147	1/75	Kidderminster	Tyseley	1/31	-	-
1148	1/75	Exeter	Exeter	7/34	-	-
1149	2/75	Bristol	Bristol SPM	9/28	-	-
1150	2/75	Aberdare	Pill	10/28	-	-
1151	2/75	Cardiff	Aberdare	7/32	-	-
1152	2/75	Tondu	Southall	5/35	-	-
1153	2/75	Aberbeeg	Tondu	5/32	-	-
1154	10/75	Tetbury	Ebbw Junction	11/34	-	-
1155	11/75	Tyseley	Neath	2/36	-	-
1156	11/75	Croes Newydd	Aylesbury	10/30	-	-
1157	12/75	Banbury	Leamington	11/35	-	-
1158	12/75	Newton Abbot	Swindon	3/31	-	-
1159	1/76	St. Blazey	Oxford	8/47	-	-
1160	1/76	Gloucester	Exeter	10/32	-	-
1161	2/76	Goodwick	Swindon	10/45	-	-
1162	2/76	Kidderminster	Newton Abbot	12/34	-	-
1163	3/76	Southall	Weymouth	5/46	-	-
1164	3/76	Swindon	Oswestry	12/37	-	-
1165	3/76	Truro	Tiverton Jct	3/30	-	-
1166	8/75	Aberbeeg	Yeovil	6/34	-	-
1167	8/75	Danygraig	Landore	12/36	-	-
1168	8/75	Laira	Millbay	10/28	-	-
1169	8/75	Swindon	Kidderminster	11/34	-	-
1170	8/75	Landore	Neath N&B	6/27	-	-
1171	8/75	Southall	Tondu	1/34	-	-
1172	9/75	Reading	Bristol SPM	4/30	-	-
1173	9/75	Pontypool	Ebbw Junction	8/28	-	-
1174	9/75	Newport	Tondu	7/34	-	-
1175	9/75	Paddington	Oxford	11/33	-	-
1176	10/75	Penzance	Exeter	3/31	-	-
1177	10/75	Yeovil	Bristol SPM	4/29	-	-
1178	10/75	Landore	Landore	4/30	-	-
1179	11/75	Slough	Yeovil	8/35	-	-
1180	11/75	Frome	Yeovil	11/34	-	-

No.	Date	1922 Shed	Last GWR Shed	Wdn GWR	Last BR shed	Wdn BR
1181	11/75	Taunton	Tondu	12/35	-	-
1182	11/75	Chippenham	Swindon	9/29	-	-
1183	11/75	Worcester	Kidderminster	4/30	-	-
1184	5/85	Cathays	Cathays	12/25	-	-
1185	12/75	Tondu	Tondu	1/31	-	-
1186	3/76	Trowbridge	Salisbury	5/23	-	-
1187	3/76	Llanelly	Neath	1/29	-	-
1188	3/76	Truro	Bristol SPM	11/27	-	-
1189	9/90	Duffryn Yard	Swindon	11/26	-	-
1190*	3/64	CONDEMNED	-	7/22	-	-
1191*	12/65	CONDEMNED	-	10/22	-	-
1192	5/66	Oswestry	Hemyock	8/29	-	-
1193	4/76	Trowbridge	Gloucester	1/29	-	-
1194	4/76	Leamington	Tyseley	6/25	-	-
1195	4/76	Wolverhampton	Oxley	10/34	-	-
1196	5/66	Oswestry	Oswestry	-	Oswestry	4/48
1197	5/66	Oswestry	Oswestry	-	Oswestry	4/48
1198	2/97	Aber'wyth VoR	Aber'wyth VoR	7/24	-	-
1199	5/92	Pill	Pill	3/31	-	-
1200	5/76	Llanelly	Whitland	10/28	-	-
1201	1/88	Pill	Pill	6/29	-	-
1202	6/76	Trowbridge	Cathays	11/27	-	-
1203	6/76	Oxford	Oxford	8/25	-	-
1204	1/88	Pill	Pill	9/29	-	-
1205	1/21	Pill	Pill	-	Cardiff	1/56
1206	1/21	Pill	Hereford	-	Hereford	1/51
1207	12/87	Pill	Pill	5/32	-	-
1208	12/87	Pill	Pill	12/29	-	-
1209	1/88	Pill	Pill	12/29	-	-
1210	9/76	Trowbridge	Bristol SPM	12/26	-	-
1211	12/87	Pill	Pill	9/29	-	-
1212	1/02	Aber'wyth VoR	Aber'wyth VoR	12/32	-	-
1213+	1/02	Aber'wyth Vor	Aber'wyth VoR	-	(STILL IN SERVICE)	
1214	10/76	Banbury	Oxley	9/29	-	-
1215	10/76	Newport	Severn Tunnel	10/26	-	-
1216	9/76	Worcester	Worcester	6/30	-	-
1217	10/76	Paddington	Paddington	4/31	-	-
1218	12/76	Carmarthen	Carmarthen	7/29	-	-
1219	12/76	Paddington	Llanelly	7/32	-	-
1220	1/77	Worcester	Worcester	11/38	-	-
1221	12/76	Hereford	Whitland	1/31	-	-
1222	1/77	Whitland	Sandy	10/33	-	-
1223	2/77	Plymouth	St. Blazey	7/35	-	-
1224	2/77	Bristol	Bath Road	3/32	-	-
1225	3/77	Bristol	St. Blazey	7/38	-	-
1226	3/77	Paddington	Pill	1/35	-	-
1227	4/77	Cardiff	St. Blazey	4/38	-	-
1228	10/76	Frome	Weymouth	9/31	-	-
1229	10/76	Swansea E Dock	Ebbw Junction	2/31	-	-
1230	11/76	Paddington	Neyland	5/32	-	-
1231	11/76	Worcester	Gloucester	9/30	-	-
1232	11/76	Yeovil	Bristol SPM	8/30	-	-
1233	11/76	Llantrisant	Westbury	3/31	-	-
1234	11/76	Tondu	Cathays	12/35	-	-
1235	12/76	Plymouth	Aberdare	10/37	-	-
1236	12/76	Duffryn Yard	Garnant	2/31	-	-
1237	12/76	Llantrisant	Banbury	12/34	-	-
1238	12/76	Llantrisant	Duffryn Yard	8/32	-	-
1239	12/76	Wells	Bridport	10/35	-	-
1240	1/77	Bristol	Reading	10/35	-	-

No.	Date	1922 Shed	Last GWR Shed	Wdn GWR	Last BR shed	Wdn BR
1241	1/77	Reading	Paddington	9/28	-	-
1242	1/77	Yeovil	Bridport	3/31	-	-
1243	1/77	Ross-on-Wye	Worcester	6/32	-	-
1244	2/77	Exeter	Exeter	5/30	-	-
1245	2/77	Pontypool	Aberdare	9/35	-	-
1246	2/77	Tondu	Kidderminster	12/34	-	-
1247	2/77	Cardiff	Cathays	5/39	-	-
1248	3/77	Paddington	Reading	5/31	-	-
1249	3/77	Oxford	Oxford	10/29	-	-
1250	3/77	Yeovil	Oxford	3/35	-	-
1251	3/77	Newport	Bath Road	12/29	-	-
1252	3/77	Laira	Laira	10/32	-	-
1253	4/77	Banbury	Chester	6/32	-	-
1254	4/77	Tondu	Kidderminster	2/37	-	-
1255	4/77	Tondu	Ebbw Junction	3/31	-	-
1256	5/77	Banbury	Ebbw Junction	1/36	-	-
1257	5/77	Didcot	Paddington	5/30	-	-
1258	5/77	Tyseley	Tyseley	3/29	-	-
1259	5/77	St. Blazey	Taunton	7/32	-	-
1260	5/77	Wolverhampton	Leamington	8/34	-	-
1261	5/77	Pontypool	Pontypool	8/28	-	-
1262	5/77	Newport	Ebbw Junction	4/30	-	-
1263	6/77	Tondu	Didcot	1/34	-	-
1264	6/77	Duffryn Yard	Llanelly	4/28	-	-
1265	6/77	Laira	Ebbw Junction	9/37	-	-
1266	6/77	Tondu	Aberbeeg	10/30	-	-
1267	7/77	Glyn Neath	Tondu	5/30	-	-
1268	8/77	Aberbeeg	Ebbw Junction	11/34	-	-
1269	8/77	Laira	Cathays	4/38	-	-
1270	8/77	Llanelly	Llanelly	3/31	-	-
1271	9/77	Plymouth	Landore	4/37	-	-
1272	9/77	Newport	Aberdare	8/34	-	-
1273	9/77	Abergavenny	Ebbw Junction	11/33	-	-
1274	9/77	Pontypool	Ebbw Junction	5/30	-	-
1275	10/77	Cardiff	Merthyr	9/28	-	-
1276	10/77	Banbury	Leamington	9/34	-	-
1278	11/77	Didcot	Abercynon	2/36	-	-
1279	11/77	Llanelly	Llanelly	9/35	-	-
1280	11/77	Paddington	Dock Street	4/29	-	-
1281	11/77	Bristol	Bridport	5/35	-	-
1282	12/77	Southall	Newton Abbot	12/36	-	-
1283	12/77	Bristol	Bristol SPM	8/29	-	-
1284	1/78	Laira	Abercynon	5/36	-	-
1285	1/78	Duffryn Yard	Cardiff	12/34	-	-
1286	1/78	Exeter	Newton Abbot	4/30	-	-
1287	1/78	Llanelly	Leamington	4/46	-	-
1288	2/78	Pontypool	Aberbeeg	7/30	-	-
1289	2/78	Swindon	Weymouth	7/34	-	-
1290	2/78	Tondu	Tondu	8/32	-	-
1291	2/78	Bristol	Westbury	5/29	-	-
1292	3/78	Newport	Ebbw Junction	12/34	-	-
1293	3/78	Basingstoke	Reading	6/28	-	-
1294	3/78	Oxford	Oxford	3/31	-	-
1295	4/78	Paddington	Reading	9/30	-	-
1296	4/78	Tondu	Tondu	9/36	-	-
1297	4/78	Wells	Chippenham	7/32	-	-
1298	12/78	Exeter	Exeter	10/26	-	-
1299	12/78	CME (Swindon)	CME (Swindon)	9/36	-	-
1300	12/78	Hemyock	Exeter	5/34	-	-
1301	8/88	Coke Ovens	Cathays	8/26	-	-

No.	Date	1922 Shed	Last GWR Shed	Wdn GWR	Last BR shed	Wdn BR
1302	8/88	Cathays	Cathays	8/26	-	-
1303	9/88	Penarth Dock	Penarth Dock	10/26	-	-
1304	8/91	Coke Ovens	Cathays	11/27	-	-
1305	7/91	Cathays	Cathays	11/26	-	-
1306	7/91	Cathays	Llantrisant	6/25	-	-
1307	4/95	Danygraig	Duffryn Yard	8/26	-	-
1308	11/02	Garnant	Oswestry	-	Oswestry	5/48
1309	4/95	Landore	Danygraig	12/26	-	-
1310	5/95	Landore	Barry	9/28	-	-
1311	3/97	Barry	Barry	12/26	-	-
1312	4/97	Barry	Barry	7/30	-	-
1313	5/97	Barry	Barry	5/28	-	-
1314	5/98	Barry	Barry	10/26	-	-
1315	5/98	Barry	Barry	10/28	-	-
1316	5/98	Barry	Barry	10/28	-	-
1317	6/99	Barry	Barry	3/28	-	-
1318	6/99	Barry	Barry	11/26	-	-
1319	6/99	Barry	Barry	3/30	-	-
1320	6/99	Barry	Barry	6/30	-	-
1321	6/99	Barry	Barry	2/26	-	-
1322	9/89	Barry	Aberdare	5/28	-	-
1323	9/89	Barry	Cathays	8/26	-	-
1324	-/90	Cardiff Docks	Aberdare	10/28	-	-
1325	-/90	Cardiff Docks	Severn Tunnel	8/28	-	-
1326	9/90	Duffryn Yard	Neath	8/30	-	-
1327*	3/98	Cardiff E M	-	5/22	-	-
1328	10/83	Dolgelly	Dinas Mawddwy	3/26	-	-
1329	9/83	Machynlleth	Aberystwyth	11/27	-	-
1330*	3/64	CONDEMNED	-	7/22	-	-
1331*	3/64	CONDEMNED	-	7/22	-	-
1332*	10/65	CONDEMNED	-	8/22	-	-
1333*	12/65	CONDEMNED	-	11/22	-	-
1334	-/94	Andover Jct	Didcot	-	Didcot	9/52
1335	-/94	Andover Jct	Reading	-	Reading	9/52
1336	-/94	Swindon	Reading	-	Reading	3/54
1337	11/89	Weymouth	Weymouth	1/26	-	-
1338	5/98	Cardiff E M	Bridgwater	-	Swansea E Dock	9/63
1339	12/99	Cardiff E M	Cardiff Docks	5/32	-	-
1340	-/97	Pill	Cathays	7/32	-	-
1341	*	Pill	Radyr	7/46	-	-
1342	12/76	Cathays	Cathays	8/26	-	-
1343	3/76	Radyr	Cathays	12/25	-	-
1344*	1/86	Pill	Pill	1/23	-	-
1345*	12/85	Pill	Pill	10/23	-	-
1346	12/85	Pill	Pill	7/27	-	-
1347	5/14	Barry	Barry	10/26	-	-
1348	5/14	Barry	Barry	10/26	-	-
1349	5/14	Barry	Barry	10/26	-	-
1350	6/14	Barry	Barry	10/26	-	-
1351	6/14	Barry	Cathays	10/26	-	-
1352	7/14	Barry	Barry	9/26	-	-
1353	7/14	Barry	Barry	9/26	-	-
1354	8/14	Barry	Barry	10/26	-	-
1355	8/14	Barry	Barry	10/26	-	-
1356	11/73	Pill	Pill	8/23	-	-
1357	9/14	Barry	Barry	10/26	-	-
1358	-/01	Duffryn Yard	Danygraig	-	Danygraig	2/48
1359	-/01	Duffryn Yard	Duffryn Yard	12/35	-	-
1360	-/01	Duffryn Yard	Duffryn Yard	11/26	-	-
1361	4/10	St. Blazey	Laira	-	Swindon	5/61

No.	Date	1922 Shed	Last GWR Shed	Wdn GWR	Last BR shed	Wdn BR
1362	6/10	Plymouth	Newton Abbot	-	Taunton	5/61
1363	6/10	Plymouth	Laira	-	Laira	11/62
1364	8/10	St. Blazey	Laira	-	Swindon	2/61
1365	8/10	Plymouth	Millbay	-	Swindon	11/62
1376	9/74	Weymouth	Oswestry	1/34	-	-
1377	12/75	Weymouth	Weymouth	1/27	-	-
1378	-/00	Duffryn Yard	Duffryn Yard	12/28	-	-
1379	-/00	Duffryn Yard	Duffryn Yard	4/29	-	-
1380	9/96	Barry	Barry	8/30	-	-
1381	9/96	Barry	Barry	6/25	-	-
1382	10/96	Barry	Barry	9/28	-	-
1383	10/96	Barry	Barry	8/30	-	-
1384	10/96	Barry	Barry	7/30	-	-
1385	10/96	Barry	Barry	6/28	-	-
1386	10/96	Barry	Barry	5/27	-	-
1387	-/88	Trehafod	Barry	11/27	-	-
1388	-/88	Barry	Barry	8/28	-	-
1389	-/86	Barry	Barry	12/27	-	-
1390	-/87	Barry	Barry	4/30	-	-
1393	-/73	Neath	Danygraig	3/36	-	-
1394	-/73	Swindon	Swindon	9/33	-	-
1395	-/73	Swindon	Swindon	3/34	-	-
1396	-/73	St. Blazey	Swindon	3/34	-	-
1397	-/73	Bridgwater	Swindon	11/33	-	-
1398	-/73	Penzance	Swindon	10/36	-	-
1399	-/73	Swindon	Swindon	3/34	-	-
1401	5/78	Croes Newydd	Ludlow	1/34	-	-
1403	5/78	Slough	Slough	11/30	-	-
1404	6/78	Southall	Marlow	2/34	-	-
1405	6/78	Bala	Bala	6/29	-	-
1406	6/78	Southall	Slough	6/29	-	-
1407	6/78	Worcester	Slough	4/31	-	-
1408	7/78	Slough	Llantrisant	12/34	-	-
1409	7/78	Marlow	Southall	7/30	-	-
1410	7/78	Birkenhead	Bala	11/34	-	-
1411	8/78	Southall	Worcester	7/29	-	-
1413	8/78	Shrewsbury	Oxford	3/31	-	-
1414	9/78	Bridgend	Aberdare	1/29	-	-
1415	9/78	Southall	Newton Abbot	4/38	-	-
1416	9/78	Southall	Southall	11/30	-	-
1417	9/78	Bristol	Southall	4/28	-	-
1418	10/78	Slough	Southall	4/29	-	-
1419	10/78	Leamington	Pontypool	7/32	-	-
1420	10/78	Southall	Aberdare	4/36	-	-
1421	4/77	Newport	Hereford	9/28	-	-
1422	5/77	Ross-on-Wye	Worcester	7/28	-	-
1423	5/77	Croes Newydd	Chippenham	2/30	-	-
1424	5/77	Much Wenlock	Leominster	9/33	-	-
1425	6/77	Tyseley	Leamington	8/32	-	-
1426	6/77	Pill	Swindon	11/34	-	-
1427	7/77	Llanelly	Weymouth	6/37	-	-
1428	7/77	Bristol	Malmesbury	10/32	-	-
1429	7/77	Worcester	Reading	10/32	-	-
1430	8/77	Weymouth	Weymouth	10/36	-	-
1431	8/77	Exeter	Exeter	1/33	-	-
1432	9/77	Kidderminster	Swindon	10/32	-	-
1433	10/77	Exeter	Weymouth	4/36	-	-
1434	10/77	Newport	Llantrisant	8/28	-	-
1435	11/77	Tyseley	Oswestry	9/32	-	-
1436	11/77	Wellington	Swindon	11/44	-	-

No.	Date	1922 Shed	Last GWR Shed	Wdn GWR	Last BR shed	Wdn BR
1437	11/77	Kidderminster	Leominster	9/28	-	-
1438	11/77	Swindon	Aylesbury	8/34	-	-
1439	12/77	Exeter	Swindon	9/30	-	-
1440	12/77	Ashburton	Weymouth	9/35	-	-
1441	1/78	Trowbridge	Gloucester	8/29	-	-
1442	1/78	Tyseley	Swindon	5/45	-	-
1443	2/78	Woodstock	Newton Abbot	12/34	-	-
1444	2/78	Southall	Chippenham	11/34	-	-
1445	9/81	Stourbridge	Abercynon	12/37	-	-
1446	10/81	Cheltenham	Gloucester	2/36	-	-
1447	10/81	Bristol	Swindon	11/28	-	-
1448	11/81	Tiverton Jct	Westbury	7/29	-	-
1449	11/81	Gloucester	Llanelly	9/28	-	-
1450	12/81	Fairford	Machynlleth	4/28	-	-
1451	12/81	Landore	Llanelly	9/28	-	-
1453	1/82	Bristol	Truro	7/33	-	-
1454	2/82	Worcester	Aberdare	11/34	-	-
1455	2/82	Reading	Ross-on-Wye	6/35	-	-
1456	2/82	Yatton	Goodwick	10/30	-	-
1457	3/82	Yatton	Yeovil	2/29	-	-
1459	3/82	Reading	Truro	3/36	-	-
1461	4/82	Fairford	Oxford	8/30	-	-
1462	4/82	Pontypool	Oxford	3/29	-	-
1463	5/82	Swindon	Truro	12/34	-	-
1464	5/82	Cheltenham	Newton Abbot	1/36	-	-
1465	1/83	Newport	Weymouth	1/36	-	-
1466	1/83	Newton Abbot	Weymouth	6/35	-	-
1467	2/83	Penzance	Penzance	3/29	-	-
1468	2/83	Southall	Neath	11/36	-	-
1469	3/83	Pontrilas	Didcot	2/31	-	-
1470	4/83	Southall	Merthyr	12/34	-	-
1471	4/83	Bridgwater	Ebbw Junction	4/32	-	-
1472	5/83	Bristol	Bristol SPM	11/34	-	-
1473	5/83	Woodstock	Oxford	8/35	-	-
1474	6/83	Worcester	Wellington	4/31	-	-
1475	7/83	Chester	Worcester	8/32	-	-
1476	7/83	Swindon	Worcester	6/29	-	-
1477	10/84	Shrewsbury	Swindon	12/37	-	-
1478	10/84	Wolverhampton	Aberystwyth	5/36	-	-
1479	11/84	Swindon	Wallingford	4/32	-	-
1481	12/84	Exeter	Worcester	2/30	-	-
1482	1/85	Stourbridge	Croes Newydd	6/35	-	-
1483	6/85	Wellington	Helston	8/32	-	-
1484	7/85	Wallingford	Southall	5/30	-	-
1485	8/85	Wellington	Weymouth	12/36	-	-
1486	9/85	Aberdare	Merthyr	1/36	-	-
1487	10/85	Leominster	Ashburton	8/36	-	-
1488	11/85	Swindon	Croes Newydd	11/37	-	-
1491	6/92	Oxford	Bath Road	3/30	-	-
1492	6/92	Bristol	Chester	2/35	-	-
1493	7/92	Fairford	Gloucester	5/35	-	-
1494	7/92	Cheltenham	St. Blazey	12/37	-	-
1495	8/92	Stourbridge	Merthyr	12/38	-	-
1496	8/92	Barnstaple	Penzance	10/35	-	-
1497	8/92	Cheltenham	Bristol SPM	6/38	-	-
1498	8/92	Oxford	Swindon	9/44	-	-
1499	9/92	Bristol	Swindon	5/46	-	-
1500	9/92	Barnstaple	St. Blazey	9/37	-	-
1501	3/78	Tyseley	Tyseley	12/34	-	-
1502	4/78	Wolverhampton	Ross-on-Wye	7/35	-	-

No.	Date	1922 Shed	Last GWR Shed	Wdn GWR	Last BR shed	Wdn BR
1503	5/78	Birkenhead	Ebbw Junction	5/32	-	-
1504	5/78	Stourbridge	Stourbridge	11/29	-	-
1505	6/78	Tyseley	Aberdare	1/37	-	-
1506	7/78	Leamington	Croes Newydd	12/37	-	-
1507	7/78	Chester	Stafford Road	6/34	-	-
1508	8/78	Birkenhead	Croes Newydd	9/36	-	-
1509	8/78	Gloucester	Hereford	10/38	-	-
1510	9/78	Wolverhampton	Stourbridge	8/29	-	-
1511	9/78	Birkenhead	Croes Newydd	1/36	-	-
1512	10/78	Tyseley	Tyseley	3/29	-	-
1513	11/78	Wolverhampton	Hereford	7/30	-	-
1514	11/78	Stourbridge	Wellington	12/34	-	-
1515	12/78	Croes Newydd	Bala	6/29	-	-
1516	12/78	Bala	Croes Newydd	6/31	-	-
1517	12/78	Tyseley	Tyseley	8/28	-	-
1518	12/78	Corwen	Stourbridge	9/34	-	-
1519	1/79	Corwen	Wellington	7/35	-	-
1520	2/79	Bullo Pill	Gloucester	5/29	-	-
1521	2/79	Wolverhampton	Ebbw Junction	12/32	-	-
1522	3/79	Stourbridge	Stourbridge	8/34	-	-
1523	4/79	Wolverhampton	Croes Newydd	11/37	-	-
1524	4/79	Stourbridge	Croes Newydd	5/39	-	-
1525	5/79	Worcester	Weymouth	7/35	-	-
1526	5/79	Wolverhampton	Stafford Road	7/33	-	-
1527	6/79	Wolverhampton	Croes Newydd	7/45	-	-
1528	7/79	Wolverhampton	Stourbridge	2/36	-	-
1529	7/79	Chester	Wellington	6/33	-	-
1530	8/79	Carmarthen	Aberdare	12/38	-	-
1531	8/79	Tyseley	Oxford	-	Oxford	12/49
1532	9/79	Birkenhead	Croes Newydd	-	Bala	7/48
1533	10/79	Croes Newydd	Oswestry	7/36	-	-
1534	10/79	Pontypool	Pontypool	8/34	-	-
1535	10/79	Wolverhampton	Aberdare	4/34	-	-
1536	11/79	St. Blazey	Stafford Road	6/34	-	-
1537	11/79	Birkenhead	Chester	2/31	-	-
1538	12/79	Gloucester	Bristol SPM	-	Bristol SPM	11/48
1539	12/79	Croes Newydd	Croes Newydd	8/38	-	-
1540	1/80	Trawsfynydd	Tyseley	8/35	-	-
1541	2/80	Croes Newydd	Yeovil	12/37	-	-
1542	2/80	Stourbridge	Swindon	-	Swindon	5/51
1543	3/80	Tyseley	Worcester	8/32	-	-
1544	3/80	Wolverhampton	Merthyr	2/30	-	-
1545	4/80	Stourbridge	Birkenhead	11/35	-	-
1546	4/80	Duffryn Yard	Goodwick	12/35	-	-
1547	5/80	Stourbridge	Stafford Road	11/34	-	-
1548	6/80	Birkenhead	Oxford	2/37	-	-
1549	8/80	Wolverhampton	Tyseley	7/35	-	-
1550	7/80	Stourbridge	Worcester	7/29	-	-
1551	8/80	Birkenhead	Tyseley	9/32	-	-
1552	9/80	Tondu	Ebbw Junction	10/36	-	-
1553	9/80	Wolverhampton	Kidderminster	4/29	-	-
1554	10/80	Crewe	Croes Newydd	11/44	-	-
1555	10/80	Stourbridge	Pontypool	7/29	-	-
1556	10/80	Wolverhampton	Croes Newydd	9/32	-	-
1557	11/80	Croes Newydd	Bristol SPM	1/38	-	-
1558	11/80	Birkenhead	Croes Newydd	9/38	-	-
1559	12/80	Stourbridge	Wellington	8/29	-	-
1560	12/80	Worcester	Wellington	1/29	-	-
1561	11/78	Pontypool	Tondu	7/34	-	-
1562	11/78	Truro	Taunton	7/38	-	-

No.	Date	1922 Shed	Last GWR Shed	Wdn GWR	Last BR shed	Wdn BR
1564	12/78	Plymouth	Newton Abbot	2/31	-	-
1565	12/78	Marlborough	Didcot	12/38	-	-
1566	12/78	Trowbridge	Weymouth	5/35	-	-
1567	12/78	Basingstoke	Stafford Road	8/36	-	-
1568	12/78	Neath	Danygraig	4/38	-	-
1569	1/79	Swansea E Dock	Oxford	7/32	-	-
1570	1/79	Worcester	Pontypool	10/35	-	-
1571	1/79	Llantrisant	Cardiff	2/31	-	-
1572	1/79	Tondu	Tondu	1/33	-	-
1573	2/79	Pontypool	Taunton	6/35	-	-
1574	2/79	Wells	Oxford	12/37	-	-
1575	2/79	Taunton	Laira	6/30	-	-
1576	2/79	Salisbury	Swindon	3/28	-	-
1577	3/79	Newport	Tondu	12/34	-	-
1578	3/79	Truro	Bridgwater	9/33	-	-
1579	3/79	Merthyr	Llantrisant	6/30	-	-
1580	3/79	Aberdare	Aberbeeg	8/35	-	-
1581	4/79	Worcester	Leominster	7/29	-	-
1582	5/79	Neath	Danygraig	2/31	-	-
1583	5/79	Neath	Neath	11/28	-	-
1584	5/79	Landore	Neath	9/33	-	-
1585	5/79	Pontypool	Aberdare	4/46	-	-
1586	6/79	Swansea E Dock	Westbury	2/31	-	-
1587	6/79	Bristol	St. Blazey	1/34	-	-
1588	6/79	Neath	Llanelly	8/29	-	-
1589	6/79	Yeovil	Bath Road	1/28	-	-
1590	7/79	Goodwick	Pembroke Dock	2/31	-	-
1592	7/79	Aberbeeg	Ebbw Junction	2/28	-	-
1593	8/79	Tondu	Ebbw Junction	7/34	-	-
1594	8/79	Bristol	Swindon	6/30	-	-
1595	8/79	Worcester	Worcester	12/27	-	-
1596	9/79	Tondu	Tondu	10/28	-	-
1597	9/79	Goodwick	Goodwick	12/34	-	-
1598	10/79	Frome	Yeovil	5/35	-	-
1599	10/79	Pontypool	Danygraig	12/36	-	-
1600	11/79	Laira	Pontypool	12/37	-	-
1601	11/79	Didcot	Didcot	4/31	-	-
1602	12/79	Newport	Dock Street	1/28	-	-
1603	12/79	Neath	Bath Road	1/31	-	-
1604	12/79	Bristol	Bristol SPM	9/29	-	-
1605	12/79	Glyn Neath	Aberbeeg	3/31	-	-
1606	12/79	Newport	Abergavenny	3/31	-	-
1607	1/80	Newport	Merthyr	9/30	-	-
1608	1/80	Llanelly	Neath	5/35	-	-
1609	1/80	Hereford	Duffryn Yard	3/31	-	-
1610	2/80	Salisbury	Didcot	10/35	-	-
1611	2/80	Duffryn Yard	Duffryn Yard	9/35	-	-
1612	2/80	Tondu	Tondu	12/28	-	-
1613	2/80	St. Blazey	St. Blazey	12/29	-	-
1614	3/80	Pontypool	Abergavenny	3/28	-	-
1615	3/80	Bridgend	Yeovil	4/39	-	-
1616	3/80	Aberbeeg	Dock Street	1/29	-	-
1617	3/80	Newton Abbot	Laira	12/34	-	-
1618	4/80	Newport	Ebbw Junction	5/29	-	-
1619	4/80	Salisbury	Bristol SPM	3/29	-	-
1620	4/80	Pontypool	Swindon	11/37	-	-
1621	5/80	Duffryn Yard	Cardiff	1/32	-	-
1622	5/80	Tondu	Tondu	6/29	-	-
1623	5/80	Llanelly	Whitland	12/36	-	-
1624	5/80	Westbury	Yeovil	4/46	-	-

No.	Date	1922 Shed	Last GWR Shed	Wdn GWR	Last BR shed	Wdn BR
1625	6/80	Newport	Dock Street	8/28	-	-
1626	6/80	Yeovil	Bristol SPM	8/28	-	-
1627	6/80	Hereford	Worcester	5/30	-	-
1628	6/80	Llantrisant	Llantrisant	10/32	-	-
1629	7/80	Swansea E Dock	Swansea E Dock	9/37	-	-
1630	7/80	Neath	Neath N&B	11/34	-	-
1631	7/80	Neath	Llanelly	6/29	-	-
1632	8/80	Wolverhampton	Stafford Road	4/39	-	-
1633	8/80	Swansea E Dock	Llantrisant	1/29	-	-
1634	8/80	Southall	Oswestry	7/29	-	-
1635	9/80	Merthyr	Aberbeeg	4/32	-	-
1636	9/80	Cardiff	Merthyr	4/29	-	-
1637	10/80	Swansea E Dock	Neath	10/35	-	-
1638	10/80	Whitland	Llanelly	7/38	-	-
1639	10/80	Paddington	St. Blazey	3/31	-	-
1640	11/80	Newport	Pontypool	3/31	-	-
1641	11/80	Duffryn Yard	Neath	11/28	-	-
1642	11/80	Newport	Llantrisant	10/35	-	-
1643	12/80	Llantrisant	Pill	2/31	-	-
1644	12/80	Gloucester	Frome	6/35	-	-
1645	12/80	Hereford	Bridgend	12/34	-	-
1646	12/80	Yeovil	Bristol SPM	12/34	-	-
1647	1/81	Newport	Aberdare	11/36	-	-
1648	1/81	Duffryn Yard	Llanelly	4/30	-	-
1649	2/81	Tondu	Bridgend	9/35	-	-
1650	2/81	Newton Abbot	Exeter	3/38	-	-
1651	3/81	Paddington	Paddington	4/30	-	-
1652	3/81	Danygraig	Weymouth	6/32	-	-
1653	3/81	Newport	Severn Tunnel	12/30	-	-
1654	4/81	Plymouth	Exeter	1/32	-	-
1655	4/81	Neath	Swansea E Dock	7/31	-	-
1656	4/81	Aberbeeg	Pontypool	5/30	-	-
1657	5/81	Swindon	Bristol SPM	2/31	-	-
1658	5/81	St. Blazey	Laira	7/34	-	-
1659	5/81	Paddington	Southall	10/30	-	-
1660	6/81	Aberavon	Swindon	6/38	-	-
1661	6/86	Swansea E Dock	Duffryn Yard	2/30	-	-
1662	6/86	Glyn Neath	Neath	2/30	-	-
1663	6/86	Westbury	Pill	10/30	-	-
1664	6/86	Oxford	Oxford	8/29	-	-
1665	7/86	Oxford	Tondu	3/31	-	-
1666	7/86	Newport	Westbury	10/27	-	-
1667	7/86	Cardiff E M	Cardiff Docks	8/31	-	-
1669	7/86	Swansea E Dock	Swansea E Dock	3/29	-	-
1671	8/86	Bristol	Neath N&B	10/27	-	-
1672	8/86	Pontypool	Cardiff	5/28	-	-
1673	9/86	Tyseley	Tyseley	11/28	-	-
1675	9/86	Cardiff	Ebbw Junction	1/29	-	-
1676	10/86	Cardiff E M	Swansea E Dock	10/26	-	-
1678	10/86	Southall	Southall	10/26	-	-
1679	11/86	Pill	Dock Street	10/26	-	-
1680	11/86	Southall	Southall	10/26	-	-
1681	11/86	Reading	Reading	7/27	-	-
1682	11/86	Paddington	Paddington	12/26	-	-
1683	12/86	Pill	Llantrisant	9/26	-	-
1684	12/86	Newport	Llantrisant	5/29	-	-
1686	12/86	Cardiff	Ebbw Junction	1/30	-	-
1687	1/87	Southall	Southall	4/29	-	-
1688	1/87	Cardiff	Oxford	4/30	-	-
1689	1/87	Cardiff E M	Cardiff	4/31	-	-

No.	Date	1922 Shed	Last GWR Shed	Wdn GWR	Last BR shed	Wdn BR
1690	2/87	Wolverhampton	Tyseley	3/28	-	-
1691	2/87	Cardiff	Pill	11/27	-	-
1695	4/87	Glyn Neath	Danygraig	10/27	-	-
1696	5/87	Landore	Southall	4/30	-	-
1697	6/87	Cardiff	Ebbw Junction	5/29	-	-
1698	6/87	Cardiff	Cardiff	11/27	-	-
1699	6/87	Newport	Pill	9/26	-	-
1700	6/87	Swindon	Danygraig	3/30	-	-
1701	2/91	Cardiff	Cardiff	7/28	-	-
1702	3/91	Truro	Truro	9/38	-	-
1703	3/91	Aberdare	Swindon	7/45	-	-
1704	3/91	Swindon	Duffryn Yard	1/35	-	-
1705	3/91	Newport	Cathays	-	Cardiff E Dock	11/50
1706	3/91	Llanelly	Trawsfynydd	-	Croes Newydd	6/48
1707	4/91	Pontypool	Cardiff E Dock	7/47	-	-
1708	4/91	Tondu	Severn Tunnel	1/35	-	-
1709	4/91	Llanelly	Pill	-	Pill	11/50
1710	4/91	Danygraig	Yeovil	2/37	-	-
1711	4/91	Aberdare	Pontypool	6/28	-	-
1712	5/91	Stourbridge	Birkenhead	1/46	-	-
1713	5/91	Exeter	Ebbw Junction	-	Ebbw Junction	6/48
1714	5/91	Truro	Cardiff	7/46	-	-
1716	5/91	Aberavon	Aberdare	5/47	-	-
1717	6/91	St. Blazey	Exeter	12/36	-	-
1718	6/91	Branches Fork	Merthyr	10/36	-	-
1719	6/91	Danygraig	Llanelly	11/45	-	-
1720	6/91	Pontypool	Ebbw Junction	-	Ebbw Junction	12/49
1721	5/92	Bristol	Merthyr	7/46	-	-
1722	5/92	Cardiff	Llantrisant	11/46	-	-
1723	5/92	Branches Fork	Ebbw Junction	1/35	-	-
1724	5/92	Tondu	Worcester	11/29	-	-
1725	6/92	Paddington	Tondu	10/46	-	-
1726	6/92	Tondu	Pill	-	Pill	4/48
1727	6/92	Llanelly	Neath	5/35	-	-
1728	6/92	Stourbridge	Newton Abbot	2/32	-	-
1729	7/92	Paddington	Yeovil	9/42	-	-
1730	7/92	Tondu	Tondu	-	Tondu	8/48
1731	7/92	Aberavon	Swindon	-	Swindon	6/49
1732	7/92	Duffryn Yard	Llanelly	7/46	-	-
1733	8/92	Neath	Duffryn Yard	6/36	-	-
1734	8/92	Newport	Ebbw Junction	2/46	-	-
1735	8/92	Newport	Ebbw Junction	7/46	-	-
1736	8/92	Stourbridge	Newton Abbot	4/45	-	-
1737	8/92	Aberdare	Ebbw Junction	11/45	-	-
1738	9/92	Severn Tunnel	Truro	2/39	-	-
1739	9/92	Tyseley	Stafford Road	3/31	-	-
1740	9/92	Llanelly	Danygraig	11/44	-	-
1741	3/92	Chester	Kington	12/38	-	-
1742	3/92	Tyseley	Oxford	-	Oxford	2/50
1743	4/92	Garnant	Oxford	2/45	-	-
1744	4/92	Corwen	Trawsfynydd	7/38	-	-
1745	5/92	Chester	Stourbridge	-	Stourbridge	8/48
1746	6/92	Stourbridge	Leominster	6/46	-	-
1747	6/92	Leamington	Croes Newydd	-	Croes Newydd	5/50
1748	7/92	Much Wenlock	Wellington	2/46	-	-
1749	7/92	Wolverhampton	Stourbridge	-	Stourbridge	10/48
1750	8/92	Wolverhampton	Tyseley	6/46	-	-
1751	10/92	Laira	Newton Abbot	11/45	-	-
1752	10/92	Bristol	Stourbridge	-	Ebbw Junction	3/50
1753	10/92	Stourbridge	Truro	-	Truro	4/48

No.	Date	1922 Shed	Last GWR Shed	Wdn GWR	Last BR shed	Wdn BR
1754	10/92	Glyn Neath	Duffryn Yard	-	Duffryn Yard	12/49
1755	11/92	Laira	Laira	4/36	-	-
1756	11/92	Danygraig	Danygraig	7/46	-	-
1757	11/92	Truro	Truro	3/31	-	-
1758	11/92	Swindon	Swindon	-	Swindon	3/49
1759	11/92	Aberdare	Ebbw Junction	11/45	-	-
1760	12/92	Bristol	Taunton	-	St. Blazey	7/50
1761	12/92	Taunton	Newton Abbot	12/46	-	-
1762	12/92	Glyn Neath	Oxley	-	Oxley	4/48
1763	12/92	Chester	Stourbridge	11/46	-	-
1764	12/92	Llantrisant	Pill	-	Pill	10/49
1765	1/93	Worcester	Southall	11/38	-	-
1766	1/93	Banbury	Stafford Road	6/38	-	-
1767	1/93	Tondu	Yeovil	12/47	-	-
1768	1/93	Swansea E Dock	Duffryn Yard	12/47	-	-
1769	2/93	Southall	Aberdare	-	Aberdare	4/48
1770	2/93	Reading	Tyseley	5/36	-	-
1771	12/92	Wolverhampton	Stourbridge	1/37	-	-
1772	12/92	Stourbridge	Wellington	3/29	-	-
1773	1/93	Gobowen	Bala	-	Croes Newydd	3/50
1774	2/93	Tyseley	Tyseley	10/32	-	-
1775	2/93	Severn Tunnel	Weymouth	10/38	-	-
1776	3/93	Wolverhampton	Stourbridge	6/38	-	-
1777	3/93	Stourbridge	St. Blazey	7/41	-	-
1778	4/93	Wellington	Croes Newydd	10/28	-	-
1779	5/93	Wellington	Croes Newydd	1/45	-	-
1780	5/93	Wellington	Croes newydd	-	Trawsfynydd	8/48
1781	6/93	Newport	Reading	4/29	-	-
1782	8/93	Wolverhampton	Truro	-	Truro	11/50
1783	9/93	Gobowen	Wellington	1/40	-	-
1784	10/93	Wolverhampton	Croes Newydd	2/45	-	-
1785	10/93	Wolverhampton	Oxford	9/46	-	-
1786	11/93	Wolverhampton	Weymouth	11/45	-	-
1787	11/93	Tyseley	Oswestry	1/47	-	-
1788	2/94	Tyseley	Croes Newydd	11/45	-	-
1789	5/94	Paddington	Weymouth	-	Weymouth	10/50
1790	6/94	Tyseley	Croes Newydd	6/46	-	-
1791	5/95	Frome	Cardiff	1/35	-	-
1792	5/95	Paddington	St. Blazey	11/45	-	-
1793	5/95	Paddington	Stourbridge	11/45	-	-
1794	6/95	Taunton	Truro	12/46	-	-
1795	6/95	Llantrisant	Newton Abbot	6/47	-	-
1796	6/95	Cardiff	Evesham	2/47	-	-
1797	6/95	Bristol	St. Blazey	7/46	-	-
1798	6/95	Bristol	Neath	7/46	-	-
1799	6/95	Stourbridge	Laira	-	Truro	12/49
1800	7/95	Bristol	Ebbw Junction	12/47	-	-
1801	1/81	Severn Tunnel	Aberbeeg	9/36	-	-
1802	2/81	Stourbridge	Chester	8/37	-	-
1803	2/81	Wellington	Stourbridge	12/44	-	-
1804	3/81	Bristol	Swindon	2/31	-	-
1805	3/81	Croes Newydd	Croes Newydd	7/30	-	-
1806	4/81	Duffryn Yard	Swansea E Dock	11/38	-	-
1807	4/81	Gloucester	Gloucester	9/28	-	-
1808	5/81	Chester	Stafford Road	6/45	-	-
1809	5/81	Tyseley	Stafford Road	5/36	-	-
1810	6/81	Wolverhampton	Tyseley	12/44	-	-
1811	7/81	Neath	Duffryn Yard	6/28	-	-
1812	7/81	Wolverhampton	Tyseley	4/37	-	-
1813	9/82	Severn Tunnel	Bassaleg	7/28	-	-

No.	Date	1922 Shed	Last GWR Shed	Wdn GWR	Last BR shed	Wdn BR
1814	9/82	Frome	Duffryn Yard	9/37	-	-
1815	9/82	Laira	Laira	12/36	-	-
1816	10/82	Bristol	Bridport	11/35	-	-
1817	10/82	Frome	Westbury	10/32	-	-
1818	10/82	Duffryn Yard	Pantyffynon	8/34	-	-
1819	11/82	Swansea E Dock	Laira	1/38	-	-
1820	11/82	Pontypool	Ebbw Junction	4/37	-	-
1821	11/82	Severn Tunnel	Carmarthen	5/33	-	-
1822	11/82	Bristol	Bath Road	3/30	-	-
1823	11/82	Newport	Duffryn Yard	8/44	-	-
1824	12/82	Croes Newydd	Tyseley	6/38	-	-
1825	12/82	Danygraig	Landore	9/28	-	-
1826	12/82	Llanelly	Carmarthen	10/36	-	-
1827	12/82	Bristol	Merthyr	12/34	-	-
1828	1/83	Llanelly	Duffryn Yard	12/34	-	-
1829	1/83	Aberdare	Aberdare	12/28	-	-
1830	1/83	Bristol	Severn Tunnel	10/33	-	-
1831	2/83	Newton Abbot	Swindon	5/45	-	-
1832	2/83	Landore	Reading	4/31	-	-
1834	3/83	Danygraig	Danygraig	8/29	-	-
1835	3/83	Bristol	Stourbridge	-	Stourbridge	1/49
1836	3/83	Oxford	Oxford	3/39	-	-
1837	3/83	Bristol	Landore	8/34	-	-
1838	4/83	Exeter	Merthyr	5/47	-	-
1839	4/83	Bristol	Worcester	9/38	-	-
1840	4/83	Severn Tunnel	Laira	3/37	-	-
1841	4/83	Bristol	Weymouth	12/34	-	-
1842	5/83	Bristol	Bristol SPM	5/28	-	-
1843	5/83	Southall	Southall	7/29	-	-
1844	5/83	Glyn Neath	Duffryn Yard	12/35	-	-
1845	6/83	Taunton	Kidderminster	8/30	-	-
1846	9/83	Tondu	Llantrisant	7/46	-	-
1847	10/83	Severn Tunnel	Pontypool	5/47	-	-
1848	10/83	Paddington	Didcot	6/35	-	-
1849	11/83	Pontypool	Pontypool	8/34	-	-
1850	11/83	Paddington	Oxford	9/38	-	-
1851	12/83	Swindon	Yeovil	9/31	-	-
1852	12/83	Bristol	Andover Jct	4/29	-	-
1853	1/84	Bristol	Leominster	12/34	-	-
1854	1/90	Oxford	Duffryn Yard	12/44	-	-
1855	2/90	Llanelly	Neath	-	Neath	12/50
1856	2/90	Westbury	Pill	11/45	-	-
1857	2/90	Branches Fork	Paddington	1/31	-	-
1858	3/90	Llanelly	Neath	-	Neath	10/50
1859	3/90	Swansea E Dock	Neath	7/46	-	-
1860	3/90	Swindon	Yeovil	12/34	-	-
1861	3/90	Llanelly	Didcot	-	Didcot	11/51
1862	3/90	Llantrisant	Ebbw Junction	-	Ebbw Junction	12/50
1863	3/90	Chester	Stafford Road	-	Stafford Road	9/49
1864	4/90	Swindon	Worcester	8/35	-	-
1865	4/90	Llanelly	Llanelly	11/33	-	-
1866	4/90	Oxford	Duffryn Yard	11/46	-	-
1867	4/90	Cardiff	Duffryn Yard	-	Duffryn Yard	11/48
1868	4/90	Aberdare	Cardiff	2/39	-	-
1869	4/90	Southall	Stafford Road	12/34	-	-
1870	5/90	Paddington	Severn Tunnel	-	Bridgend	10/50
1871	5/90	Llanelly	Llanelly	9/28	-	-
1872	5/90	Southall	Oxford	4/35	-	-
1873	5/90	Aberdare	Cathays	11/45	-	-
1874	10/90	St. Blazey	St. Blazey	11/31	-	-

No.	Date	1922 Shed	Last GWR Shed	Wdn GWR	Last BR shed	Wdn BR
1875	11/90	Aberdare	Pill	11/45	-	-
1876	11/90	Taunton	Newton Abbot	2/36	-	-
1877	11/90	Paddington	Southall	11/35	-	-
1878	11/90	Landore	Merthyr	-	Merthyr	11/49
1879	11/90	Neath	Llanelly	9/28	-	-
1880	11/90	Southall	Reading	4/35	-	-
1881	12/90	Cardiff	Yeovil	12/36	-	-
1883	12/90	Duffryn Yard	Ebbw Junction	12/47	-	-
1884	12/90	Tondu	Cardiff E Dock	-	Cardiff E Dock	8/49
1885	12/90	Worcester	Reading	11/28	-	-
1886	12/90	Llantrisant	Pill	6/45	-	-
1887	12/90	Tondu	Merthyr	7/45	-	-
1888	1/91	Aberdare	Cardiff E Dock	-	Cardiff E Dock	12/49
1889	1/91	Paddington	Cardiff	-	Cardiff	12/48
1890	1/91	Cardiff	Pill	6/47	-	-
1891	2/91	Llantrisant	Cardiff	-	Cardiff	12/49
1892	2/91	Wolverhampton	Stafford Road	4/36	-	-
1893	2/91	Taunton	Neath	11/45	-	-
1894	7/95	Pontypool	Ebbw Junction	-	Ebbw Junction	2/49
1895	7/95	Laira	Yeovil	5/47	-	-
1896	7/95	Reading	Pill	-	Pill	12/49
1897	7/95	Laira	Cardiff E Dock	-	Cardiff E Dock	1/49
1898	7/95	Paddington	Hereford	4/45	-	-
1899	8/95	Bristol	Taunton	1/45	-	-
1900	8/95	St. Blazey	St. Blazey	-	St. Blazey	4/48
1901	8/81	Paddington	Paddington	9/29	-	-
1902	9/81	Bristol	Swindon	10/43	-	-
1903	9/81	Cardigan	Carmarthen	-	Carmarthen	6/52
1904	10/81	Shrewsbury	Birkenhead	12/34	-	-
1905	10/81	Exeter	Laira	10/36	-	-
1906	11/81	Bristol	Frome	7/32	-	-
1907	11/81	Llanelly	Burry Port	-	Burry Port	1/50
1908	12/81	Paddington	Bristol SPM	11/36	-	-
1909	1/82	Laira	Taunton	-	Bridgwater	11/49
1910	4/82	Whitland	Whitland	7/44	-	-
1911	4/82	Paddington	Duffryn Yard	5/36	-	-
1912	4/82	Didcot	Paddington	-	Paddington	12/49
1913	6/82	Worcester	Worcester	11/29	-	-
1914	7/82	Birkenhead	Birkenhead	6/31	-	-
1915	7/82	Pontypool	Westbury	8/35	-	-
1916	7/82	Whitland	Burry Port	7/28	-	-
1917	8/82	Birkenhead	Birkenhead	-	Birkenhead	3/51
1918	8/82	Paddington	Llanelly	6/29	-	-
1919	9/82	Bristol	Worcester	-	Worcester	11/49
1920	10/82	Neath	Sandy	9/33	-	-
1921	10/82	Swansea E Dock	Didcot	5/35	-	-
1922	11/82	Laira	Penzance	4/36	-	-
1923	11/82	Tyseley	Whitland	7/39	-	-
1924	12/82	Tyseley	Machynlleth	8/38	-	-
1925	12/83	Laira	Southall	-	Southall	4/51
1926	1/84	Llanelly	Llanelly	11/28	-	-
1927	2/84	Taunton	Laira	10/29	-	-
1928	3/84	Bristol	Cardiff E Dock	12/34	-	-
1929	4/84	Whitland	Burry Port	7/35	-	-
1930	5/84	Laira	St. Blazey	-	St. Blazey	8/49
1931	6/84	Newport	Cardiff	12/38	-	-
1932	7/84	Exeter	Laira	4/30	-	-
1933	7/84	Swindon	Swindon	3/28	-	-
1934	8/84	Bristol	Swindon	4/28	-	-
1935	8/84	Didcot	Oxford	-	Birkenhead	11/53

No.	Date	1922 Shed	Last GWR Shed	Wdn GWR	Last BR shed	Wdn BR
1936	9/84	Worcester	Burry Port	2/35	-	-
1937	9/86	Hereford	Llanelly	12/34	-	-
1938	10/86	Paddington	Whitland	7/36	-	-
1939	10/86	Worcester	Worcester	11/29	-	-
1940	11/86	Swindon	Llanelly	4/33	-	-
1941	11/86	Plymouth	Llanelly	-	Llanelly	2/51
1942	12/86	Birkenhead	Sandy	12/34	-	-
1943	1/87	Tyseley	Gloucester	-	Gloucester	3/51
1944	1/87	Wolverhampton	Bristol SPM	10/34	-	-
1945	2/87	Paddington	Danygraig	-	Danygraig	11/49
1946	3/87	Paddington	Paddington	1/35	-	-
1947	3/87	Stourbridge	Tyseley	11/38	-	-
1948	3/87	Swindon	Sandy	8/34	-	-
1949	2/88	Shrewsbury	Birkenhead	-	Birkenhead	4/50
1950	2/88	Birkenhead	Tyseley	6/34	-	-
1951	3/88	Birkenhead	Stafford Road	10/37	-	-
1952	3/88	Bridgwater	Millbay	11/28	-	-
1953	4/88	Didcot	Didcot	4/29	-	-
1954	5/88	Taunton	Swindon	9/44	-	-
1955	5/88	Tyseley	Tyseley	8/39	-	-
1956	6/88	Laira	Exeter	12/38	-	-
1957	6/88	Aberayron	Burry Port	-	Burry Port	4/51
1958	7/88	Landore	Oxford	9/36	-	-
1959	8/88	Croes Newydd	Birkenhead	11/35	-	-
1960	9/88	Worcester	Croes Newydd	4/35	-	-
1961	10/89	Paddington	Sandy	9/36	-	-
1962	10/89	Gloucester	Kidderminster	2/29	-	-
1963	11/89	Birkenhead	Worcester	1/43	-	-
1964	11/89	Bristol	Whitland	-	Birkenhead	2/52
1965	12/89	Tyseley	Machynlleth	-	Machynlleth	1/50
1966	12/89	Pontypool	Whitland	7/39	-	-
1967	1/90	Bristol	Burry Port	-	Burry Port	6/51
1968	2/90	Birkenhead	Birkenhead	-	Birkenhead	9/51
1969	2/90	Paddington	Southall	-	Southall	8/49
1970	3/90	Paddington	Cardiff E Dock	7/38	-	-
1971	3/90	Newport	Cardiff	12/38	-	-
1972	6/90	Bristol	Bristol SPM	10/29	-	-
1973	7/90	Newton Abbot	Truro	-	Laira	12/49
1974	7/90	Neath	Pantyffynon	8/35	-	-
1975	9/90	Paddington	Llanelly	9/35	-	-
1976	9/90	Stourbridge	Paddington	11/36	-	-
1977	10/90	Newport	Salisbury	2/31	-	-
1978	10/90	Swansea E Dock	Burry Port	2/36	-	-
1979	11/90	Birkenhead	Whitland	-	Whitland	8/50
1980	11/90	Bristol	Bristol SPM	11/34	-	-
1981	12/90	Llanelly	Llanelly	7/28	-	-
1982	1/91	Swansea E Dock	Llanelly	1/36	-	-
1983	1/91	Paddington	Whitland	12/35	-	-
1984	2/91	Pontypool	Pontypool	6/28	-	-
1985	2/91	Laira	Laira	7/34	-	-
1986	3/91	Carmarthen	Llanelly	10/29	-	-
1987	3/91	Paddington	Paddington	1/29	-	-
1988	5/91	Shrewsbury	Llanelly	1/47	-	-
1989	6/91	Worcester	Gloucester	-	Gloucester	9/50
1990	6/91	Cardiff	Laira	-	Laira	11/49
1991	7/91	Bristol	Llanelly	-	Burry Port	1/53
1992	7/91	Moorswater	Exeter	3/35	-	-
1993	8/91	St. Blazey	Barry	-	Barry	4/51
1994	8/91	Paddington	Whitland	10/38	-	-
1995	9/91	Bristol	Bristol SPM	8/30	-	-

No.	Date	1922 Shed	Last GWR Shed	Wdn GWR	Last BR shed	Wdn BR
1996	10/91	Paddington	Whitland	-	Gloucester	1/53
1997	10/91	Truro	Penzance	5/29	-	-
1998	11/91	Cardiff	Cardiff	11/34	-	-
1999	11/91	Plymouth	Exeter	6/46	-	-
2000	12/91	Paddington	Abercynon	-	Abercynon	12/49
2001	12/91	Worcester	Worcester	-	Worcester	8/52
2002	1/92	Whitland	Burry Port	-	Burry Port	2/52
2003	1/92	Didcot	Paddington	5/29	-	-
2004	9/92	Birkenhead	Birkenhead	-	Birkenhead	1/52
2005	9/92	Wolverhampton	Tyseley	12/44	-	-
2006	10/92	Birkenhead	Birkenhead	-	Birkenhead	12/49
2007	10/92	Tyseley	Worcester	-	Worcester	12/49
2008	11/92	Paddington	Abercynon	-	Birkenhead	3/58
2009	4/94	Landore	Gloucester	-	Gloucester	1/51
2010	5/94	Birkenhead	Whitland	-	Whitland	3/53
2011	11/94	Bristol	Whitland	-	Birkenhead	8/56
2012	11/94	Chippenham	Burry Port	-	Birkenhead	6/58
2013	12/94	Neath	Whitland	-	Whitland	5/50
2014	12/94	Swansea E Dock	Swindon	-	Swindon	11/51
2015	1/95	Bristol	Bristol SPM	11/45	-	-
2016	2/95	Worcester	Worcester	-	Worcester	1/52
2017	2/95	Bristol	Swindon	-	Swindon	3/51
2018	3/95	Tyseley	Cardigan	-	Whitland	12/49
2019	3/95	Hereford	Burry Port	-	Burry Port	12/49
2020	5/95	Newton Abbot	Worcester	12/38	-	-
2021	2/97	Newport	Pontypool	-	Pontypool	6/51
2022	3/97	Lydney	Cardiff E Dock	-	Cardiff E Dock	12/49
2023	4/97	Bristol	Frome	-	Westbury	1/52
2024	4/97	Lydney	Hereford	3/39	-	-
2025	5/97	Lydney	Lydney	-	Gloucester	5/52
2026	5/97	Paddington	Hereford	-	Hereford	4/51
2027	6/97	Tondu	Llanelly	-	Burry Port	2/57
2028	6/97	Stourbridge	Stafford Road	9/38	-	-
2029	7/97	Oxford	Hereford	-	Llanelly	11/49
2030	8/97	Stourbridge	Wellington	-	Wellington	2/52
2031	9/97	Bristol	Bristol SPM	-	Bristol SPM	1/53
2032	10/97	Swindon	Oswestry	-	Oswestry	6/51
2033	11/97	Newport	Pill	-	Pill	3/51
2034	11/97	Pontypool	Lydney	-	Kidderminster	9/55
2035	1/98	Pontypool	Pontypool	-	Ebbw Junction	3/55
2036	1/98	Abingdon	Birkenhead	2/39	-	-
2037	1/98	Paddington	Worcester	-	Worcester	7/50
2038	2/98	Lydney	Bridgwater	-	Laira	4/52
2039	2/98	Lydney	Lydney	-	Lydney	4/50
2040	3/98	Swindon	Hereford	-	Birkenhead	10/56
2041	4/98	Lydney	Pontypool	3/47	-	-
2042	4/98	Gloucester	Llanelly	-	Birkenhead	4/53
2043	5/98	Lydney	Lydney	-	Birkenhead	1/55
2044	6/98	Bristol	Lydney	-	Ebbw Junction	7/51
2045	6/98	Duffryn Yard	Lydney	-	Lydney	12/49
2046	6/98	Paddington	Llanelly	6/47	-	-
2047	7/98	Pontypool	Carmarthen	-	Carmarthen	12/49
2048	8/98	Cardiff	Cardiff E Dock	-	Ebbw Junction	5/52
2049	9/98	Tondu	Pill	11/44	-	-
2050	9/98	Bodmin	St. Blazey	-	St. Blazey	10/51
2051	10/98	Hereford	Worcester	-	Kidderminster	7/51
2052	10/98	Wolverhampton	Birkenhead	-	Birkenhead	5/50
2053	11/98	Gloucester	Westbury	-	Bristol SPM	4/54
2054	12/98	Southall	Oswestry	-	Moat Lane	3/51
2055	1/99	Shrewsbury	Slough	-	Danygraig	1/51

No.	Date	1922 Shed	Last GWR Shed	Wdn GWR	Last BR shed	Wdn BR
2056	1/99	Stourbridge	Carmarthen	-	Carmarthen	3/51
2058	3/99	Aberdare	Cardiff E Dock	9/45	-	-
2059	3/99	Weymouth	Llanelly	-	Llanelly	11/49
2060	4/99	Swindon	Swindon	-	Swindon	12/54
2061	4/99	Stourbridge	Stafford Road	-	Wellington	4/55
2062	5/99	Cardiff	St. Blazey	8/30	-	-
2063	6/99	Newport	Ebbw Junction	-	Ebbw Junction	5/51
2064	6/99	Bristol	Bristol SPM	-	Bristol SPM	11/49
2065	6/99	Tyseley	Banbury	-	Tyseley	11/49
2066	7/99	Pontypool	Cathays	-	Birkenhead	9/51
2067	8/99	Tyseley	Stafford Road	-	Birkenhead	11/52
2068	8/99	Swindon	Moat Lane	-	Oswestry	1/53
2069	9/99	Lydney	Carmarthen	-	Birkenhead	4/59
2070	10/99	Lydney	Bristol SPM	-	Bristol SPM	8/55
2071	10/99	Shrewsbury	Tyseley	-	Croes Newydd	6/50
2072	11/99	Southall	Bath Road	-	Birkenhead	7/56
2073	11/99	Newport	Ebbw Junction	-	Ebbw Junction	6/51
2074+	12/99	Laira	Croes Newydd	-	Croes Newydd	5/55
2075	1/00	Paddington	Oswestry	-	Oswestry	3/51
2076	2/00	Tyseley	Reading	-	Oxford	10/51
2077	2/00	Croes Newydd	Pontypool	2/47	-	-
2078	3/00	Stourbridge	Slough	12/44	-	-
2079	6/00	Bristol	Duffryn Yard	-	Duffryn Yard	11/52
2080	9/00	Gloucester	Lydney	-	Lydney	3/52
2081	10/00	Paddington	Llanelly	-	Llanelly	10/54
2082	10/00	Cheltenham	Danygraig	-	Birkenhead	6/55
2083	11/00	Paddington	Llanelly	-	Llanelly	11/51
2084	11/00	Bullo Pill	Lydney	2/46	-	-
2085	12/00	Gloucester	Llanelly	-	Birkenhead	8/53
2086	1/01	Lydney	Cardiff E Dock	-	Pill	5/52
2087+	1/01	Swindon	Croes Newydd	-	Croes Newydd	2/52
2088	3/01	Lydney	Exeter	-	Taunton	8/55
2089	3/01	Stourbridge	Birkenhead	-	Birkenhead	9/51
2090	3/01	Stourbridge	Stourbridge	-	Ebbw Junction	3/55
2091	4/01	Landore	Lydney	-	Lydney	4/50
2092	4/01	Shrewsbury	Stourbridge	-	Birkenhead	8/55
2093	5/01	Lydney	Kidderminster	-	Worcester	1/52
2094	5/01	Pontypool	Pontypool	-	Pontypool	5/52
2095	6/01	Tyseley	Stafford Road	-	Stafford Road	4/51
2096	7/01	Stourbridge	Hereford	-	Lydney	5/50
2097	7/01	Paddington	Penzance	-	Laira	3/55
2098	8/01	Merthyr	Llanelly	-	Llanelly	5/51
2099	9/01	Kidderminster	Hereford	-	Birkenhead	6/54
2100	11/01	Kidderminster	Worcester	-	Worcester	6/52
2101	1/02	Southall	Worcester	-	Birkenhead	3/56
2102	1/02	Stourbridge	Ross-on-Wye	-	Hereford	11/49
2103	2/02	Southall	Laira	3/46	-	-
2104	3/02	Shrewsbury	Birkenhead	-	Birkenhead	5/51
2105+	4/02	Stourbridge	Stourbridge	-	Stourbridge	10/50
2106	4/02	Stourbridge	Birkenhead	-	Birkenhead	8/52
2107	5/02	Stourbridge	Stourbridge	-	Birkenhead	6/56
2108	5/02	Stourbridge	Birkenhead	-	Birkenhead	12/54
2109	6/02	Shrewsbury	Stafford Road	-	Worcester	2/52
2110	7/02	Stourbridge	Stafford Road	-	Stafford Road	7/50
2111	7/02	Llanelly	Carmarthen	-	Carmarthen	3/53
2112	8/02	Reading	Slough	-	Birkenhead	9/54
2113	9/02	Merthyr	Pill	-	Pill	3/50
2114	9/02	Lydney	Lydney	-	Lydney	12/49
2115	10/02	Swindon	Worcester	-	Worcester	6/52
2116	10/02	Newton Abbot	Taunton	2/46	-	-

No.	Date	1922 Shed	Last GWR Shed	Wdn GWR	Last BR shed	Wdn BR
2117	11/02	Southall	Aberdare	-	Pontypool	6/51
2118+	11/02	St. Blazey	Stourbridge	-	Croes Newydd	4/55
2119	12/02	Weymouth	Swindon	8/46	-	-
2120	1/03	Croes Newydd	Stourbridge	8/46	-	-
2121	2/03	Watlington	Lydney	-	Gloucester	6/52
2122	3/03	Merthyr	Ebbw Junction	-	Ebbw Junction	11/52
2123	3/03	Cardiff	Cardiff E Dock	-	Cardiff E Dock	10/52
2124	5/03	Croes Newydd	Cardiff E Dock	-	Cardiff E Dock	7/50
2125+	5/03	Bodmin	St. Blazey	-	St. Blazey	8/55
2126	5/03	Bridport	Llanelly	-	Burry Port	9/50
2127	5/03	Bodmin	Taunton	-	Taunton	9/52
2129	7/03	Croes Newydd	Birkenhead	-	Birkenhead	3/53
2130	7/03	Gobowen	Cardiff E Dock	-	Cardiff E Dock	5/50
2131	8/03	Bristol	Lydney	-	Lydney	11/51
2132	9/03	Bristol	Lydney	-	Lydney	8/50
2133+	9/03	Pontypool	St. Blazey	-	St. Blazey	2/52
2134	10/03	Llanelly	Danygraig	-	Birkenhead	5/57
2135	10/03	Croes Newydd	Bristol SPM	-	Bristol SPM	1/53
2136	11/03	Aberdare	Pill	-	Birkenhead	4/55
2137	11/03	Watlington	Llanelly	-	Llanelly	12/49
2138	12/03	Lydney	Hereford	-	Hereford	5/56
2139	1/04	Neath	Danygraig	2/45	-	-
2140	1/04	Pontypool	Cardiff E Dock	-	Cardiff E Dock	5/52
2141	3/04	Merthyr	Cardiff E Dock	-	Cardiff E Dock	10/50
2142	3/04	Kidderminster	Cathays	2/47	-	-
2143+	4/04	Tondu	Stourbridge	-	Stourbridge	2/52
2144	5/04	Croes Newydd	Lydney	-	Worcester	5/55
2145+	5/04	Tyseley	Croes Newydd	-	Croes Newydd	10/50
2146	6/04	Hereford	Gloucester	-	Danygraig	3/53
2147	6/04	Lydney	Cardiff E Dock	-	Cardiff E Dock	3/53
2148	7/04	St. Blazey	Penzance	-	Laira	2/52
2149+	8/04	Lydney	Stourbridge	-	Croes Newydd	12/52
2150	9/04	Lydney	Llanelly	-	Llanelly	1/52
2151	9/04	Leominster	Machynlleth	-	Danygraig	5/52
2152	10/04	Stourbridge	Tyseley	-	Birkenhead	11/51
2153	10/04	Worcester	Lydney	-	Lydney	12/50
2154	11/04	Newport	Pill	-	Pill	2/52
2155	11/04	Gloucester	Lydney	-	Lydney	11/50
2156	12/04	Kington	Stafford Road	-	Birkenhead	2/53
2157+	1/05	Lydney	Croes Newydd	-	Croes Newydd	4/51
2158	2/05	Stourbridge	Lydney	11/45	-	-
2159	2/05	Slough	Pontypool	-	Pill	8/51
2160	3/05	Pontypool	Lydney	-	Birkenhead	3/57
2221	9/05	Slough	Reading	9/33	-	-
2222	8/06	Slough	Didcot	12/34	-	-
2223	9/06	Slough	Slough	4/32	-	-
2224	9/06	Slough	Paddington	8/33	-	-
2225	9/06	Aylesbury	Oxford	12/34	-	-
2226	10/06	Slough	Paddington	6/34	-	-
2227	10/06	Paddington	Slough	6/31	-	-
2228	10/06	Slough	Slough	1/31	-	-
2229	10/06	Slough	Reading	1/32	-	-
2230	10/06	Aylesbury	Reading	9/32	-	-
2231	9/08	Reading	Slough	5/31	-	-
2232	9/08	Slough	Paddington	10/33	-	-
2233	10/08	Paddington	Paddington	10/32	-	-
2234	10/08	Slough	Paddington	1/32	-	-
2235	11/08	Aylesbury	Reading	1/35	-	-
2236	11/08	Reading	Paddington	1/32	-	-
2237	11/08	Slough	Reading	5/31	-	-

No.	Date	1922 Shed	Last GWR Shed	Wdn GWR	Last BR shed	Wdn BR
2238	12/08	Paddington	Slough	6/31	-	-
2239	1/09	Aylesbury	Didcot	12/34	-	-
2240	1/09	Reading	Slough	10/31	-	-
2241	6/12	Paddington	Slough	4/31	-	-
2242	6/12	Slough	Raeding	9/35	-	-
2243	6/12	Slough	Reading	12/34	-	-
2244	6/12	Reading	Reading	8/33	-	-
2245	6/12	Paddington	Slough	3/31	-	-
2246	6/12	Slough	Reading	11/35	-	-
2247	7/12	Paddington	Reading	10/33	-	-
2248	7/12	Slough	Paddington	10/31	-	-
2249	8/12	Slough	Reading	8/32	-	-
2250	8/12	Slough	Paddington	1/34	-	-
2301	5/83	Worcester	Moat Lane	1/34	-	-
2302	6/83	Bristol	Westbury	3/31	-	-
2303	6/83	Cardiff	Winchester	2/35	-	-
2304	6/83	Reading	Reading	1/29	-	-
2305	7/83	Exeter	Reading	6/35	-	-
2306	7/83	Worcester	Worcester	4/29	-	-
2307	7/83	Tyseley	Pontypool	4/32	-	-
2309	8/83	Newport	Ebbw Junction	4/31	-	-
2310	8/83	Worcester	Worcester	11/34	-	-
2311	8/83	Carmarthen	Bristol SPM	11/34	-	-
2312	9/83	Reading	Slough	2/31	-	-
2313	9/83	Exeter	Oxford	12/38	-	-
2314	9/83	Newport	Ebbw Junction	5/32	-	-
2315	10/83	Carmarthen	Moat Lane	11/45	-	-
2316	10/83	Taunton	Oswestry	11/34	-	-
2317	10/83	Hereford	Bristol SPM	11/34	-	-
2318	11/83	Bristol	Bristol SPM	8/29	-	-
2319	11/83	Neyland	Swindon	2/31	-	-
2320	11/83	Taunton	Builth Wells	11/44	-	-
2321	1/84	Kidderminster	Aberystwyth	10/38	-	-
2322	1/84	Aberystwyth	Bristol SPM	-	Bath Road	6/51
2323	1/84	Neyland	Portmadoc	-	Oswestry	6/53
2324	2/84	Worcester	Worcester	2/31	-	-
2325	2/84	Tyseley	Hereford	11/45	-	-
2326	2/84	Worcester	Worcester	5/32	-	-
2327	2/84	Barnstaple	Oswestry	-	Oswestry	4/53
2328	3/84	Worcester	Carmarthen	9/38	-	-
2329	3/84	Swindon	Gloucester	12/33	-	-
2330	3/84	Weymouth	Swindon	7/31	-	-
2331	3/84	Hereford	Worcester	3/31	-	-
2332	4/84	Worcester	Oxford	7/38	-	-
2333	4/84	Hereford	Worcester	11/31	-	-
2335	5/84	Bristol	Llanidloes	9/33	-	-
2336	5/84	Carmarthen	Brecon	11/34	-	-
2337	5/84	Westbury	Oswestry	11/35	-	-
2338	6/84	Swindon	Westbury	4/31	-	-
2339	6/84	Wells	Worcester	-	Worcester	3/52
2340	6/84	Reading	Bristol SPM	-	Westbury	6/54
2341	7/84	Hereford	Machynlleth	11/34	-	-
2342	8/84	Tyseley	Brecon	8/38	-	-
2343	8/84	Oxford	Brecon	-	Brecon	2/53
2344	8/84	Worcester	Oxford	2/31	-	-
2345	9/84	Carmarthen	Portmadoc	5/47	-	-
2346	9/84	Machynlleth	Reading	9/36	-	-
2347	10/84	Worcester	Swindon	4/38	-	-
2348	10/84	Gloucester	Shrewsbury	7/45	-	-
2349	11/84	Paddington	Hereford	-	Hereford	3/52

No.	Date	1922 Shed	Last GWR Shed	Wdn GWR	Last BR shed	Wdn BR
2350	11/84	Neyland	Lydney	-	Hereford	1/53
2351	11/84	Carmarthen	Brecon	-	Brecon	2/53
2352	11/84	Worcester	Machynlleth	5/35	-	-
2353	11/84	Weymouth	Machynlleth	12/36	-	-
2354	11/84	Llandovery	Moat Lane	-	Brecon	4/53
2355	12/84	Hereford	Taunton	3/30	-	-
2356	12/84	Stourbridge	Machynlleth	-	Portmadoc	7/48
2357	12/84	Swindon	Westbury	7/35	-	-
2358	12/84	Aberdare	Reading	11/34	-	-
2359	12/84	Birkenhead	Stafford Road	12/35	-	-
2360	12/84	Landore	Swindon	2/45	-	-
2361	9/85	Reading	Didcot	5/32	-	-
2362	10/85	Duffryn Yard	Oswestry	11/46	-	-
2363	11/85	Gloucester	Honeybourne	10/32	-	-
2364	11/85	Neath	Westbury	8/34	-	-
2365	11/85	Lydney	Hereford	10/28	-	-
2366	12/85	Duffryn Yard	Westbury	2/33	-	-
2367	12/85	Llanelly	Landovery	5/32	-	-
2368	12/85	Banbury	Banbury	3/38	-	-
2369	12/85	Reading	Reading	11/37	-	-
2370	12/85	Duffryn Yard	Southall	3/35	-	-
2371	1/86	Llanelly	Goodwick	3/30	-	-
2372	1/86	Neath	Llanelly	2/38	-	-
2373	2/86	Llanelly	Bristol SPM	11/34	-	-
2374	3/86	Lydney	Southall	11/30	-	-
2375	3/86	Lydney	Worcester	5/38	-	-
2376	3/86	Gloucester	Paddington	12/38	-	-
2377	4/86	Didcot	Bristol SPM	2/31	-	-
2378	4/86	Banbury	Bristol SPM	3/45	-	-
2379	4/86	Shrewsbury	Worcester	2/31	-	-
2380	5/86	Westbury	Duffryn Yard	9/43	-	-
2381	6/90	Taunton	Southall	11/45	-	-
2382	6/90	Bristol	Llanidloes	-	Llanidloes	4/50
2383	6/90	Gloucester	Machynlleth	8/46	-	-
2384	6/90	Bristol	Evesham	12/38	-	-
2385	7/90	Wolverhampton	Pontypool	-	Pontypool	10/51
2386	7/90	Wolverhampton	Brecon	-	Oswestry	12/50
2388	7/90	Aberdare	Moat Lane	6/38	-	-
2389	8/90	Bristol	Stourbridge	7/44	-	-
2390	8/90	Aberdare	Llanelly	8/37	-	-
2391	8/90	Bristol	Oswestry	12/33	-	-
2392	8/90	Tyseley	Oxley	10/40	-	-
2393	9/90	Llanelly	Didcot	10/40	-	-
2394	9/90	Bristol	Westbury	5/38	-	-
2395	9/90	Bristol	Fairford	11/45	-	-
2396	9/90	Cardiff	Carmarthen	6/34	-	-
2397	10/90	Southall	Didcot	4/35	-	-
2398	10/90	Bristol	Kingham	12/38	-	-
2399	10/90	Cardiff	Bristol SPM	12/40	-	-
2400	10/90	Carmarthen	Worcester	10/40	-	-
2401	10/91	Wolverhampton	Brecon	-	Brecon	1/53
2402	10/91	Banbury	Brecon	10/40	-	-
2403	10/91	Bristol	Gloucester	10/40	-	-
2404	10/91	Banbury	Carmarthen	10/40	-	-
2405	11/91	Taunton	Reading	10/40	-	-
2406	11/91	Tyseley	Oxley	11/45	-	-
2407	11/91	Pembroke Dock	Ebbw Junction	-	Oswestry	1/52
2408	11/91	Stourbridge	Banbury	-	Oswestry	1/53
2409	11/91	Paddington	Carmarhten	-	Oswestry	4/53
2410	11/91	Bristol	Taunton	10/40	-	-

No.	Date	1922 Shed	Last GWR Shed	Wdn GWR	Last BR shed	Wdn BR
2411	11/91	Wells	Carmarthen	-	Bristol SPM	4/54
2412	11/91	Kingham	Westbury	10/40	-	-
2413	11/91	Carmarthen	Stourbridge	10/40	-	-
2414	12/91	Newport	Severn Tunnel	-	Svern Tunnel	3/53
2415	12/91	Bristol	Swindon	12/40	-	-
2416	12/91	Evesham	Severn Tunnel	10/40	-	-
2417	12/91	Paddington	Aberystwyth	2/39	-	-
2418	12/91	Whitland	Taunton	10/40	-	-
2419	12/91	Chester	Stafford Road	10/40	-	-
2421	12/91	Wolverhampton	Aberystwyth	5/38	-	-
2422	1/92	Bristol	Kingham	10/40	-	-
2423	1/92	Chester	Didcot	10/40	-	-
2424	1/92	Cardiff	Llanidloes	7/46	-	-
2425	1/92	Wolverhampton	Shrewsbury	12/40	-	-
2426	2/92	Stourbridge	Bristol SPM	-	Westbury	12/53
2427	2/92	Oxford	Shrewsbury	10/40	-	-
2428	2/92	Gloucester	Oswestry	10/40	-	-
2429	3/92	Oxford	Oxford	10/40	-	-
2430	3/92	Reading	Reading	10/40	-	-
2431	2/93	Corwen	Carmarthen	-	Carmarthen	11/51
2432	3/93	Llanelly	Brecon	10/40	-	-
2433	3/93	Llanelly	Cardiff	12/40	-	-
2434	3/93	Neath	Severn Tunnel	10/40	-	-
2435	3/93	Taunton	Oswestry	10/40	-	-
2436	3/93	Tyseley	Chester	10/40	-	-
2437	4/93	Tyseley	Machynlleth	10/40	-	-
2438	4/93	Bristol	Aberystwyth	10/40	-	-
2439	4/93	Pontypool	Tyseley	10/40	-	-
2440	4/93	Bristol	Whitland	10/40	-	-
2441	4/93	Cardiff	Southall	10/40	-	-
2442	5/93	Tyseley	Stourbridge	12/40	-	-
2443	5/93	Bristol	Paddington	10/40	-	-
2444	5/93	Banbury	Bath Road	-	Westbury	11/52
2445	5/93	Worcester	Westbury	-	Bristol SPM	3/53
2446	6/93	Pontypool	Carmarthen	10/40	-	-
2447	6/93	Neath	Brecon	10/40	-	-
2449	7/93	Trowbridge	Moat Lane	-	Oswestry	1/53
2450	7/93	Westbury	Portmadoc	5/38	-	-
2451	10/95	Pontypool	Oxley	10/40	-	-
2452	10/95	Croes Newydd	Brecon	-	Brecon	10/52
2454	11/95	Southall	Bristol SPM	10/40	-	-
2455	11/95	Paddington	Oswestry	10/40	-	-
2456	12/95	Exeter	Westbury	10/40	-	-
2457	12/95	Aberystwyth	Bristol SPM	10/40	-	-
2458	12/95	Bristol	Worcester	-	Brecon	5/54
2459	12/95	Leamington	Worcester	10/40	-	-
2460	12/95	Taunton	Severn Tunnel	-	Severn Tunnel	4/54
2461	12/95	Banbury	Bristol SPM	10/40	-	-
2462	12/95	Stourbridge	Bath Road	-	Bristol SPM	1/53
2463	12/95	Reading	Reading	10/40	-	-
2464	1/96	Tyseley	Portmadoc	-	Portmadoc	12/49
2465	1/96	Carmarthen	Portmadoc	10/40	-	-
2466	1/96	Birkenhead	Aberystwyth	10/40	-	-
2467	1/96	Taunton	Wells	10/40	-	-
2468	2/96	Westbury	Brecon	-	Brecon	1/53
2469	2/96	Leamington	Portmadoc	10/40	-	-
2470	2/96	Oxford	Bristol SPM	12/40	-	-
2471	2/96	Reading	Hereford	10/40	-	-
2472	3/96	Southall	Taunton	10/40	-	-
2473	3/96	Bristol	Swindon	10/40	-	-

No.	Date	1922 Shed	Last GWR Shed	Wdn GWR	Last BR shed	Wdn BR
2474	2/96	Southall	Carmarthen	-	Reading	4/55
2475	3/96	Neath	Shrewsbury	10/40	-	-
2476	4/96	Tyseley	Brecon	10/40	-	-
2477	4/96	Chester	Oxford	10/40	-	-
2478	4/96	Hereford	Llanidloes	10/40	-	-
2479	4/96	Reading	Cardiff	10/40	-	-
2480	5/96	Bristol	Gloucester	10/40	-	-
2481	5/96	Lambourn	Oswestry	10/40	-	-
2482	5/96	Honeybourne	Moat Lane	-	Brecon	12/52
2483	5/96	Exeter	Oswestry	-	Oswestry	9/52
2484	6/96	Paddington	Cardiff	-	Oswestry	5/54
2485	6/96	Bristol	Bristol SPM	10/40	-	-
2486	6/96	Landore	Salisbury	10/40	-	-
2487	7/96	Llanelly	Banbury	10/40	-	-
2488	7/96	Reading	Croes Newydd	10/40	-	-
2489	7/96	Basingstoke	Reading	10/40	-	-
2490	7/96	Cardiff	Portmadoc	10/40	-	-
2511	2/97	Wolverhampton	Llanfyllin	10/40	-	-
2512	2/97	Reading	Reading	10/40	-	-
2513	2/97	Shrewsbury	Chester	-	Brecon	7/55
2514	3/97	Paddington	Moat Lane	10/40	-	-
2515	3/97	Paddington	Gloucester	-	Hereford	2/53
2516	3/97	Carmarthen	Llanidloes	-	Oswestry	5/56
2517	4/97	Westbury	Evesham	10/40	-	-
2518	4/97	Bristol	Weymouth	10/40	-	-
2519	4/97	Salisbury	Worcester	10/40	-	-
2520	4/97	Bristol	Paddington	10/40	-	-
2521	5/97	Wolverhampton	Severn Tunnel	12/40	-	-
2522	5/97	Barnstaple	Brecon	10/40	-	-
2523	5/97	Didcot	Oswestry	-	Brecon	9/49
2524	5/97	Llanelly	Pontypool	10/40	-	-
2525	5/97	Chippenham	Oswetry	12/45	-	-
2526	5/97	Chippenham	Westbury	10/40	-	-
2527	6/97	Bristol	Taunton	10/40	-	-
2528	6/97	Aberystwyth	Westbury	12/40	-	-
2529	6/97	Carmarthen	Carmarthen	10/40	-	-
2530	6/97	Newport	Carmarthen	10/45	-	-
2531	6/97	Southall	Banbury	10/40	-	-
2532	6/97	Wolverhampton	Didcot	-	Didcot	5/54
2533	7/97	Carmarthen	Didcot	10/40	-	-
2534	7/97	Worcester	Bristol SPM	-	Westbury	1/53
2535	7/97	Bristol	Oswestry	6/46	-	-
2536	7/97	Pontypool	Moat Lane	10/40	-	-
2537	7/97	Southall	Cardiff	-	Cardiff	1/53
2538	8/97	Shrewsbury	Cardiff	-	Oswestry	5/57
2539	8/97	Neath	Worcester	10/40	-	-
2540	8/97	Reading	Ebbw Junction	10/40	-	-
2541	8/97	Salisbury	Hereford	-	Hereford	6/54
2543	8/97	Taunton	Oswestry	-	Oswestry	2/53
2544	9/97	Neath	Portmadoc	10/40	-	-
2545	9/97	Swindon	Llanidloes	10/40	-	-
2546	9/97	Reading	Llanelly	10/40	-	-
2547	9/97	Bristol	Southall	10/40	-	-
2548	9/97	Wolverhampton	Croes Newydd	10/40	-	-
2549	9/97	Cardiff	Westbury	10/40	-	-
2550	9/97	Wolverhampton	Stafford Road	10/40	-	-
2551	11/97	Slough	Worcester	-	Kingham	9/53
2552	11/97	Wolverhampton	Newbury	10/40	-	-
2553	11/97	Shrewsbury	Reading	10/40	-	-
2554	11/97	Wolverhampton	Portmadoc	11/45	-	-

No.	Date	1922 Shed	Last GWR Shed	Wdn GWR	Last BR shed	Wdn BR
2555	11/97	Shrewsbury	Chester	10/40	-	-
2556	11/97	Southall	Builth Wells	-	Hereford	6/53
2557	11/97	Neath	Worcester	10/40	-	-
2558	12/97	Worcester	Machynlleth	10/40	-	-
2559	12/97	Oxford	Reading	10/40	-	-
2560	12/97	Wolverhampton	Weymouth	10/40	-	-
2561	1/98	Reading	Brecon	10/40	-	-
2562	1/98	Cardiff	Oswestry	10/40	-	-
2564	1/98	Bristol	Reading	9/45	-	-
2565	2/98	Westbury	Moat Lane	10/40	-	-
2566	2/98	Westbury	Stafford Road	10/40	-	-
2567	2/98	Pontypool	Bristol SPM	10/40	-	-
2568	2/98	Pontypool	Swindon	-	Swindon	5/53
2569	4/98	Shrewsbury	Llanidloes	-	Brecon	10/48
2570	4/98	Bristol	Cardiff	-	Cardiff	1/49
2571	10/98	Swindon	Oxford	10/40	-	-
2572	11/98	Reading	Portmadoc	-	Oswestry	12/52
2573	11/98	Cardiff	Reading	-	Reading	2/53
2574	11/98	Gloucester	Bristol SPM	10/40	-	-
2575	12/98	Stourbridge	Brecon	2/47	-	-
2576	12/98	Stourbridge	Builth Wells	10/40	-	-
2577	1/99	Pontypool	Honeybourne	10/40	-	-
2578	1/99	Didcot	Bristol SPM	-	Bristol SPM	9/53
2579	1/99	Barnstaple	Oxford	-	Brecon	1/54
2580	1/99	Reading	Gloucester	10/40	-	-
2600	8/00	Weymouth	Llanelly	2/36	-	-
2601	2/06	Newton Abbot	Cardiff	9/36	-	-
2602	1/07	Paddington	Laira	8/34	-	-
2603	2/06	Bristol	Pontypool	9/36	-	-
2604	4/06	Llantrisant	Oxley	4/36	-	-
2605	2/06	Pontypool	Gloucester	10/34	-	-
2606	5/06	Neath	Banbury	10/34	-	-
2607	1/06	Pontypool	Banbury	8/34	-	-
2608	1/06	Bristol	Stourbridge	10/45	-	-
2609	12/06	Cardiff	Oxford	9/34	-	-
2610	4/06	Aberdare	Banbury	8/35	-	-
2611	7/03	Aberdare	Paddington	9/34	-	-
2612	7/03	Llantrisant	Banbury	-	Banbury	1/48
2613	7/03	Llanelly	Worcester	10/38	-	-
2614	7/03	Aberdare	Severn Tunnel	12/35	-	-
2615	8/03	Wolverhampton	Stourbridge	11/34	-	-
2616	8/03	Pontypool	Pontypool	11/34	-	-
2617	8/03	Salisbury	Croes Newydd	9/34	-	-
2618	8/03	Aberdare	Ebbw Junction	12/38	-	-
2619	8/03	Bristol	Croes Newydd	1/46	-	-
2620	8/03	St. Blazey	Stourbridge	-	Stourbridge	8/49
2621	3/01	Croes Newydd	Stourbridge	9/34	-	-
2622	3/01	Neyland	Pontypool	6/46	-	-
2623	4/01	Tyseley	Oxley	-	Oxley	2/48
2624	4/01	Gloucester	Severn Tunnel	12/34	-	-
2625	4/01	Pontypool	Duffryn Yard	9/45	-	-
2626	4/01	Cardiff	Cheltenham	4/37	-	-
2627	4/01	Swansea E Dock	Gloucester	9/34	-	-
2628	4/01	Chester	Llanelly	8/35	-	-
2629	5/01	Hereford	Oxford	12/35	-	-
2630	5/01	Salisbury	Duffryn Yard	10/34	-	-
2631	5/01	Tyseley	Worcester	9/47	-	-
2632	5/01	Llantrisant	Bristol SPM	9/35	-	-
2633	5/01	Hereford	Oxley	9/35	-	-
2634	5/01	Pontypool	Ebbw Junction	11/35	-	-

No.	Date	1922 Shed	Last GWR Shed	Wdn GWR	Last BR shed	Wdn BR
2635	6/01	Wolverhampton	Severn Tunnel	8/47	-	-
2636	6/01	Carmarthen	Ebbw Junction	7/46	-	-
2637	6/01	Cardiff	Severn Tunnel	9/47	-	-
2638	6/01	Cardiff	Chester	10/45	-	-
2639	6/01	St. Blazey	Gloucester	5/46	-	-
2640	6/01	Newport	Banbury	6/47	-	-
2641	10/01	Wolverhampton	Oxley	9/34	-	-
2642	10/01	Laira	Ebbw Junction	6/39	-	-
2643	10/01	Landore	Banbury	-	Banbury	7/48
2644	10/01	Pontypool	Severn Tunnel	10/35	-	-
2645	11/01	Bristol	Llanelly	12/35	-	-
2646	11/01	Tyseley	Cardiff	10/34	-	-
2647	11/01	Tyseley	Exeter	12/34	-	-
2648	11/01	Aberdare	Pontypool	2/44	-	-
2649	11/01	Tondu	Pontypool	8/46	-	-
2650	11/01	Aberdare	Stourbridge	5/36	-	-
2651	12/01	Neyland	Hereford	-	Gloucester	6/49
2652	12/01	Oxford	Ebbw Junction	8/45	-	-
2653	12/01	Croes Newydd	Cardiff	11/34	-	-
2654	1/02	Croes Newydd	Ebbw Junction	1/36	-	-
2655	1/02	Aberdare	Stourbridge	-	Stourbridge	6/49
2656	1/02	Hereford	Gloucester	-	Gloucester	3/48
2657	1/02	Croes Newydd	Pontypool	10/46	-	-
2658	2/02	Laira	Chester	11/34	-	-
2659	2/02	Chester	Severn Tunnel	10/35	-	-
2660	2/02	Pontypool	Llanelly	1/46	-	-
2661	9/02	Tondu	Hereford	11/35	-	-
2662	9/02	Oxford	Chester	-	Chester	7/48
2663	8/02	Bristol	Neyland	10/35	-	-
2664	8/02	Chester	Severn Tunnel	1/36	-	-
2665	9/02	Pontypool	Oxley	-	Oxley	1/48
2666	9/02	Tyseley	Swansea E Dock	10/34	-	-
2667	9/02	Newport	Cardiff	-	Pontypool	10/49
2668	9/02	Tyseley	Oxford	12/36	-	-
2669	10/02	Wolverhampton	Pontypool	-	Pontypool	5/48
2670	10/02	Wolverhampton	Gloucester	8/38	-	-
2671	10/02	Salisbury	Cardiff	11/34	-	-
2672	10/02	Gloucester	Cardiff	11/35	-	-
2673	11/02	Leamington	Banbury	5/46	-	-
2674	11/02	St. Blazey	Chester	9/35	-	-
2675	11/02	Penzance	Bristol SPM	9/35	-	-
2676	11/02	Bristol	Cheltenham	9/46	-	-
2677	11/02	Cardiff	Aberdare	4/46	-	-
2678	11/02	Croes Newydd	Bristol SPM	6/36	-	-
2679	12/02	Tyseley	Worcester	2/46	-	-
2680	12/02	Cardiff	Hereford	-	Hereford	6/48
2700	3/01	Llanelly	Neath	11/45	-	-
2701	1/96	Severn Tunnel	Tondu	11/45	-	-
2702	2/96	Wolverhampton	Bristol SPM	-	Bristol SPM	1/50
2703	2/96	Wolverhampton	Bristol SPM	11/45	-	-
2704	3/96	Leamington	Croes Newydd	-	Croes Newydd	3/50
2705	3/96	Wolverhampton	Cardiff E Dock	2/45	-	-
2706	4/96	Tyseley	Stourbridge	-	Stourbridge	10/48
2707	5/96	Landore	Pantyffynnon	-	Pantyffynnon	7/50
2708	6/96	St. Balzey	Taunton	-	Taunton	7/49
2709	6/96	Stourbridge	Bristol SPM	-	Bath Road	9/48
2710	6/96	Wellington	Swindon	11/45	-	-
2711	7/96	Stourbridge	Trawsfynydd	11/45	-	-
2712	8/96	Tyseley	Stourbridge	-	Stourbridge	3/50
2713	9/96	Wolverhampton	Oxley	-	Croes Newydd	9/49

No.	Date	1922 Shed	Last GWR Shed	Wdn GWR	Last BR shed	Wdn BR
2714	9/96	Leamington	Leominster	-	Leominster	5/48
2715	10/96	Wolverhampton	Duffryn Yard	-	Duffryn Yard	7/50
2716	11/96	Leamington	Croes Newydd	-	Croes Newydd	9/50
2717	11/96	Tyseley	Croes Newydd	-	Croes Newydd	10/48
2718	12/96	Wolverhampton	Tyseley	10/45	-	-
2719	12/96	Tyseley	Tyseley	-	Croes Newydd	11/50
2720	2/97	Chester	Shrewsbury	11/45	-	-
2721	11/97	Landore	Duffryn Yard	-	Duffryn Yard	8/50
2722	3/98	Neath	Neath	-	Neath	11/50
2723	3/98	Llantrisant	Swindon	8/47	-	-
2724	3/98	Tondu	Cardiff E Dock	-	Cardiff E Dock	2/49
2725	3/98	Newton Abbot	Truro	1/46	-	-
2726	4/98	Pontypool	Ebbw Junction	8/47	-	-
2727	4/98	Pontypool	Cardiff E Dock	11/45	-	-
2728	5/98	Pontypool	Pontypool	-	Pontypool	4/48
2729	5/98	Pontypool	Tondu	3/46	-	-
2730	5/98	Llanelly	Llanelly	-	Llanelly	4/48
2731	5/98	Cardiff	Swindon	10/47	-	-
2732	6/98	Tondu	Merthyr	10/45	-	-
2733	6/98	Chester	Swindon	11/47	-	-
2734	7/98	Tondu	Pill	-	Pill	8/48
2735	7/98	Westbury	Swindon	8/47	-	-
2736	8/98	Tondu	Llanelly	8/47	-	-
2737	8/98	St. Blazey	Penzance	11/47	-	-
2738	8/98	Newport	Pill	-	Pill	12/49
2739	9/98	Cardiff	Pontypool	-	Pontypool	5/48
2740	9/98	Llantrisant	Truro	11/45	-	-
2741	4/99	Danygraig	Swansea E Dock	10/45	-	-
2742	4/99	Aberbeeg	Merthyr	10/45	-	-
2743	4/99	Cardiff	Worcester	-	Worcester	10/50
2744	4/99	Bristol	Croes Newydd	-	Shrewsbury	11/50
2745	5/99	Llantrisant	Shrewsbury	-	Croes Newydd	1/50
2746	5/99	Duffryn Yard	Llanelly	-	Llanelly	7/48
2747	5/99	Pontypool	Slough	11/45	-	-
2748	5/99	Exeter	Taunton	-	Taunton	4/48
2749	5/99	Branches Fork	Pontypool	-	Pontypool	4/48
2750	6/99	Cardiff	Treherbert	11/45	-	-
2751	6/99	Bristol	Llanelly	-	Llanelly	4/48
2752	6/99	Southall	Penznance	-	Penzance	3/48
2753	7/99	Wolverhampton	Banbury	11/45	-	-
2754	7/99	Cardiff	Cardiff E Dock	-	Cardiff E Dock	10/50
2755	7/99	Newton Abbot	Taunton	-	Taunton	8/48
2756	8/99	Danygraig	Gloucester	-	Gloucester	5/49
2757	8/99	Reading	Southall	-	Southall	2/50
2758	8/99	Newport	Wellington	12/45	-	-
2759	8/99	Newport	Cardiff	10/45	-	-
2760	8/99	Newport	Merthyr	-	Merthyr	10/50
2761	3/00	Danygraig	Tondu	-	Bridgend	3/50
2762	3/00	Aberdare	Tondu	7/46	-	-
2763	3/00	Tondu	Danygraig	11/45	-	-
2764	3/00	Tondu	Pill	-	Pill	7/48
2765	3/00	Aberdare	Slough	11/45	-	-
2766	8/00	Cardiff	Merthyr	11/45	-	-
2767	8/00	Llantrisant	Pontypool	-	Pontypool	1/49
2768	9/00	Reading	St. Blazey	10/47	-	-
2769	9/00	Frome	Bridgend	-	Bridgend	3/49
2770	9/00	Cardiff	Cardiff	11/45	-	-
2771	9/00	Aberdare	Stourbridge	-	Stourbridge	6/50
2772	9/00	Aberdare	Leamington	-	Leamington	11/49
2773	9/00	Tondu	Bridgend	10/46	-	-

No.	Date	1922 Shed	Last GWR Shed	Wdn GWR	Last BR shed	Wdn BR
2774	9/00	Tondu	Evesham	-	Evesham	4/48
2775	10/00	Newport	Duffryn Yard	7/46	-	-
2776	10/00	Laira	Laira	-	Laira	4/48
2777	10/00	Ledbury	Neath	5/47	-	-
2778	10/00	Chester	Chester	7/47	-	-
2779	10/00	Glyn Neath	Yeovil	11/45	-	-
2780	10/00	Trowbridge	St. Blazey	-	St. Blazey	7/50
2781	1/01	Branches Fork	Cardiff E Dock	-	Cardiff E Dock	6/48
2782	1/01	Neath	Pantyffynnon	2/46	-	-
2783	1/01	Newport	Didcot	11/47	-	-
2784	1/01	Newport	Didcot	5/47	-	-
2785	1/01	St. Blazey	Newton Abbot	-	Newton Abbot	4/48
2786	1/01	Cardiff	Bristol SPM	-	Bristol SPM	12/49
2787	1/01	Llanelly	Pantyffynnon	-	Pantyffynnon	12/49
2788	1/01	Pontypool	Tondu	6/47	-	-
2789	2/01	Neath	Swansea E Dock	-	Swansea E Dock	5/49
2790	2/01	Llanelly	Southall	-	Southall	6/50
2791	2/01	Wolverhampton	Stafford Road	-	Stafford Road	4/50
2792	2/01	Paddington	Duffryn Yard	-	Duffryn Yard	6/50
2793	2/01	Tondu	Pill	-	Pill	1/48
2794	2/01	Pontypool	Ebbw Junction	-	Ebbw Junction	11/49
2795	2/01	Trowbridge	Ebbw Junction	-	Ebbw Junction	1/49
2796	2/01	Pembroke Dock	Whitland	11/45	-	-
2797	3/01	Tondu	Neath	-	Neath	6/48
2798	3/01	Neath	Danygraig	-	Duffryn Yard	11/49
2799	3/01	Swindon	Worcester	-	Worcester	3/50
2800	6/03	Swansea E Dock	Pontypool	-	Ebbw Junction	4/58
2801	9/05	Aberdare	Aberdare	-	Pontypool	12/58
2802	9/05	Paddington	Pontypool	-	Didcot	12/58
2803	10/05	Laira	Westbury	-	Pontypool	4/59
2804	10/05	Bristol	Severn Tunnel	-	Banbury	7/59
2805	10/05	Pontypool	Banbury	-	Westbury	5/60
2806	10/05	Cardiff	Aberdare	-	Severn Tunnel	3/60
2807	10/05	Bristol	Hereford	-	Severn Tunnel	3/63
2808	10/05	Severn Tunnel	Aberdare	-	Llanelly	9/59
2809	10/05	Severn Tunnel	Severn Tunnel	-	Laira	1/60
2810	10/05	Aberdare	Aberdare	-	Aberdare	9/59
2811	11/05	Bristol	Aberdare	-	Westbury	10/59
2812	11/05	Pontypool	Chester	-	Stourbridge	1/59
2813	11/05	Severn Tunnel	Pontypool	-	Aberdare	11/60
2814	11/05	Aberdare	Taunton	-	Ebbw Junction	8/58
2815	11/05	Reading	Severn Tunnel	-	Severn Tunnel	12/59
2816	12/05	Severn Tunnel	Banbury	-	Banbury	10/59
2817	12/05	Reading	Weymouth	-	Pontypool	3/59
2818	12/05	Pontypool	Westbury	-	Neath	10/63
2819	12/05	Severn Tunnel	Severn Tunnel	-	Didcot	1/61
2820	12/05	Pontypool	Cardiff	-	Ebbw Junction	11/58
2821	1/07	Aberdare	Cardiff	-	Cardiff	9/60
2822	1/07	Aberdare	Aberdare	-	Barrow Road	11/64
2823	1/07	Aberdare	Aberdare	-	Stourbridge	4/59
2824	1/07	Cardiff	Severn Tunnel	-	Llanelly	8/59
2825	1/07	Aberdare	Oxley	-	Pontypool	3/59
2826	1/07	Severn Tunnel	Paddington	-	Severn Tunnel	9/59
2827	2/07	Oxford	Oxford	-	Westbury	8/58
2828	2/07	Pontypool	Aberdare	-	Aberdare	1/59
2829	2/07	Aberdare	Severn Tunnel	-	Gloucester	2/59
2830	2/07	Aberdare	Oxley	-	Oxley	1/59
2831	3/11	Pontypool	Aberdare	-	Aberdare	1/60
2832+	4/11	Severn Tunnel	Llanelly	-	Severn Tunnel	11/59
2833	4/11	Salisbury	Birkenhead	-	Aberdare	3/59

No.	Date	1922 Shed	Last GWR Shed	Wdn GWR	Last BR shed	Wdn BR
2834+	5/11	Severn Tunnel	Laira	-	Didcot	11/62
2835	5/11	Gloucester	Paddington	-	Swindon	5/60
2836	9/12	Aberdare	Aberdare	-	Severn Tunnel	6/64
2837	9/12	Severn Tunnel	Cardiff	-	Severn Tunnel	5/60
2838	9/12	Aberdare	Severn Tunnel	-	Severn Tunnel	8/59
2839+	10/12	Pontypool	Bristol SPM	-	Pontypool	6/64
2840	10/12	Salisbury	Paddington	-	Croes Newydd	6/59
2841	10/12	Bristol	Aberdare	-	Southall	12/63
2842	10/12	Tyseley	Ebbw Junction	-	Didcot	9/63
2843	11/12	Gloucester	Southall	-	Laira	6/59
2844	11/12	Paddington	Bristol SPM	-	Didcot	3/60
2845+	11/12	Aberdare	Paddington	-	Banbury	5/63
2846	12/12	Southall	Bristol SPM	-	Westbury	11/60
2847+	12/12	Severn Tunnel	Laira	-	Severn Tunnel	3/60
2848+	12/12	Severn Tunnel	Laira	-	Pontypool	6/59
2849+	12/12	Salisbury	Severn Tunnel	-	Ebbw Junction	8/62
2850	1/13	Westbury	Paddington	-	Oxley	2/60
2851	2/13	Tyseley	Ebbw Junction	-	Southall	6/63
2852	2/13	Pontypool	Stourbridge	-	Didcot	10/63
2853+	2/13	Aberdare	Bristol SPM	-	Reading	5/62
2854+	2/13	Aberdare	Severn Tunnel	-	Severn Tunnel	10/63
2855	2/13	Pontypool	Paddington	-	Banbury	12/62
2856	4/18	Pontypool	Paddington	-	Southall	4/64
2857	5/18	Paddington	Laira	-	Neath	4/63
2858	5/18	Tyseley	Southall	-	Severn Tunnel	1/63
2859	5/18	Paddington	Bristol SPM	-	Southall	12/64
2860	5/18	Paddington	Cardiff	-	Severn Tunnel	4/62
2861	6/18	Stourbridge	Oxford	-	Severn Tunnel	3/63
2862+	6/18	Paddington	Llanelly	-	Severn Tunnel	4/64
2863+	6/18	Westbury	Llanelly	-	Aberdare	6/59
2864	7/18	Chester	Cardiff	-	Cardiff	6/59
2865	7/18	Salisbury	Ebbw Junction	-	Severn Tunnel	1/63
2866	9/18	Pontypool	Ebbw Junction	-	Southall	3/63
2867	10/18	Aberdare	Laira	-	Ebbw Junction	7/63
2868	10/18	Chester	Paddington	-	Ebbw Junction	6/59
2869	10/18	Southall	Banbury	-	Severn Tunnel	6/59
2870	10/18	Cardiff	Aberdare	-	Aberdare	6/59
2871	11/18	Llanelly	Banbury	-	Taunton	5/63
2872+	11/18	Llanelly	Llanelly	-	Severn Tunnel	8/63
2873	11/18	Bristol	Exeter	-	Southall	12/64
2874	11/18	Reading	Banbury	-	Neath	5/63
2875	12/18	Aberdare	Paddington	-	Cardiff E Dock	4/64
2876	1/19	Banbury	Ebbw Junction	-	Ebbw Junction	1/65
2877	1/19	Aberdare	Cardiff	-	Cardiff	1/60
2878	2/19	Llanelly	Banbury	-	Tyseley	6/59
2879	3/19	Pontypool	Ebbw Junction	-	Southall	8/64
2880	3/19	Bristol	Aberdare	-	Oxford	6/59
2881	3/19	Paddington	Oxford	-	Newton Abbot	1/60
2882	4/19	Southall	Banbury	-	Taunton	12/63
2883	4/19	Wolverhampton	Chester	-	Westbury	11/62
2900	2/02	Pontypool	Chester	6/32	-	-
2901	5/06	Swindon	Cardiff	4/33	-	-
2902	5/06	Wolverhampton	Leamington	-	Leamington	8/49
2903	5/06	Pontypool	Tyseley	-	Tyseley	11/49
2904	5/06	Wolverhampton	Shrewsbury	10/32	-	-
2905	5/06	Pontypool	Cardiff	-	Cardiff	4/48
2906	5/06	Bristol	Cardiff	-	Cardiff	8/52
2907	5/06	Bristol	Ebbw Junction	7/33	-	-
2908	5/06	Newton Abbot	Swindon	-	Swindon	12/50
2909	5/06	Cardiff	Stafford Road	11/31	-	-

No.	Date	1922 Shed	Last GWR Shed	Wdn GWR	Last BR shed	Wdn BR
2910	5/06	Wolverhampton	Westbury	10/31	-	-
2911	8/07	Cardiff	Taunton	3/35	-	-
2912	8/07	Bristol	Weymouth	-	Weymouth	2/51
2913	8/07	Cardiff	Swindon	-	Swindon	5/48
2914	8/07	Goodwick	Weymouth	1/46	-	-
2915	8/07	Paddington	Chester	-	Chester	10/50
2916	8/07	Bristol	Tyseley	-	Tyseley	7/48
2917	8/07	Newton Abbot	Cardiff	10/34	-	-
2918	8/07	Worcester	Leamington	2/35	-	-
2919	9/07	Paddington	Taunton	2/32	-	-
2920	9/07	Wolverhampton	Hereford	-	Hereford	10/53
2921	9/07	Bristol	Banbury	12/45	-	-
2922	9/07	Bristol	Severn Tunnel	12/44	-	-
2923	9/07	Wolverhampton	Bath Road	10/34	-	-
2924	9/07	Shrewsbury	Hereford	-	Hereford	3/50
2925	9/07	Worcester	Shrewsbury	8/24	-	-
2926	9/07	Shrewsbury	Swindon	-	Chester	9/51
2927	9/07	Bristol	Swindon	-	Swindon	12/51
2928	9/07	Paddington	Westbury	-	Westbury	8/48
2929	9/07	Cardiff	Weston-S-Mare	-	Bath Road	12/49
2930	9/07	Bristol	Chester	-	Chester	10/49
2931	10/11	Exeter	Bath Road	-	Bath Road	2/51
2932	10/11	Bristol	Hereford	-	Tyseley	6/51
2933	11/11	Wolverhampton	Leamington	-	Leamington	1/53
2934	11/11	Exeter	Swindon	-	Swindon	6/52
2935	11/11	Cardiff	Swindon	-	Swindon	12/48
2936	11/11	Cardiff	Ebbw Junction	-	Ebbw Junction	4/51
2937	12/11	Exeter	Hereford	-	Hereford	6/53
2938	12/11	Bristol	Gloucester	-	Gloucester	8/52
2939	12/11	Paddington	Bath Road	-	Bath Road	12/50
2940	12/11	Cardiff	Cardiff	-	Cardiff	1/52
2941	5/12	Worcester	Westbury	-	Westbury	12/49
2942	5/12	Paddington	Bath Road	-	Bath Road	12/49
2943	5/12	Paddington	Cardiff	-	Cardiff	1/51
2944	5/12	Paddington	Hereford	-	Hereford	11/51
2945	6/12	Cardiff	Swindon	-	Cardiff	6/53
2946	6/12	Cardiff	Westbury	-	Westbury	11/49
2947	6/12	Paddington	Swindon	-	Swindon	4/51
2948	6/12	Pontypool	Hereford	-	Bath Road	11/51
2949	5/12	Bristol	Swindon	-	Swindon	1/52
2950	5/12	Bristol	Weston-S-Mare	-	Bath Road	9/52
2951	3/13	Cardiff	Hereford	-	Gloucester	6/52
2952	3/13	Paddington	Severn Tunnel	-	Severn Tunnel	9/51
2953	3/13	Pontypool	Chester	-	Chester	2/52
2954	3/13	Cardiff	Swindon	-	Swindon	7/52
2955	4/13	Goodwick	Weymouth	-	Weymouth	3/50
2971	12/03	Paddington	Swindon	1/46	-	-
2972	2/05	Bristol	Cardiff	3/35	-	-
2973	3/05	Exeter	Bath Road	7/33	-	-
2974	3/05	Paddington	Pontypool	8/33	-	-
2975	3/05	Swindon	Banbury	11/44	-	-
2976	4/05	Cardiff	Stafford Road	1/34	-	-
2977	4/05	Cardiff	Westbury	2/35	-	-
2978	4/05	Newton Abbot	Swindon	8/46	-	-
2979	4/05	Paddington	Ebbw Junction	-	Ebbw Junction	1/51
2980	5/05	Bristol	Gloucester	-	Gloucester	5/48
2981	6/05	Exeter	Banbury	-	Banbury	3/51
2982	6/05	Bristol	Ebbw Junction	6/34	-	-
2983	7/05	Bristol	Tyseley	3/46	-	-
2984	7/05	Paddington	Bath Road	5/33	-	-

No.	Date	1922 Shed	Last GWR Shed	Wdn GWR	Last BR shed	Wdn BR
2985	7/05	Paddington	Landore	8/31	-	-
2986	7/05	Paddington	Swindon	11/32	-	-
2987	8/05	Exeter	Hereford	-	Hereford	10/49
2988	8/05	Worcester	Tyseley	-	Tyseley	5/48
2989	9/05	Paddington	Chester	-	Chester	9/48
2990	9/05	Wolverhampton	Cardiff	1/39	-	-
2998	3/03	Exeter	Cardiff	6/33	-	-
3000	3/19	Neath	Pontypool	8/46	-	-
3001	3/19	Llanelly	Oxley	8/46	-	-
3002	3/19	Neath	Pontypool	-	Pontypool	4/48
3003	3/19	Landore	Chester	9/47	-	-
3004	3/19	Carmarthen	Carmarthen	-	Carmarthen	2/48
3005	3/19	Gloucester	Tyseley	-	Tyseley	8/48
3006	3/19	Llanelly	Carmarthen	-	Carmarthen	6/48
3007	3/19	Paddington	Worcester	11/47	-	-
3008	3/19	Llanelly	Stafford Road	-	Stafford Road	9/48
3009	4/19	Gloucester	Carmarthen	-	Carmarthen	6/48
3010	4/19	Cardiff	Carmarthen	-	Carmarthen	3/56
3011	4/19	Llanelly	Carmarthen	-	Carmarthen	10/58
3012	4/19	Swindon	Pontypool	-	Tyseley	5/56
3013	4/19	Landore	Bristol SPM	-	Bristol SPM	9/48
3014	4/19	Llanelly	Westbury	-	Carmarthen	10/55
3015	4/19	Landore	Carmarthen	-	Carmarthen	10/58
3016	4/19	Swindon	Oxley	-	Oxley	10/56
3017	4/19	Neath	Bristol SPM	-	Tyseley	10/56
3018	4/19	Llanelly	Pontypool	-	Swindon	1/57
3019	4/19	Llanelly	Westbury	-	Westbury	4/48
3100+	9/03	Newport	Tyseley	5/38	-	-
3111+	1/05	Severn Tunnel	Stafford Road	-	Stafford Road	10/48
3112+	1/05	Cardiff	Worcester	-	Stafford Road	10/55
3113+	1/05	Severn Tunnel	Newton Abbot	-	Newton Abbot	10/55
3114+	2/05	Brimscombe	Worcester	-	Gloucester	9/50
3115+	2/05	Brimscombe	Tyseley	6/39	-	-
3116+	2/05	Croes Newydd	Tyseley	3/39	-	-
3117+	2/05	Paddington	Tyseley	-	Chester	5/49
3118+	2/05	Paddington	Bristol SPM	3/38	-	-
3119+	2/05	Cardiff	Southall	-	Southall	6/48
3120+	3/05	Wolverhampton	Chester	2/39	-	-
3121+	5/05	Severn Tunnel	Tyseley	-	Wellington	10/48
3122+	5/05	Newport	Stourbridge	-	Stourbridge	9/50
3123+	5/05	Severn Tunnel	Kidderminster	7/37	-	-
3124+	5/05	Newton Abbot	Tyseley	9/38		
3125+	5/05	Newport	Tyseley	-	Wellington	7/52
3126+	6/05	Duffryn Yard	Bristol SPM	10/38	-	-
3127+	6/05	Cardiff	Wellington	-	Wellington	5/48
3128+	6/05	Swansea E Dock	Rhymney	-	Barry	11/48
3129+	6/05	Barry	Tyseley	-	Chester	7/51
3130+	6/05	Llanelly	Leamington	-	Leamington	8/48
3131+	1/06	Birkenhead	Stourbridge	-	Stourbridge	10/48
3132+	1/06	Duffryn Yard	Newton Abbot	-	Newton Abbot	8/51
3133+	1/06	Birkenhead	Worcester	4/39	-	-
3134+	2/06	Newton Abbot	Stourbridge	-	Stourbridge	4/51
3135+	2/06	Severn Tunnel	Wellington	-	Wellington	8/49
3136+	2/06	Severn Tunnel	Stourbridge	-	Stourbridge	10/51
3137+	2/06	Cardiff	Wellington	-	Banbury	10/51
3138+	2/06	Severn Tunnel	Stourbridge	-	Wellington	11/52
3139+	3/06	Llanelly	Crewe	-	Wellington	11/52
3140+	3/06	Severn Tunnel	St. Blazey	-	Taunton	7/53
3141+	3/06	Slough	Stourbridge	-	Chester	10/52
3142+	3/06	Newport	Newton Abbot	-	Newton Abbot	5/52

No.	Date	1922 Shed	Last GWR Shed	Wdn GWR	Last BR shed	Wdn BR
3143+	3/06	Birkenhead	Stafford Road	-	Stafford Road	12/51
3144+	3/06	Newport	Leamington	-	Leamington	1/52
3145+	3/06	Paddington	Stourbridge	5/38	-	-
3146+	3/06	Paddington	Stourbridge	-	Stourbridge	5/48
3147+	3/06	Cardiff	Stourbridge	-	Truro	1/53
3148+	3/06	Brimscombe	Laira	-	Laira	12/59
3149+	3/06	Severn Tunnel	Stafford Road	4/47	-	-
3150	4/06	Severn Tunnel	Stourbridge	-	Stourbridge	9/57
3151	4/07	Basingstoke	Stratford	-	Tyseley	3/52
3152	4/07	Severn Tunnel	Severn Tunnel	1/47	-	-
3153	4/07	Chester	Gloucester	-	Gloucester	2/53
3154	4/07	Severn Tunnel	Severn Tunnel	-	Severn Tunnel	10/50
3155	4/07	Llanelly	Newton Abbot	3/39	-	-
3156	4/07	Laira	Severn Tunnel	10/38	-	-
3157	4/07	Newport	Severn Tunnel	-	Severn Tunnel	10/52
3158	4/07	Newton Abbot	Tyseley	-	Tyseley	4/48
3159	5/07	Newport	Severn Tunnel	-	Severn Tunnel	11/49
3160	5/07	Leamington	Stafford Road	-	Stafford Road	7/53
3161	5/07	Newport	Severn Tunnel	-	Severn Tunnel	12/52
3162	5/07	Duffryn Yard	Severn Tunnel	1/47	-	-
3163	6/07	Barry	Severn Tunnel	-	Gloucester	6/57
3164	6/07	Barry	Gloucester	-	Gloucester	3/56
3165	6/07	Newton Abbot	Severn Tunnel	-	Severn Tunnel	7/48
3166	6/07	Llanelly	Severn Tunnel	1/47	-	-
3167	6/07	Leamington	Severn Tunnel	-	Severn Tunnel	8/52
3168	6/07	Cardiff	Severn Tunnel	-	Severn Tunnel	9/50
3169	6/07	Cardiff	Birkenhead	-	Birkenhead	7/50
3170	6/07	Swansea E Dock	Severn Tunnel	-	Ebbw Junction	8/58
3171	10/07	Cardiff	Brimscombe	-	Gloucester	7/57
3172	10/07	Cardiff	Severn Tunnel	-	Severn Tunnel	10/57
3173	10/07	Bristol	Gloucester	6/38	-	-
3174	10/07	Severn Tunnel	Severn Tunnel	-	Severn Tunnel	3/58
3175	10/07	Newport	Brimscombe	-	Severn Tunnel	3/49
3176	11/07	Newton Abbot	Severn Tunnel	-	Severn Tunnel	11/57
3177	11/07	Cardiff	Severn Tunnel	-	Severn Tunnel	10/57
3178	11/07	Cardiff	Severn Tunnel	-	Laira	1/51
3179	11/07	Leamington	Severn Tunnel	4/39	-	-
3180	11/07	Severn Tunnel	Tyseley	-	Gloucester	10/57
3181	12/07	Reading	Newton Abbot	11/38	-	-
3182	12/07	Severn Tunnel	Severn Tunnel	-	Severn Tunnel	8/49
3183	12/07	Severn Tunnel	Severn Tunnel	-	Severn Tunnel	10/57
3184	12/07	Severn Tunnel	Severn Tunnel	-	Severn Tunnel	7/48
3185	1/08	Cardiff	Severn Tunnel	-	Severn Tunnel	2/56
3186	1/08	Leamington	Laira	-	Laira	6/57
3187	1/08	Laira	Laira	-	Laira	11/57
3188	1/08	Bristol	Severn Tunnel	-	Severn Tunnel	11/52
3189	1/08	Severn Tunnel	Severn Tunnel	-	Severn Tunnel	9/49
3190	1/08	Newport	Severn Tunnel	-	Severn Tunnel	3/58
3201	12/84	Chester	Wellington	10/33	-	-
3202	6/85	Chester	Stourbridge	2/29	-	-
3203	6/85	Llandovery	Oswestry	8/25	-	-
3204	6/85	Crewe	Shrewsbury	5/29	-	-
3206	2/89	Trowbridge	Aberystwyth	6/34	-	-
3207	3/89	Salisbury	Machynlleth	6/33	-	-
3208	4/89	Salisbury	Swindon	9/26	-	-
3209	5/89	Trowbridge	Machynlleth	3/29	-	-
3210	5/89	Machynlleth	Wellington	3/37	-	-
3211	5/89	Trowbridge	Croes Newydd	11/34	-	-
3212	6/89	Westbury	Westbury	4/29	-	-
3213	7/89	Westbury	Machynlleth	6/36	-	-

No.	Date	1922 Shed	Last GWR Shed	Wdn GWR	Last BR shed	Wdn BR
3214	7/89	Didcot	Stourbridge	4/31	-	-
3215	7/89	Salisbury	Salisbury	7/27	-	-
3216	8/89	Aberystwyth	Croes Newydd	5/36	-	-
3217	9/89	Swindon	Machynlleth	10/35	-	-
3218	10/89	Crewe	Stafford Road	11/30	-	-
3219	10/89	Westbury	Croes Newydd	1/35	-	-
3220	11/89	Wellington	Wellington	10/27	-	-
3221	11/89	Westbury	Croes Newydd	8/33	-	-
3222	11/89	Trowbridge	Wellington	3/37	-	-
3223	12/89	Swindon	Wellington	5/36	-	-
3224	12/89	Gloucester	Welshpool	2/26	-	-
3225	12/89	Trowbridge	Pwllheli	4/35	-	-
3231	6/89	Wellington	Wellington	9/22	-	-
3232	9/92	Southall	Llanidloes	8/29	-	-
3233	9/92	Didcot	Didcot	10/25	-	-
3234	9/92	Wolverhampton	Wellington	12/26	-	-
3235	10/92	Westbury	Bath Road	3/36	-	-
3236	10/92	Reading	Reading	10/26	-	-
3237	10/92	Wolverhampton	Crewe	3/28	-	-
3238	11/92	Slough	Slough	5/28	-	-
3239	11/92	Pwllheli	Machynlleth	3/25	-	-
3240	11/92	Wolverhampton	Wellington	11/28	-	-
3241	11/92	Pwllheli	Winchester	4/27	-	-
3242	5/93	Reading	Reading	10/26	-	-
3244	6/93	Wolverhampton	Wellington	9/26	-	-
3245	6/93	Winchester	Winchester	9/26	-	-
3246	6/93	Wellington	Crewe	9/26	-	-
3247	6/93	Oxford	Fairford	10/25	-	-
3248	7/93	Didcot	Didcot	9/23	-	-
3249	7/93	Winchester	Didcot	2/25	-	-
3250	7/93	Reading	Reading	2/25	-	-
3251	7/93	Reading	Whitchurch	4/30	-	-
3252	5/95	Swindon	Aberystwyth	8/37	-	-
3253	7/95	Falmouth	Welshpool	1/39	-	-
3254+	7/95	Westbury	Machynlleth	-	Portmadoc	6/50
3255	8/95	Truro	Machynlleth	6/36	-	-
3256	8/95	Reading	Didcot	8/39	-	-
3257	8/95	Truro	Machynlleth	5/37	-	-
3258	9/95	Truro	Stratford	2/38	-	-
3259	9/95	Bristol	Machynlleth	10/38	-	-
3260	9/95	Bristol	Gloucester	4/38	-	-
3261	5/96	Newport	Cheltenham	4/37	-	-
3262	5/96	Penzance	Oswestry	9/37	-	-
3263	6/96	Didcot	Pwllheli	3/36	-	-
3264+	6/96	Leamington	Gloucester	-	Gloucester	12/49
3265+	7/96	Wolverhampton	Machynlleth	-	Oswestry	12/49
3266	8/96	Truro	Didcot	3/38	-	-
3267	11/96	Wellington	Didcot	10/36	-	-
3268	11/96	Wolverhampton	Oswestry	3/39	-	-
3269	12/96	Aberystwyth	Oswestry	3/37	-	-
3270	1/97	Bristol	Pwllheli	4/39	-	-
3271	1/97	Cambrian line	Machynlleth	6/36	-	-
3272+	1/97	Truro	Machynlleth	-	Aberystwyth	6/49
3273	2/97	Bristol	Shrewsbury	-	Oxley	12/49
3274	2/97	Truro	Stratford	11/36	-	-
3275	2/97	Plymouth	Oswestry	5/36	-	-
3276+	3/97	Reading	Shrewsbury	-	Shrewsbury	11/49
3277	3/97	Truro	Machynlleth	1/37	-	-
3278	3/97	Didcot	Swindon	12/38	-	-
3279	3/97	Exeter	Swindon	10/38	-	-

No.	Date	1922 Shed	Last GWR Shed	Wdn GWR	Last BR shed	Wdn BR
3280	3/97	Bristol	Didcot	5/39	-	-
3281	3/99	Banbury	Tyseley	2/37	-	-
3282	3/99	Shrewsbury	Didcot	10/37	-	-
3283+	3/99	Plymouth	Didcot	-	Winchester	12/50
3284+	4/99	Reading	Stourbridge	-	Oswestry	4/51
3285	4/99	Swindon	Cheltenham	1/37	-	-
3286	4/99	Reading	Reading	4/36	-	-
3287+	4/99	Bristol	Aberystwyth	-	Aberystwyth	7/49
3288	6/99	Reading	Aberystwyth	3/36	-	-
3289+	7/99	Bristol	Gloucester	-	Oswestry	7/51
3290	7/99	Wolverhampton	Oswestry	1/39	-	-
3291+	7/99	Bristol	Welshpool	-	Welshpool	2/49
3300	5/95	Hereford	Bath Road	1/36	-	-
3301	4/96	Bristol	Worcester	4/31	-	-
3302	4/96	Carmarthen	Paddington	7/32	-	-
3303	5/96	Neyland	Gloucester	5/32	-	-
3304	6/96	Carmarthen	Reading	11/34	-	-
3305	6/96	Hereford	Andover Jct	9/36	-	-
3306	11/96	Hereford	Swindon	1/39	-	-
3307	1/97	Taunton	Hereford	2/34	-	-
3308	1/97	Crewe	Bath Road	8/38	-	-
3309	2/97	Reading	Crewe	6/34	-	-
3310	2/97	Tyseley	Carmarthen	5/32	-	-
3311	10/98	Carmarhten	Pontypool	5/32	-	-
3312	3/99	Plymouth	Bath Road	4/31	-	-
3313	4/99	Hereford	Newton Abbot	4/46	-	-
3314	4/99	Shrewsbury	Gloucester	11/34	-	-
3315	6/99	Tondu	Worcester	6/31	-	-
3316	6/99	Shrewsbury	Hereford	4/38	-	-
3317	7/99	Chester	Bath Road	3/31	-	-
3318	7/99	Taunton	Hereford	2/34	-	-
3319	8/99	Salisbury	Reading	5/32	-	-
3320	11/99	Oxford	Hereford	8/29	-	-
3321	11/99	Reading	Tyseley	4/35	-	-
3322	11/99	Wolverhampton	Gloucester	3/35	-	-
3323	11/99	Hereford	Reading	8/35	-	-
3324	12/99	Bristol	Didcot	6/35	-	-
3325	12/99	Wolverhampton	Oxford	9/35	-	-
3326	1/00	Didcot	Paddington	11/33	-	-
3327	1/00	Landore	Croes Newydd	3/36	-	-
3328	1/00	Chester	Hereford	4/34	-	-
3329	1/00	Bristol	Salisbury	5/32	-	-
3330	2/00	Southall	Bristol SPM	8/38	-	-
3331	2/00	Oxford	Taunton	2/34	-	-
3332	2/00	Landore	Landore	10/31	-	-
3333	2/00	Neyland	Hereford	10/32	-	-
3334	2/00	Didcot	Gloucester	4/30	-	-
3335	2/00	Plymouth	Exeter	-	Exeter	10/48
3336	3/00	Landore	Newton Abbot	1/36	-	-
3337	3/00	Severn Tunnel	Hereford	5/34	-	-
3338	3/00	Shrewsbury	Chester	11/33	-	-
3339	3/00	Gloucester	Worcester	4/36	-	-
3340	10/99	Westbury	Swindon	6/34	-	-
3341	5/00	Cardiff	Newton Abbot	-	Exeter	11/49
3342	6/00	Swindon	Chester	10/38	-	-
3343	6/00	Worcester	Westbury	3/34	-	-
3344	6/00	Goodwick	Carmarthen	1/34	-	-
3345	6/00	Cardiff	Weymouth	1/36	-	-
3346	10/00	Wolverhampton	Westbury	1/34	-	-
3347	10/00	Carmarthen	Hereford	8/36	-	-

No.	Date		Last GWR Shed	Wdn GWR	Last BR shed	Wdn BR
3348	10/00	Truro	Worcester	11/34	-	-
3349	10/00	Newport	Hereford	11/34	-	-
3350	11/00	Cardiff	Hereford	7/35	-	-
3351	11/00	Oxford	Worcester	3/31	-	-
3352	11/00	Cardiff	Shrewsbury	11/33	-	-
3353	11/00	Bristol	Worcester	12/46	-	-
3354	11/00	Bristol	Westbury	11/34	-	-
3355	11/00	Carmarthen	Hereford	1/34	-	-
3356	11/00	Westbury	Reading	1/36	-	-
3357	11/00	Plymouth	Newton Abbot	11/34	-	-
3358	12/00	Worcester	Wellington	11/45	-	-
3359	12/00	Shrewsbury	Chester	9/36	-	-
3360	12/00	Reading	Tyseley	11/34	-	-
3361	12/02	Oxford	Taunton	9/47	-	-
3362	12/02	Tyseley	Hereford	4/37	-	-
3363	1/03	Taunton	Westbury	-	Westbury	10/49
3364	2/03	Tyseley	Westbury	-	Westbury	6/49
3365	2/03	Wolverhampton	Hereford	1/30	-	-
3366	2/03	Leamington	Chester	-	Chester	4/48
3367	2/03	Goodwick	Bristol SPM	9/35	-	-
3368	2/03	Bristol	Worcester	3/35	-	-
3369	2/03	Wolverhampton	Tyseley	7/36	-	-
3370	3/03	Swindon	Bath Road	2/39	-	-
3371	3/03	Plymouth	Westbury	11/44	-	-
3372	3/03	Carmarthen	Hereford	10/36	-	-
3373	5/03	Weymouth	Stratford	2/39	-	-
3374	5/03	Bristol	Exeter	6/37	-	-
3375	5/03	Neyland	Newton Abbot	9/47	-	-
3376	5/03	Penzance	Didcot	-	Didcot	9/48
3377	5/03	Worcester	Shrewsbury	-	Worcester	3/51
3378	5/03	Neyland	Swindon	11/45	-	-
3379	5/03	Truro	Gloucester	-	Gloucester	6/48
3380	5/03	Plymouth	Bath Road	3/38	-	-
3381	9/03	Cardiff	Crewe	11/35	-	-
3382	9/03	Neyland	Worcester	-	Worcester	11/49
3383	9/03	Hereford	Newton Abbot	-	Newton Abbot	12/49
3384	9/03	Reading	Andover Jct	5/36	-	-
3385	9/03	Swindon	Didcot	11/34	-	-
3386	9/03	Reading	Reading	-	Reading	11/49
3387	10/03	Bristol	Banbury	12/34	-	-
3388	10/03	Hereford	Hereford	10/35	-	-
3389	10/03	Worcester	Hereford	11/45	-	-
3390	10/03	Paddington	Carmarthen	3/39	-	-
3391	1/04	Bristol	Laira	-	Laira	5/48
3392	1/04	Yeovil	Cardiff	1/37	-	-
3393	1/04	Cardiff	Worcester	-	Worcester	11/49
3394	1/04	Paddington	Worcester	11/34	-	-
3395	1/04	Pontypool	Exeter	-	Exeter	8/48
3396	1/04	Landore	Didcot	-	Didcot	3/48
3397	2/04	Neyland	Neyland	11/34	-	-
3398	2/04	Newton Abbot	Laira	9/35	-	-
3399	2/04	Oxford	Chester	10/47	-	-
3400	2/04	Wolverhampton	Newton Abbot	-	Exeter	5/49
3401	3/04	Weymouth	Laira	-	Hereford	11/49
3402	3/04	Neyland	Hereford	3/37	-	-
3403	3/04	Shrewsbury	Croes Newydd	1/37	-	-
3404	3/04	Wolverhampton	Reading	9/37	-	-
3405	3/04	Hereford	Crewe	4/37	-	-
3406	3/04	Penzance	Pontypool	-	Hereford	1/51
3407	4/04	Reading	Newton Abbot	-	Newton Abbot	12/49

No.	Date	1922 Shed	Last GWR Shed	Wdn GWR	Last BR shed	Wdn BR
3408	4/04	Bristol	Didcot	-	Didcot	4/48
3409	4/04	Hereford	Hereford	1/39	-	-
3410	4/04	Paddington	Stratford	11/36	-	-
3411	4/06	Weymouth	Cardiff	10/38	-	-
3412	4/06	Worcester	Bath Road	3/36	-	-
3413	4/06	Swindon	Westbury	8/36	-	-
3414	5/06	Tyseley	Crewe	10/38	-	-
3415	5/06	Hereford	Oxford	2/37	-	-
3416	6/06	Penzance	Taunton	5/36	-	-
3417	6/06	Plymouth	Wellington	-	Wellington	4/48
3418	6/06	Newport	Reading	-	Southall	8/49
3419	6/06	Newport	Winchester	-	Didcot	8/49
3420	6/06	Shrewsbury	Didcot	9/37	-	-
3421	6/06	Weymouth	Swindon	-	Swindon	4/48
3422	6/06	Newport	Severn Tunnel	3/36	-	-
3423	6/06	Reading	Chester	1/39	-	-
3424	7/06	Plymouth	Laira	5/36	-	-
3425	6/06	Worcester	Andover Jct	2/38	-	-
3426	7/06	Weymouth	Reading	-	Reading	12/49
3427	7/06	Worcester	Laira	4/38	-	-
3428	7/06	Weymouth	Worcester	10/36	-	-
3429	7/06	Paddington	Swindon	9/39	-	-
3430	8/06	Landore	Newton Abbot	-	Newton Abbot	12/48
3431	8/06	Weymouth	Laira	-	Laira	12/48
3432	8/06	Westbury	Hereford	-	Worcester	12/49
3433	8/06	Worcester	Swindon	4/39	-	-
3434	8/06	Reading	Didcot	4/37	-	-
3435	8/06	Hereford	Laira	11/45	-	-
3436	8/06	Gloucester	Gloucester	12/38	-	-
3437	9/06	Gloucester	Shrewsbury	3/39	-	-
3438	9/06	Swindon	Westbury	-	Westbury	10/49
3439	9/06	Westbury	Cardiff	7/36	-	-
3440	9/06	Swindon	Worcester	-	Worcester	6/48
3441	5/09	Hereford	Laira	-	Worcester	2/49
3442	5/09	Gloucester	Shrewsbury	-	Shrewsbury	7/48
3443	5/09	Swindon	Taunton	-	Taunton	5/49
3444	5/09	Hereford	Taunton	-	Taunton	6/51
3445	5/09	Banbury	Laira	-	Gloucester	10/48
3446	11/09	Oxford	Laira	-	Worcester	12/48
3447	12/09	Hereford	Neyland	-	Worcester	4/51
3448	12/09	Westbury	Didcot	-	Winchester	1/49
3449	12/09	Weymouth	Cheltenham	-	Cheltenham	6/51
3450	12/09	Hereford	Stourbridge	-	Stourbridge	12/49
3451	1/10	Weymouth	Exeter	-	Exeter	4/51
3452	1/10	Leamington	Swindon	-	Swindon	4/48
3453	1/10	Neyland	Pontypool	-	Reading	11/51
3454	1/10	Neyland	Hereford	-	Reading	11/51
3455	1/10	Chester	Neath	-	Worcester	6/50
3500	11/99	Southall	Southall	4/42	-	-
3501	3/85	Wolverhampton	Oswestry	8/30	-	-
3502	3/85	Oxford	Oxford	6/25	-	-
3503	3/85	Chester	Croes Newydd	7/29	-	-
3504	3/85	Worcester	Worcester	12/25	-	-
3505	4/85	Croes Newydd	Wellington	4/31	-	-
3506	4/85	Didcot	Didcot	9/26	-	-
3507	4/85	Chester	Stourbridge	10/26	-	-
3508	4/85	Corwen	Chester	9/29	-	-
3510	5/85	Banbury	Wellington	9/28	-	-
3511	5/85	Chester	Croes Newydd	4/31	-	-
3512	5/85	Shrewsbury	Chester	3/29	-	-

No.	Date	1922 Shed	Last GWR Shed	Wdn GWR	Last BR shed	Wdn BR
3513	6/85	Didcot	Oxford	7/27	-	-
3514	6/85	Stourbridge	Wellington	4/29	-	-
3515	7/85	Didcot	Brecon	12/30	-	-
3516	7/85	Stratford	Brecon	12/30	-	-
3518	8/85	Corwen	Wellington	9/32	-	-
3519	8/85	Oswestry	Machynlleth	3/25	-	-
3521	8/87	Machynlleth	Machynlleth	10/31	-	-
3522	8/87	Stratford	Worcester	1/25	-	-
3523	8/87	Gloucester	Cheltenham	5/27	-	-
3524	9/87	Gloucester	Gloucester	10/27	-	-
3525	9/87	Gloucester	Gloucester	5/29	-	-
3526	10/87	Gloucester	Evesham	10/27	-	-
3527	10/87	Worcester	Worcester	4/27	-	-
3528	11/87	Reading	Reading	7/27	-	-
3529	11/87	Carmarthen	Kidderminster	4/31	-	-
3531	11/87	Worcester	Worcester	11/27	-	-
3532	12/87	Taunton	Taunton	3/23	-	-
3533	12/87	Malvern Wells	Worcester	2/29	-	-
3534	12/87	Carmarthen	Carmarthen	10/27	-	-
3535	1/88	Worcester	Kidderminster	12/28	-	-
3536	1/88	Gloucester	Gloucester	11/26	-	-
3537	2/88	Cheltenham	Gloucester	9/28	-	-
3538	2/88	Worcester	Worcester	7/27	-	-
3539	3/88	Bristol	Bristol SPM	10/27	-	-
3540	3/88	Gloucester	Gloucester	11/27	-	-
3542	9/88	Aberystwyth	Aberystwyth	3/26	-	-
3543	9/88	Didcot	Winchester	9/28	-	-
3544	10/88	Stratford	Aberystwyth	2/27	-	-
3545	10/88	Gloucester	Machynlleth	4/31	-	-
3546	10/88	Machynlleth	Machynlleth	3/27	-	-
3547	10/88	Taunton	Barnstaple	10/27	-	-
3548	11/88	Landore	Gloucester	8/29	-	-
3549	11/88	Didcot	Winchester	9/28	-	-
3550	11/88	Bristol	Westbury	10/29	-	-
3551	11/88	Landore	Didcot	6/29	-	-
3552	12/88	Yeovil	Bath Road	11/29	-	-
3553	12/88	Bristol	Bath Road	11/28	-	-
3554	1/89	Gloucester	Aberystwyth	6/30	-	-
3555	1/89	Cheltenham	Kidderminster	9/29	-	-
3556	2/89	Gloucester	Cheltenham	10/27	-	-
3557	3/89	Carmarthen	Kidderminster	5/34	-	-
3558	6/89	Gloucester	Evesham	9/23	-	-
3559	7/89	Malvern Wells	Worcester	11/31	-	-
3560	7/89	Didcot	Didcot	10/23	-	-
3561	1/94	Slough	Swindon	-	Swindon	10/49
3562	2/94	Southall	Southall	-	Oxford	2/49
3563	2/94	Slough	Cheater	12/44	-	-
3564	3/94	Aylesbury	Fairford	1/44	-	-
3565	3/94	Slough	Oxford	6/36	-	-
3566	3/94	Slough	Bala	4/38	-	-
3567	4/94	Paddington	Croes Newydd	9/36	-	-
3568	4/94	Paddington	Fairford	11/45	-	-
3569	4/94	Slough	Birkenhead	3/36	-	-
3570	4/94	Paddington	Ebbw Junction	12/38	-	-
3571	5/95	Birkenhead	Swindon	10/42	-	-
3572	6/95	Birkenhead	Birkenhead	10/28	-	-
3573	7/95	Birkenhead	Swindon	12/45	-	-
3574	8/95	Birkenhead	Worcester	-	Worcester	12/49
3575	9/95	Chester	Severn Tunnel	-	Swindon	10/49
3576	9/95	Yeovil	Birkenhead	6/29	-	-

No.	Date	1922 Shed	Last GWR Shed	Wdn GWR	Last BR shed	Wdn BR
3577	10/95	Birkenhead	Swansea E Dock	-	Swansea E Dock	5/49
3578	11/95	Wolverhampton	Tyseley	7/45	-	-
3579	12/95	Birkenhead	Chester	9/42	-	-
3580	1/97	Birkenhead	Swindon	12/45	-	-
3581	1/99	Slough	St. Blazey	11/45	-	-
3582	2/99	Southall	Taunton	-	St. Blazey	11/49
3583	2/99	Southall	Fairford	12/47	-	-
3584	4/99	Southall	Llantrisant	11/45	-	-
3585	5/99	Southall	Oxford	-	Oxford	1/48
3586	6/99	Paddington	Llantrisant	-	Llantrisant	11/49
3587	8/99	Slough	Taunton	12/44	-	-
3588	8/99	Minehead	Oxford	-	Oxford	12/49
3589	9/99	Aberdare	Oxford	-	Oxford	8/48
3590	10/99	Southall	Taunton	4/44	-	-
3591	9/99	Paddington	Swindon	5/36	-	-
3592	9/99	Paddington	Carmarthen	-	Carmarthen	4/49
3593	9/99	Reading	Reading	11/27	-	-
3594	10/99	Slough	Llantrisant	5/47	-	-
3595	10/99	Aberdare	Abercynon	1/45	-	-
3596	10/99	Slough	Southall	3/45	-	-
3597	10/99	Southall	Cathays	-	Cathays	8/48
3598	10/99	Southall	Swindon	10/36	-	-
3599	11/99	Southall	Radyr	-	Radyr	10/49
3600	12/00	Paddington	Reading	6/33	-	-
3601	2/02	Slough	Slough	3/31	-	-
3602	2/02	Kidderminster	Goodwick	2/31	-	-
3603	3/02	Kidderminster	Chester	5/31	-	-
3604	3/02	Chester	Chester	11/34	-	-
3605	4/02	Kidderminster	Paddington	12/30	-	-
3606	4/02	Leamington	Leamington	3/31	-	-
3607	5/02	Wolverhampton	Birkenhead	4/31	-	-
3608	5/02	Stourbridge	Birkenhead	4/31	-	-
3609	5/02	Leamington	Chester	3/32	-	-
3610	5/02	Kidderminster	Reading	11/34	-	-
3611	6/02	Tyseley	Cardiff	2/31	-	-
3612	6/02	Newport	Cardiff	10/30	-	-
3613	6/02	Stourbridge	Cardiff	5/32	-	-
3614	6/02	Paddington	Paddington	4/31	-	-
3615	7/02	Paddington	Bridgend	4/31	-	-
3616	7/02	Leamington	Stratford	9/33	-	-
3617	7/02	Paddington	Aylesbury	3/33	-	-
3618	7/02	Leamington	Chester	11/34	-	-
3619	8/02	Paddington	Paddington	4/31	-	-
3620	8/02	Leamington	Chester	5/31	-	-
3621	10/03	Wellington	Llanelly	1/33	-	-
3622	10/03	Wolverhampton	Cardiff	5/32	-	-
3623	11/03	Wolverhampton	Stratford	4/31	-	-
3624	11/03	Stourbridge	Birkenhead	10/31	-	-
3625	11/03	Stourbridge	Leamington	7/31	-	-
3626	12/03	Slough	Paddington	8/32	-	-
3627	12/03	Stourbridge	Birkenhead	1/34	-	-
3628	12/03	Wolverhampton	Leamington	11/34	-	-
3629	12/03	Tyseley	Chester	11/31	-	-
3630	12/03	Leamington	Birkenhead	10/30	-	-
3700	8/01	Oxford	Stafford Road	11/29	-	-
3701	8/01	Wolverhampton	Stafford Road	8/28	-	-
3702	8/01	Oxford	Leamington	4/29	-	-
3703	9/01	Wolverhampton	Banbury	8/29	-	-
3704	9/01	Banbury	Chester	9/28	-	-
3705	9/01	Oxford	Oxford	9/28	-	-

No.	Date	1922 Shed	Last GWR Shed	Wdn GWR	Last BR shed	Wdn BR
3706	9/01	Wolverhampton	Banbury	6/29	-	-
3707	9/01	Shrewsbury	Bath Road	4/29	-	-
3708	10/01	Worcester	Shrewsbury	10/29	-	-
3709	10/01	Chester	Oxford	9/29	-	-
3710	3/03	Barnstaple	Stafford Road	9/28	-	-
3711	5/03	Chester	Leamington	7/30	-	-
3712	5/03	Westbury	Reading	5/31	-	-
3713	5/03	Wolverhampton	Chester	12/29	-	-
3714	5/03	Oxford	Oxford	11/29	-	-
3715	5/03	Reading	Chester	10/29	-	-
3716	5/03	Westbury	Bath Road	4/29	-	-
3717	5/03	Bristol	Shrewsbury	3/31	-	-
3718	5/03	Reading	Paddington	10/27	-	-
3719	5/03	Chester	Oxford	4/29	-	-
3800	5/04	Wolverhampton	Weston-S-Mare	3/31	-	-
3801	10/06	Bristol	Westbury	4/31	-	-
3802	10/06	Worcester	Swindon	5/31	-	-
3803	10/06	Hereford	Oxford	1/32	-	-
3804	10/06	Banbury	Oxford	3/31	-	-
3805	10/06	Shrewsbury	Oxford	5/33	-	-
3806	11/06	Westbury	Tyseley	2/31	-	-
3807	11/06	Worcester	Didcot	12/30	-	-
3808	11/06	Pontypool	Bath Road	10/31	-	-
3809	11/06	Swindon	Leamington	9/31	-	-
3810	11/06	Swindon	Swindon	3/31	-	-
3811	11/06	Taunton	Oxford	1/31	-	-
3812	11/06	Swindon	Hereford	7/32	-	-
3813	11/06	Shrewsbury	Didcot	11/31	-	-
3814	11/06	Wolverhampton	Reading	6/33	-	-
3815	11/06	Westbury	Leamington	1/32	-	-
3816	12/06	Landore	Swindon	9/31	-	-
3817	12/06	Bristol	Swindon	1/31	-	-
3818	12/06	Pontypool	Swindon	8/31	-	-
3819	12/06	Wolverhampton	Weston-S-Mare	5/31	-	-
3820	12/06	Reading	Reading	5/31	-	-
3821	12/11	Paddington	Leamington	9/31	-	-
3822	12/11	Shrewsbury	Pontypool	4/33	-	-
3823	12/11	Bristol	Swindon	4/31	-	-
3824	12/11	Shrewsbury	Hereford	3/31	-	-
3825	12/11	Worcester	Reading	3/31	-	-
3826	1/12	Worcester	Oxford	8/31	-	-
3827	1/12	Bristol	Reading	12/31	-	-
3828	1/12	Bristol	Oxford	3/33	-	-
3829	1/12	Swindon	Oxford	2/32	-	-
3830	2/12	Swindon	Reading	2/31	-	-
3831	6/04	Pontypool	Tyseley	11/30	-	-
3832	6/04	Westbury	Oxford	5/30	-	-
3833	8/04	Wolverhampton	Swindon	2/30	-	-
3834	8/04	Bristol	Tyseley	11/33	-	-
3835	8/04	Wolverhampton	Paddington	1/31	-	-
3836	10/04	Bristol	Leamington	11/31	-	-
3837	10/04	Wolverhampton	Swindon	3/31	-	-
3838	10/04	Reading	Swindon	8/30	-	-
3839	10/04	Swindon	Weston-S-Mare	3/30	-	-
3901	8/96	Tyseley	Neath	5/34	-	-
3902	9/96	Tyseley	Southall	8/32	-	-
3903	10/96	Wolverhampton	Swansea E Dock	5/32	-	-
3904	10/96	Tyseley	Neath	10/33	-	-
3905	9/96	Neath	Barry	8/30	-	-
3906	8/96	Tyseley	Slough	5/32	-	-

No.	Date	1922 Shed	Last GWR Shed	Wdn GWR	Last BR shed	Wdn BR
3907	11/96	Gloucester	Landore	9/34	-	-
3908	9/96	Westbury	Barry	9/30	-	-
3909	11/96	Wolverhampton	Barry	10/30	-	-
3910	9/96	Westbury	Ledbury	4/31	-	-
3911	11/96	Tyseley	Neath	7/32	-	-
3912	12/96	Tyseley	Neyland	4/31	-	-
3913	10/96	Tyseley	Barry	6/31	-	-
3914	11/96	Landore	Llandovery	6/31	-	-
3915	8/96	Neath	Reading	5/32	-	-
3916	8/96	Tyseley	Landore	11/34	-	-
3917	10/96	Landore	Ledbury	10/32	-	-
3918	8/96	Tyseley	Llanelly	5/32	-	-
3919	9/96	Tyseley	Slough	3/34	-	-
3920	10/96	Tyseley	Hereford	2/31	-	-
4000	4/06	Wolverhampton	Bath Road	11/29	-	-
4001	2/07	Paddington	Bath Road	1/34	-	-
4002	3/07	Laira	Paddington	6/33	-	-
4003	2/07	Paddington	Landore	-	Landore	7/51
4004	2/07	Laira	Oxford	-	Oxford	4/48
4005	2/07	Laira	Paddington	11/34	-	-
4006	4/07	Paddington	Landore	11/32	-	-
4007	4/07	Worcester	Worcester	-	Worcester	9/51
4008	5/07	Paddington	Paddington	6/35	-	-
4009	5/07	Newton Abbot	Laira	4/25	-	-
4010	5/07	Paddington	Bath Road	11/34	-	-
4011	3/08	Paddington	Worcester	11/32	-	-
4012	3/08	Laira	Newton Abbot	-	Newton Abbot	10/49
4013	3/08	Paddington	Chester	-	Chester	5/50
4014	3/08	Laira	Stafford Road	6/46	-	-
4015	3/08	Paddington	Swindon	-	Swindon	2/51
4016	4/08	Paddington	Paddington	10/25	-	-
4017	4/08	Paddington	Swindon	-	Swindon	11/49
4018	4/08	Laira	Stafford Road	-	Stafford Road	4/51
4019	5/08	Wolverhampton	Bath Road	-	Bath Road	10/49
4020	5/08	Exeter	Bath Road	-	Bath Road	3/51
4021	6/09	Paddington	Oxford	-	Oxford	10/52
4022	6/09	Laira	Swindon	-	Swindon	2/52
4023	6/09	Paddington	Landore	-	Landore	7/52
4024	6/09	Paddington	Taunton	2/35	-	-
4025	7/09	Laira	Stafford Road	-	Stafford Road	8/50
4026	9/09	Paddington	Taunton	-	Taunton	2/50
4027	9/09	Paddington	Landore	10/34	-	-
4028	9/09	Newton Abbot	Westbury	-	Westbury	11/51
4029	10/09	Paddington	Stafford Road	11/34	-	-
4030	10/09	Paddington	Bath Road	-	Bath Road	5/50
4031	10/10	Paddington	Stafford Road	-	Stafford Road	6/51
4032	10/10	Paddington	Exeter	4/26	-	-
4033	11/10	Paddington	Bath Road	-	Bath Road	6/51
4034	11/10	Newton Abbot	Bath Road	-	Swindon	9/52
4035	11/10	Newton Abbot	Bath Road	-	Bath Road	10/51
4036	12/10	Paddington	Swindon	-	Swindon	3/52
4037	12/10	Paddington	Paddington	6/26	-	-
4038	1/11	Newton Abbot	Westbury	-	Westbury	4/52
4039	2/11	Newton Abbot	Landore	-	Landore	11/50
4040	3/11	Exeter	Shrewsbury	-	Shrewsbury	6/51
4041	6/13	Paddington	Bath Road	-	Bath Road	4/51
4042	5/13	Paddington	Bath Road	-	Bath Road	11/51
4043	5/13	Wolverhampton	Bath Road	-	Bath Road	1/52
4044	5/13	Laira	Shrewsbury	-	Shrewsbury	2/53
4045	6/13	Wolverhampton	Westbury	-	Westbury	11/50

No.	Date	1922 Shed	Last GWR Shed	Wdn GWR	Last BR shed	Wdn BR
4046	5/14	Newton Abbot	Shrewsbury	-	Shrewsbury	11/51
4047	5/14	Laira	Bath Road	-	Bath Road	7/51
4048	5/14	Wolverhampton	Landore	-	Landore	1/53
4049	5/14	Wolverhampton	Oxford	-	Stafford Road	7/53
4050	6/14	Paddington	Landore	-	Landore	2/52
4051	6/14	Paddington	Worcester	-	Worcester	10/50
4052	6/14	Laira	Oxford	-	Shrewsbury	6/53
4053	6/14	Laira	Stafford Road	-	Stafford Road	7/54
4054	6/14	Paddington	Exeter	-	Laira	2/52
4055	7/14	Paddington	Swindon	-	Swindon	2/51
4056	7/14	Paddington	Taunton	-	Bath Road	10/57
4057	7/14	Paddington	Swindon	-	Swindon	2/52
4058	7/14	Wolverhampton	Tyseley	-	Stafford Road	4/51
4059	7/14	Laira	Gloucester	-	Gloucester	9/52
4060	7/14	Newton Abbot	Stafford Road	-	Bath Road	10/52
4061	5/22	Penzance	Shrewsbury	-	Stafford Road	3/57
4062	5/22	Laira	Swindon	-	Swindon	11/56
4100	12/97	Landore	Tyseley	9/29	-	-
4101	4/98	Worcester	Oxford	4/30	-	-
4102	5/98	Oxford	Tyseley	9/28	-	-
4103	5/98	Bristol	Bath Road	4/30	-	-
4104	5/98	Cardiff	Cardiff	6/29	-	-
4105	5/98	Taunton	Bath Road	11/27	-	-
4106	6/98	Tyseley	Bath Road	8/29	-	-
4107	6/98	Tyseley	Tyseley	2/30	-	-
4108	7/98	Bristol	Bath Road	3/30	-	-
4109	7/98	Worcester	Bath Road	3/31	-	-
4110	7/98	Cardiff	Tyseley	10/28	-	-
4111	7/98	Wolverhampton	Oxford	10/28	-	-
4112	9/98	Banbury	Tyseley	9/29	-	-
4113	9/98	Shrewsbury	Didcot	5/31	-	-
4114	9/98	Cardiff	Goodwick	7/27	-	-
4115	11/98	Landore	Tyseley	3/31	-	-
4116	12/98	Cardiff	Cardiff	7/27	-	-
4117	12/98	Swindon	Bath Road	10/27	-	-
4118	1/99	Weston-S-Mare	Bristol SPM	7/27	-	-
4119	1/99	Oxford	Didcot	7/27	-	-
4120	4/00	Tyseley	Leamington	9/29	-	-
4121	4/00	Landore	Bath Road	12/28	-	-
4122	4/00	Bristol	Landore	10/28	-	-
4123	4/00	Cardiff	Cardiff	11/28	-	-
4124	5/00	Taunton	Didcot	4/30	-	-
4125	5/00	Bristol	Bath Road	4/27	-	-
4126	5/00	Tyseley	Tyseley	10/27	-	-
4127	5/00	Bristol	Tyseley	9/29	-	-
4128	5/00	Westbury	Swindon	7/27	-	-
4129	7/00	Paddington	Shrewsbury	11/28	-	-
4130	7/00	Paddington	Didcot	4/30	-	-
4131	7/00	Goodwick	Didcot	4/29	-	-
4132	8/00	Tyseley	Tyseley	4/31	-	-
4133	8/00	Cardiff	Cardiff	7/27	-	-
4134	8/00	Shrewsbury	Cardiff	10/27	-	-
4135	8/00	Weston-S-Mare	Bristol SPM	11/27	-	-
4136	8/00	Wolverhampton	Banbury	10/27	-	-
4137	9/00	Swindon	Westbury	10/28	-	-
4138	9/00	Paddington	Paddington	11/29	-	-
4139	6/01	Bristol	Chester	9/28	-	-
4140	6/01	Tyseley	Salisbury	4/29	-	-
4141	7/01	Bristol	Salisbury	2/30	-	-
4142	7/01	Tyseley	Landore	10/28	-	-

No.	Date	1922 Shed	Last GWR Shed	Wdn GWR	Last BR shed	Wdn BR
4143	8/01	Bristol	Tyseley	4/29	-	-
4144	8/01	Bristol	Goodwick	11/27	-	-
4145	8/01	Pontypool	Westbury	12/30	-	-
4146	10/01	Chester	Salisbury	10/27	-	-
4147	10/01	Cardiff	Tyseley	7/27	-	-
4148	10/01	Worcester	Severn Tunnel	5/31	-	-
4149	5/08	Cardiff	Leamington	9/29	-	-
4150	5/08	Hereford	Didcot	4/31	-	-
4151	5/08	Tyseley	Pontypool	11/27	-	-
4152	5/08	Landore	Bristol SPM	9/28	-	-
4153	5/08	Oxford	Cardiff	7/27	-	-
4154	5/08	Landore	Tyseley	5/30	-	-
4155	6/08	Bristol	Landore	11/27	-	-
4156	6/08	Swindon	Leamington	4/29	-	-
4157	6/08	Shrewsbury	Leamington	10/28	-	-
4158	6/08	Landore	Leamington	4/29	-	-
4159	6/08	Bristol	Shrewsbury	10/29	-	-
4160	6/08	Oxford	Landore	7/27	-	-
4161	6/08	Wolverhampton	Salisbury	4/29	-	-
4162	7/08	Bristol	Salisbury	8/29	-	-
4163	7/08	Tyseley	Cardiff	4/29	-	-
4164	7/08	Bristol	Bath Road	10/30	-	-
4165	7/08	Pontypool	Salisbury	7/27	-	-
4166	7/08	Oxford	Tyseley	11/27	-	-
4167	7/08	Cardiff	Cardiff	7/29	-	-
4168	7/08	Landore	Taunton	5/30	-	-
4169	6/94	Wolverhampton	Cardiff	7/30	-	-
4170	5/94	Wolverhampton	Bath Road	8/28	-	-
4201	12/10	Newport	Pill	-	Pill	10/59
4202	1/12	Swansea E Dock	Aberdare	8/37	-	-
4203	2/12	Newport	Ebbw Junction	-	Ebbw Junction	1/61
4204	2/12	Glyn Neath	Llanelly	11/37	-	-
4205	2/12	Neath	Aberbeeg	3/38	-	-
4206	2/12	Llanelly	Ebbw Junction	-	St. Blazey	12/59
4207	2/12	Newport	Landore	-	Cardiff	10/61
4208	2/12	Newport	Llantrisant	-	Llantrisant	11/59
4209	2/12	Llanelly	Swansea E Dock	7/38	-	-
4210	3/12	Llanelly	Swansea E Dock	2/39	-	-
4211	3/12	Neath	Pill	-	Ebbw Junction	8/59
4212	8/12	Newport	Landore	-	Shrewsbury	6/59
4213	9/12	Newport	Llanelly	-	Tondu	1/64
4214	9/12	Newport	Tondu	-	Ebbw Junction	5/64
4215	9/12	Llanelly	St. Blazey	-	Severn Tunnel	10/59
4216	11/12	Llanelly	Ebbw Junction	2/38	-	-
4217	11/12	Swansea E Dock	Aberbeeg	-	Severn Tunnel	7/59
4218	11/12	Landore	Tondu	-	Tondu	10/62
4219	12/12	Llanelly	Glyn Neath	12/38	-	-
4220	12/12	Landore	Ebbw Junction	8/37	-	-
4221	12/12	Aberdare	Glyn Neath	-	Neath	9/59
4222	11/13	Newport	Aberbeeg	-	Pontypool	10/64
4223	11/13	Newport	Carmarthen	-	Llanelly	8/59
4224	11/13	Newport	Ebbw Junction	-	Tondu	2/59
4225	11/13	Barry	Ebbw Junction	-	Llanelly	1/63
4226	11/13	Aberdare	Pill	-	Cardiff	3/59
4227	11/13	Llanelly	Cardiff	-	Ebbw Junction	2/65
4228	11/13	Newport	Aberdare	-	Tondu	12/63
4229	11/13	Swansea E Dock	Pill	-	Ebbw Junction	3/61
4230	11/13	Newport	Ebbw Junction	-	Llanelly	9/62
4231	11/13	Newport	Aberbeeg	-	Cardiff	12/59
4232	8/14	Llanelly	Neath	-	Swansea E Dock	10/63

No.	Date	1922 Shed	Last GWR Shed	Wdn GWR	Last BR shed	Wdn BR
4233	8/14	Llanelly	Pill	-	Ebbw Junction	10/64
4234	9/14	Swansea E Dock	Pill	4/38	-	-
4235	9/14	Newport	Pill	-	Ebbw Junction	8/62
4236	10/14	Barry	Cardiff	-	Aberdare	9/62
4237	10/14	Landore	Pill	-	Aberdare	4/64
4238	10/14	Llanelly	Pontypool	-	Ebbw Junction	12/63
4239	10/14	Newport	Cardiff	5/37	-	-
4240	10/14	Newport	Barry	12/38	-	-
4241	10/14	Landore	Tondu	-	Ebbw Junction	4/64
4242	2/16	Newport	Ebbw Junction	-	Cardiff E Dock	3/64
4243	2/16	Newport	Aberbeeg	-	Pontypool	5/64
4244	2/16	Pontypool	Pill	5/38	-	-
4245	3/16	Newport	Ebbw Junction	8/39	-	-
4246	3/16	Newport	Pill	-	Aberbeeg	11/62
4247	3/16	Newport	Ebbw Junction	-	Tondu	4/64
4248	4/16	Newport	Ebbw Junction	-	Severn Tunnel	5/63
4249	4/16	Newport	Tondu	7/38	-	-
4250	4/16	Newport	Landore	-	Llanelly	9/62
4251	5/16	Newport	Tondu	-	Tondu	6/63
4252	3/17	Swansea E Dock	Neath	-	Neath	9/63
4253	3/17	Newport	Pill	-	Pill	4/63
4254	3/17	Newport	Llanelly	-	Ebbw Junction	4/65
4255	4/17	Swansea E Dock	Duffryn Yard	-	Ebbw Junction	4/64
4256	4/17	Newport	Landore	-	Duffryn Yard	1/64
4257	4/17	Newport	Aberdare	-	Aberdare	10/63
4258	5/17	Barry	Pill	-	Severn Tunnel	4/65
4259	5/17	Swansea E Dock	Neath	-	Ebbw Junction	3/64
4260	5/17	Tondu	Ebbw Junction	-	Llanelly	6/59
4261	5/17	Llanelly	Llantrisant	-	Llantrisant	3/59
4262	9/19	Newport	Severn Tunnel	-	Tondu	4/64
4263	10/19	Newport	Ebbw Junction	-	Tondu	2/64
4264	10/19	Newport	Aberdare	-	Swansea E Dock	7/63
4265	11/19	Newport	Landore	-	Ebbw Junction	6/63
4266	11/19	Newport	Carmarthen	-	Ebbw Junction	8/62
4267	11/19	Newport	Aberbeeg	-	Llantrisant	10/62
4268	11/19	Newport	Ebbw Junction	-	Ebbw Junction	8/65
4269	12/19	Newport	Pill	-	Tondu	11/62
4270	12/19	Newport	Ebbw Junction	-	Cardiff E Dock	9/62
4271	2/20	Newport	Ebbw Junction	-	Ebbw Junction	12/63
4272	2/20	Newport	Glyn Neath	-	Cardiff E Dock	10/63
4273	3/20	Newport	Tondu	-	Barry	10/64
4274	3/20	Newport	Neath	-	Tondu	6/62
4275	3/20	Newport	Pontypool	-	Neath	1/64
4276	4/20	Newport	Ebbw Junction	-	Pill	11/62
4277	4/20	Aberbeeg	Duffryn Yard	-	Aberbeeg	6/64
4278	4/20	Newport	Llanelly	-	Aberdare	7/64
4279	5/20	Newport	Glyn Neath	-	Aberdare	1/64
4280	5/20	Newport	Pill	-	Pill	1/63
4281	5/20	Pontypool	Llanelly	-	Neath	12/60
4282	6/20	Pontypool	Swansea E Dock	-	Neath	9/63
4283	6/20	Llanelly	Swansea E Dock	-	Barry	10/64
4284	8/20	Newport	Neath	-	Neath	9/64
4285	8/20	Newport	Aberdare	-	Pontypool	4/65
4286	8/21	Newport	Duffryn Yard	-	Neath	9/64
4287	8/21	Newport	Aberbeeg	-	Aberbeeg	1/61
4288	9/21	Newport	Neath	-	Duffryn Yard	11/60
4289	9/21	Newport	Ebbw Junction	-	Aberdare	10/62
4290	9/21	Barry	Cardiff	-	Ebbw Junction	2/63
4291	9/21	Newport	Pill	-	Aberbeeg	9/62
4292	10/21	Swansea E Dock	Duffryn Yard	-	Neath	10/64

No.	Date	1922 Shed	Last GWR Shed	Wdn GWR	Last BR shed	Wdn BR
4293	10/21	Swansea E Dock	Glyn Neath	-	Ebbw Junction	8/62
4294	11/21	Tondu	Ebbw Junction	-	Llanelly	9/64
4295	11/21	Aberdare	Landore	-	Pontypool	12/64
4300	11/16	Bristol	Severn Tunnel	7/37	-	-
4301	6/11	Shrewsbury	Bristol SPM	8/36	-	-
4302	6/11	Cardiff	Hereford	6/38	-	-
4303	6/11	Southall	Pontypool	-	Gloucester	11/52
4304	7/11	Pontypool	Ebbw Junction	10/38	-	-
4305	7/11	Tyseley	Chester	8/37	-	-
4306	7/11	Swindon	Neyland	6/37	-	-
4307	7/11	Wolverhampton	Oxley	3/38	-	-
4308	8/11	Birkenhead	Croes Newydd	4/37	-	-
4309	8/11	Chester	Neyland	2/36	-	-
4310	8/11	Landore	Westbury	8/36	-	-
4311	8/11	Carmarthen	Chester	12/36	-	-
4312	9/11	Crewe	Birkenhead	3/38	-	-
4313	9/11	Tyseley	Barnstaple	2/36	-	-
4314	9/11	Weymouth	Taunton	6/37	-	-
4315	9/11	Llanelly	Tyseley	4/37	-	-
4316	9/11	Paddington	Bristol SPM	9/37	-	-
4317	10/11	Bristol	Stourbridge	6/37	-	-
4318	10/11	Wolverhampton	Didcot	-	Cardiff	6/52
4319	10/11	Bristol	Severn Tunnel	5/38	-	-
4320	10/11	Pontypool	Cheltenham	-	Gloucester	1/49
4321	6/13	Southall	Westbury	9/37	-	-
4322	6/13	Exeter	Neyland	9/37	-	-
4323	7/13	Llanelly	Oxley	10/36	-	-
4324	7/13	Carmarthen	Oxley	12/36	-	-
4325	8/13	Leamington	Pontypool	4/38	-	-
4326	8/13	Cardiff	Didcot	-	Stourbridge	3/57
4327	8/13	Pontypool	Chester	1/37	-	-
4328	8/13	Reading	Neyland	8/38	-	-
4329	8/13	Neyland	Swindon	9/36	-	-
4330	8/13	Wolverhampton	Barnstaple	8/37	-	-
4331	9/13	Truro	Neyland	2/36	-	-
4332	9/13	Tyseley	Salisbury	2/36	-	-
4333	9/13	Wolverhampton	Chester	2/37	-	-
4334	9/13	Paddington	Ebbw Junction	10/36	-	-
4335	9/13	Banbury	Pontypool	11/36	-	-
4336	9/13	Worcester	Birkenhead	7/38	-	-
4337	10/13	Birkenhead	Birkenhead	-	Stourbridge	11/51
4338	10/13	Cardiff	Stafford Road	2/37	-	-
4339	10/13	Cardiff	Taunton	1/39	-	-
4340	10/13	Bristol	Neyland	8/36	-	-
4341	12/13	Exeter	Reading	8/36	-	-
4342	12/13	Newton Abbot	Pontypool	5/38	-	-
4343	1/14	Paddington	Tyseley	10/37	-	-
4344	1/14	Wolverhampton	Cardiff	7/36	-	-
4345	1/14	Neath	Reading	8/36	-	-
4346	1/14	Severn Tunnel	Hereford	11/36	-	-
4347	1/14	Neath	Tyseley	1/37	-	-
4348	1/14	Westbury	Taunton	2/37	-	-
4349	2/14	Wolverhampton	Taunton	6/38	-	-
4350	2/14	Newport	Worcester	3/36	-	-
4351	2/14	Oxley	Worcester	8/36	-	-
4352	2/14	Neyland	Pontypool	8/37	-	-
4353	3/14	Wolverhampton	Birkenhead	-	Birkenhead	11/48
4354	3/14	Wolverhampton	Westbury	2/39	-	-
4355	3/14	Exeter	Shrewsbury	6/38	-	-
4356	3/14	Exeter	Paddington	4/36	-	-

No.	Date	1922 Shed	Last GWR Shed	Wdn GWR	Last BR shed	Wdn BR
4357	4/14	Severn Tunnel	Worcester	11/36	-	-
4358	4/14	Southall	Neyland	-	Gloucester	8/59
4359	4/14	Cardiff	Tyseley	5/37	-	-
4360	5/14	Tyseley	Weymouth	5/38	-	-
4361	5/15	Neath	Barnstaple	2/39	-	-
4362	5/15	Bristol	Croes Newydd	8/37	-	-
4363	6/15	Paddington	Oxley	2/39	-	-
4364	6/15	Neath	Bristol SPM	2/36	-	-
4365	7/15	Westbury	Westbury	-	Westbury	4/48
4366	7/15	Wolverhampton	Cardiff	2/37	-	-
4367	7/15	Exeter	Worcester	1/39	-	-
4368	8/15	Wolverhampton	Reading	2/39	-	-
4369	9/15	Tyseley	Weymouth	2/38	-	-
4370	9/15	Wolverhampton	Landore	1/37	-	-
4371	10/15	Paddington	Bristol SPM	11/38	-	-
4372	10/15	Goodwick	Severn Tunnel	9/36	-	-
4373	10/15	Newport	Reading	8/37	-	-
4374	11/15	Laira	Worcester	1/37	-	-
4375	11/15	Bristol	Croes Newydd	-	Stourbridge	1/58
4376	12/15	Cardiff	Chester	11/36	-	-
4377	12/15	Reading	Westbury	-	Aberwystwyth	1/59
4378	12/15	Laira	Pontypool	1/37	-	-
4379	12/15	Neath	Taunton	2/38	-	-
4380	12/15	Laira	Salisbury	7/38	-	-
4381	5/16	Chester	Andover Jct	-	Southall	5/53
4382	5/16	Carmarthen	Westbury	1/38	-	-
4383	6/16	Crewe	Gloucester	3/36	-	-
4384	6/16	Laira	Westbury	8/36	-	-
4385	6/16	Bristol	Birkenhead	2/39	-	-
4386	7/16	Penzance	Birkenhead	-	Birkenhead	4/48
4387	7/16	Pontypool	Bath Road	3/36	-	-
4388	7/16	Wolverhampton	Cheltenham	1/39	-	-
4389	8/16	Hereford	Tyseley	10/37	-	-
4390	8/16	Wolverhampton	Chester	2/39	-	-
4391	8/16	Newton Abbot	Barnstaple	2/37	-	-
4392	9/16	Cardiff	Worcester	12/38	-	-
4393	9/16	Newport	Worcester	8/37	-	-
4394	9/16	Newton Abbot	Bristol SPM	8/38	-	-
4395	9/16	Leamington	Oxley	7/36	-	-
4396	9/16	Laira	Tyseley	6/37	-	-
4397	10/16	Penzance	Tyseley	3/36	-	-
4398	10/16	Bristol	Pontypool	8/36	-	-
4399	11/16	Tyseley	Carmarthen	5/38	-	-
4400	10/04	Tondu	Wellington	-	Wellington	4/51
4401	7/05	St. Blazey	Much Wenlock	-	Newton Abbot	10/54
4402	7/05	St. Ives	Princetown	-	Princetown	12/49
4403	10/05	Helston	Much Wenlock	-	Laira	1/53
4404	12/05	Princetown	Tondu	-	Tondu	3/52
4405	1/06	St. Blazey	Newton Abbot	-	Newton Abbot	9/55
4406	2/06	Tondu	Wellington	-	Newton Abbot	9/55
4407	4/06	Princetown	Princetown	-	Laira	3/53
4408	5/06	Penzance	Tondu	-	Tondu	1/53
4409	6/06	St. Ives	Much Wenlock	-	Laira	2/51
4410	6/06	Plymouth	Exeter	-	Laira	9/55
4500	10/06	Newton Abbot	St. Ives	-	Penzance	8/53
4501	11/06	Duffryn Yard	Portmadoc	-	Machynlleth	3/53
4502	12/06	Kingswear	Swindon	-	Swindon	10/51
4503	1/07	Newport	St. Blazey	-	St. Blazey	1/51
4504	2/07	Newport	Worcester	-	Truro	6/52
4505	2/07	Taunton	St. Blazey	-	Penzance	10/57

No.	Date	1922 Shed	Last GWR Shed	Wdn GWR	Last BR shed	Wdn BR
4506	4/07	Newport	Whitland	-	Whitland	3/55
4507	5/07	Laira	Swindon	-	Yeovil Town	10/63
4508	6/07	Laira	Westbury	-	Truro	10/59
4509	6/07	Newport	Penzance	-	Penzance	6/51
4510	7/07	St. Blazey	Swindon	-	Westbury	3/53
4511	9/07	Newton Abbot	Oxford	-	Bath Road	11/53
4512	9/07	Kingswear	Machynlleth	-	Machynlleth	2/53
4513	10/07	Plymouth	Machynlleth	-	Oxford	10/50
4514	11/07	Kingsbridge	Aberbeeg	-	Aberbeeg	2/53
4515	12/07	St. Blazey	Whitland	-	Whitland	4/53
4516	1/08	Laira	St. Blazey	-	Bath Road	12/52
4517	2/08	Chester	Laira	-	Bath Road	2/53
4518	3/08	Newport	Ebbw Junction	-	Laira	10/52
4519	4/08	Newport	Whitland	-	Whitland	2/59
4520	3/09	Launceston	Westbury	-	Weymouth	1/53
4521	3/09	St. Blazey	Swindon	-	Whitland	12/55
4522	3/09	Newton Abbot	Aberbeeg	-	Aberbeeg	2/55
4523	3/09	St. Blazey	Truro	-	Truro	10/55
4524	3/09	Newport	Launceston	-	Bath Road	6/58
4525	2/10	Taunton	Helston	-	Penzance	7/53
4526	2/10	Newport	Newton Abbot	-	St. Blazey	1/58
4527	3/10	Duffryn Yard	Weymouth	-	Weymouth	3/53
4258	3/10	Chester	Laira	-	Laira	12/50
4529	3/10	Newton Abbot	St. Blazey	-	St. Blazey	3/52
4530	4/13	Truro	Exeter	-	Laira	3/55
4531	4/13	Plymouth	Laira	-	Laira	2/50
4532	4/13	Newport	Truro	-	Bath Road	2/55
4533	4/13	Newport	Pontypool	-	Pontypool	3/55
4534	4/13	Newport	Cheltenham	-	Laira	2/55
4535	4/13	Kingsbridge	Bath Road	-	Bath Road	2/55
4536	5/13	Newton Abbot	Bath Road	-	Westbury	4/59
4537	5/13	Newton Abbot	Penzance	-	Penzance	2/55
4538	5/13	Moreton'stead	Andover Jct	-	Swindon	5/57
4539	5/13	Ludlow	Bath Road	-	Bath Road	10/55
4540	11/14	Plymouth	Penzance	-	Penzance	3/59
4541	11/14	Kingswear	Pontypool	-	Whitland	10/55
4542	11/14	Newton Abbot	Laira	-	Laira	9/55
4543	11/14	St. Blazey	Swindon	-	Frome	8/50
4544	11/14	Neyland	Swindon	-	Bath Road	9/52
4545	12/14	Plymouth	Penzance	-	Penzance	9/58
4546	12/14	Plymouth	Worcester	-	Oswestry	1/58
4547	12/14	Launceston	Newton Abbot	-	St. Blazey	2/60
4548	1/15	Newton Abbot	Penzance	-	Exeter	11/57
4549	1/15	Kingsbridge	Machynlleth	-	Penzance	12/61
4550	2/15	Plymouth	Swindon	-	St. Blazey	10/60
4551	2/15	Newton Abbot	Andover Jct	-	Westbury	2/58
4552	3/15	Newton Abbot	St. Blazey	-	St. Blazey	9/61
4553	3/15	St. Blazey	Whitland	-	Laira	12/58
4554	3/15	Plymouth	Truro	-	Penzance	9/58
4600	11/13	Pembroke Dock	Neath	7/25	-	-
4700	5/19	Bristol	Paddington	-	Southall	10/62
4701	1/22	Paddington	Paddington	-	Paddington	9/63
4702	2/22	Wolverhampton	Paddington	-	Southall	6/62
4703	3/22	Wolverhampton	Laira	-	Paddington	5/64
4704	4/22	Wolverhampton	Birkenhead	-	Paddington	5/64
4705	4/22	Wolverhampton	Paddington	-	Paddington	12/63
5300+	11/16	Plymouth	Oxley	-	Oxley	1/53
5301+	12/16	Laira	Bristol SPM	10/37	-	-
5302+	12/16	Worcester	Banbury	-	Banbury	5/48
5303	12/16	Worcester	Kidderminster	-	Worcester	5/51

No.	Date	1922 Shed	Last GWR Shed	Wdn GWR	Last BR shed	Wdn BR
5304+	12/16	Wolverhampton	Weymouth	5/37	-	-
5305+	1/17	Paddington	Weymouth	-	Weymouth	7/52
5306	1/17	Paddington	Westbury	-	Neath	7/64
5307+	1/17	Truro	Cardiff	-	Bristol SPM	11/56
5308+	2/17	Banbury	Bristol SPM	1/39	-	-
5309+	2/17	Shrewsbury	Wellington	-	Oxley	1/53
5310	6/17	Shrewsbury	Neyland	-	Carmarthen	7/58
5311	6/17	Cardiff	Westbury	-	Bristol SPM	10/60
5312	6/17	Banbury	Gloucester	-	Llanelly	10/58
5313+	6/17	Wolverhampton	Oxley	-	Oxley	7/58
5314+	6/17	Severn Tunnel	Weymouth	-	Weymouth	7/57
5315+	7/17	Chester	Croes Newydd	-	Bristol SPM	2/59
5316	7/17	Bristol	Birkenhead	-	Birkenhead	6/56
5317	7/17	Cardiff	Banbury	-	Tyseley	11/56
5318+	7/17	Wolverhampton	Laira	-	Pontypool	9/61
5319	8/17	Newport	Croes Newydd	-	Shrewsbury	11/59
5320+	8/17	Bristol	Reading	-	Reading	9/48
5321	8/17	Chester	Exeter	-	Pontypool	9/59
5322+	8/17	Chester	Andover Jct	-	Pontypool	4/64
5323	9/17	Birkenhead	Oxford	-	Llanelly	6/58
5324	9/17	Chester	Banbury	-	Shrewsbury	9/60
5325+	9/17	Cardiff	Bath Road	-	Tyseley	8/57
5326+	9/17	Wolverhampton	Westbury	-	Bristol SPM	3/62
5327+	9/17	Birkenhead	Bath Road	-	Westbury	7/56
5328+	9/17	Pontypool	Weymouth	-	Swindon	7/58
5329+	9/17	Newton Abbot	Weymouth	3/39	-	-
5330	9/17	Tyseley	Didcot	-	Didcot	6/64
5331+	9/17	Reading	Oxley	-	Shrewsbury	11/60
5332+	11/17	Paddington	Wellington	-	Carmarthen	10/61
5333+	11/17	Swindon	Oxley	-	Kidderminster	5/60
5334+	11/17	Bristol	Croes Newydd	-	Cardiff	11/57
5335+	12/17	Pontypool	Cardiff	-	Llanelly	10/58
5336	12/17	Chester	Gloucester	-	Taunton	9/64
5337+	12/17	Laira	Weymouth	-	Bristol SPM	10/60
5338+	1/18	Newport	Weymouth	-	Westbury	8/58
5339	1/18	Plymouth	Carmarthen	-	Severn Tunnel	11/60
5340+	1/18	Severn Tunnel	Weymouth	-	Weymouth	9/48
5341+	2/18	Penzance	Landore	-	Llanelly	7/59
5342+	2/18	Paddington	Laira	4/37	-	-
5343+	2/18	Newton Abbot	Bath Road	-	Bath Road	7/48
5344+	3/18	Swindon	Chester	-	Taunton	9/58
5345	3/18	Severn Tunnel	Cheltenham	-	Pontypool	6/59
5346	3/18	Newton Abbot	Tyseley	-	Tyseley	5/51
5347	3/18	Swindon	Gloucester	-	Shrewsbury	4/58
5348	3/18	Taunton	Hereford	-	Hereford	1/52
5349	4/18	Cardiff	Banbury	-	Banbury	5/48
5350+	7/18	Bristol	Newton Abbot	-	Hereford	12/59
5351+	8/18	Landore	Bristol SPM	-	Didcot	6/61
5352+	8/18	Bristol	Cardiff	6/37	-	-
5353+	8/18	Bristol	Neyland	-	Carmarthen	5/60
5354+	8/18	Gloucester	Swindon	5/37	-	-
5355	8/18	Worcester	Pontypool	-	Kidderminster	4/59
5356	8/18	Pontypool	Reading	-	Bristol SPM	10/59
5357+	9/18	Weymouth	Neyland	-	Neyland	9/62
5358+	9/18	Swindon	Bristol SPM	-	Machynlleth	7/62
5359+	9/18	Severn Tunnel	Weymouth	-	Weymouth	10/51
5360+	1/19	Laira	Southall	-	Cardiff	9/58
5361+	2/19	Bristol	Banbury	-	Didcot	1/60
5362+	2/19	Wolverhampton	Severn Tunnel	-	Shrewsbury	7/58
5363+	2/19	Neath	Exeter	9/37	-	-

No.	Date	1922 Shed	Last GWR Shed	Wdn GWR	Last BR shed	Wdn BR
5364+	3/19	Bristol	Ebbw Junction	-	Ebbw Junction	7/51
5365+	4/19	Wolverhampton	Croes Newydd	-	Croes Newydd	3/51
5366+	5/19	Pontypool	Newton Abbot	3/39	-	-
5367	5/19	Carmarthen	Swindon	-	Reading	9/58
5368+	6/19	Paddington	Neyland	-	Reading	9/58
5369+	6/19	Wolverhampton	Tyseley	-	Pontypool	11/63
5370	6/19	Banbury	Tyseley	-	Llanelly	9/60
5371	7/19	Chester	Andover Jct	-	Stourbridge	7/58
5372+	7/19	Chester	Neyland	-	Llanelly	7/58
5373+	7/19	Swindon	Banbury	-	Banbury	6/50
5374+	7/19	Newport	Bristol SPM	-	Bristol SPM	6/48
5375	8/19	Penzance	Reading	-	Banbury	10/59
5376+	8/19	Plymouth	Laira	-	Bristol SPM	6/62
5377	8/19	Penzance	Hereford	-	Carmarthen	9/58
5378+	8/19	Plymouth	Cardiff	-	Croes Newydd	9/59
5379+	8/19	Neath	Oxley	-	Banbury	5/58
5380	4/20	Plymouth	Didcot	-	Didcot	9/63
5381+	6/20	Newport	Didcot	-	Severn Tunnel	9/59
5382+	6/20	Newton Abbot	Cardiff	-	Severn Tunnel	4/59
5383+	7/20	Truro	Newton Abbot	5/37	-	-
5384+	7/20	Newton Abbot	Weymouth	-	Weymouth	10/60
5385	8/20	Chester	Reading	-	Landore	7/62
5386+	8/20	Chester	Oxley	-	Oswestry	10/58
5387+	8/20	Cardiff	Oxley	3/39	-	-
5388+	8/20	Southall	Cardiff	-	Pontypool	3/59
5389+	9/20	Laira	Ebbw Junction	5/37	-	-
5390+	10/20	Gloucester	Oxley	-	Didcot	8/58
5391+	10/20	Carmarthen	Newton Abbot	-	Tyseley	3/57
5392	10/20	Wolverhampton	Neyland	-	Worcester	8/58
5393+	10/20	Bristol	Birkenhead	-	Barrow Road	10/59
5394	10/20	Neyland	Gloucester	-	Barrow Road	1/59
5395	11/20	Neyland	Goodwick	-	Aberystwyth	10/55
5396	11/20	Neyland	Swindon	-	Worcester	5/60
5397	11/20	Neyland	Didcot	-	Didcot	7/58
5398	11/20	Neyland	Gloucester	-	Gloucester	2/59
5399	11/20	Reading	Chester	-	Croes Newydd	9/62
6300	11/20	Taunton	Oxford	-	Pontypool	6/60
6301	12/20	Goodwick	Exeter	-	Croes Newydd	10/62
6302	12/20	Llanelly	Reading	-	Didcot	3/62
6303	12/20	Paddington	Croes Newydd	-	Shrewsbury	4/59
6304	12/20	Bristol	Landore	-	Gloucester	1/64
6305	12/20	Bristol	Taunton	-	Carmarthen	9/59
6306	12/20	Neyland	Worcester	-	Croes Newydd	11/61
6307	1/21	Reading	Shrewsbury	-	Croes Newydd	7/60
6308	1/21	Paddington	Chester	-	Cardiff	8/59
6309	1/21	Basingstoke	Gloucester	-	Didcot	9/64
6310	1/21	Swindon	Carmarthen	-	Llanelly	7/62
6311	1/21	Paddington	Chester	-	Banbury	1/60
6312	1/21	Cardiff	Reading	-	Bristol SPM	9/62
6313	1/21	Cardiff	Reading	-	Didcot	11/61
6314	1/21	Wolverhampton	Westbury	-	Kidderminster	5/63
6315	1/21	Wolverhampton	Banbury	10/45	-	-
6316	1/21	Wolverhampton	Croes Newydd	-	Llanelly	7/62
6317	1/21	Wolverhampton	Taunton	-	Kidderminster	11/63
6318	2/21	Gloucester	Penzance	-	Pontypool	1/59
6319	2/21	Wolverhampton	Laira	-	Gloucester	8/63
6320	2/21	Gloucester	Swindon	-	Severn Tunnel	11/63
6321	2/21	Hereford	Stafford Road	-	Tyseley	3/56
6322	2/21	Swindon	Swindon	-	Newton Abbot	1/59
6323	3/21	Bristol	Taunton	-	Barrow Road	7/60

No.	Date	1922 Shed	Last GWR Shed	Wdn GWR	Last BR shed	Wdn BR
6324	3/21	Plymouth	Worcester	–	Reading	4/62
6325	3/21	Reading	Southall	–	Severn Tunnel	11/59
6326	3/21	Paddington	Cheltenham	–	Taunton	9/64
6327	3/21	Paddington	Croes Newydd	–	Taunton	9/63
6328	3/21	Weymouth	Taunton	–	Llanelly	1/59
6329	4/21	Paddington	Didcot	–	Carmarthen	10/61
6330	4/21	Paddington	St. Blazey	–	Gloucester	9/62
6331	4/21	Penzance	Carmarthen	–	Banbury	4/59
6332	4/21	Paddington	Oxley	–	Stourbridge	9/60
6333	5/21	Paddington	Pontypool	–	Shrewsbury	10/60
6334	5/21	Landore	Reading	–	Swindon	4/59
6335	5/21	Worcester	Oxley	–	Pontypool	7/63
6336	5/21	Bristol	Tyseley	–	Machynlleth	4/62
6337	6/21	Bristol	Chester	–	Didcot	7/64
6338	6/21	Plymouth	Shrewsbury	–	Pontypool	6/64
6339	6/21	Plymouth	Chester	–	Croes Newydd	7/62
6340	6/21	Exeter	Swindon	–	Taunton	7/62
6341	7/21	Bristol	Cheltenham	–	Gloucester	11/61
6370	4/21	Gloucester	Pontypool	–	Pontypool	1/63
6371	5/21	Penzance	Neyland	–	Machynlleth	9/60
6372	5/21	Bristol	Taunton	–	Taunton	12/63
6373	6/21	Hereford	Truro	–	Severn Tunnel	12/63
6374	5/21	Bristol	Swindon	–	Shrewsbury	8/62
6375	6/21	Worcester	Westbury	–	Carmarthen	9/63
6376	6/21	Bristol	Birkenhead	–	Barrow Road	5/62
6377	6/21	Bristol	Taunton	–	Carmarthen	10/60
6378	7/21	Paddington	Worcester	–	Didcot	6/64
6379	7/21	Paddington	Didcot	–	Didcot	8/63
6380	7/21	Paddington	Chester	–	Severn Tunnel	12/63
6381	7/21	Banbury	Gloucester	–	Gloucester	11/63
6382	8/21	Banbury	Worcester	–	Worcester	11/60
6383	8/21	Plymouth	Reading	–	Aberystwyth	5/56
6384	8/21	Goodwick	Swindon	–	Severn Tunnel	6/63
6385	8/21	Banbury	Worcester	–	Didcot	12/63
6386	9/21	Paddington	Severn Tunnel	–	Severn Tunnel	9/62
6387	9/21	Carmarthen	Andover Jct	–	Worcester	6/62
6388	10/21	Cardiff	Southall	–	Kidderminster	9/62
6389	10/21	Reading	Neyland	–	Neyland	9/60
6390	10/21	Bristol	Banbury	–	Taunton	5/62
6391	11/21	Swindon	Stafford Road	–	Reading	9/62
6392	11/21	Worcester	Chester	–	Machynlleth	10/61
6393	11/21	Wolverhampton	Reading	–	Pontypool	1/60
6394	11/21	Goodwick	Taunton	–	Gloucester	6/64
6395	12/21	Southall	Hereford	–	Stourbridge	11/64
6396	12/21	Westbury	Worcester	–	Llanelly	3/58
6397	12/21	Swindon	Exeter	–	St. Blazey	9/59
6398	12/21	Banbury	Taunton	–	Taunton	10/60
6399	12/21	Paddington	Westbury	–	Tyseley	11/59
7300	12/21	Paddington	Westbury	–	Shrewsbury	9/62
7301	12/21	Gloucester	Worcester	–	Bristol SPM	9/62
7302	12/21	Newton Abbot	Westbury	–	Pontypool	8/62
7303	1/22	Neath	Cheltenham	–	Taunton	9/64
7304	1/22	Cheltenham	Taunton	–	Taunton	12/63
7305	11/21	Goodwick	Croes Newydd	–	Taunton	9/62
7306	11/21	Goodwick	Neyland	–	Severn Tunnel	9/64
7307	11/21	Goodwick	Oxley	–	Gloucester	6/64
7308	11/21	Goodwick	Worcester	–	Didcot	6/64
7309	11/21	Goodwick	Westbury	–	Shrewsbury	9/62
7310	12/21	Goodwick	Croes Newydd	–	Severn Tunnel	6/64
7311	12/21	Goodwick	Oxley	–	Exeter	9/62

No.	Date	1922 Shed	Last GWR Shed	Wdn GWR	Last BR shed	Wdn BR
7312	12/21	Goodwick	Cheltenham	-	Severn Tunnel	12/63
7313	12/21	Goodwick	Chester	-	Croes Newydd	7/62
7314	12/21	Goodwick	Taunton	-	Shrewsbury	2/63
7315	12/21	Goodwick	Stafford Road	-	Neath	12/63
7316	12/21	Goodwick	Exeter	-	Exeter	9/62
7317	12/21	Hereford	Oxley	-	Taunton	12/63
7318	12/21	Hereford	Reading	-	Gloucester	11/64
7319	1/22	Hereford	Oxley	-	Gloucester	10/64

The undermentioned locomotives were withdrawn in 1938-9 from the engine sheds shown in the left hand columns and re-inststed in 1939-40 at the sheds given in the right hand columns:-

213	7/39	Barry	11/39	Barry
876	6/39	Brecon	12/39	Llanidloes
884	9/39	Pwllheli	12/39	Machynlleth
885	8/39	Oswestry	12/39	Machynlleth
1161	4/39	Weymouth	2/40	Swindon
1287	9/39	Whitland	9/39	Leamington
1624	6/39	Swindon	12/39	Swindon
1705	9/39	Pontypool	2/40	Cardiff
1709	6/39	Swansea E Dock	12/40	Pill
1722	5/39	Pontypool	2/40	Llantrisant
1740	7/39	Duffryn Yard	1/40	Danygraig
1742	10/39	Stourbridge	11/39	Didcot
1747	6/39	Tyseley	12/39	Stourbridge
1756	5/39	Duffryn Yard	1/40	Danygraig
1785	4/39	Oxford	2/40	Oxford
1789	9/39	Swindon	12/39	Weymouth
1795	4/39	Llantrisant	11/39	Newton Abbot
1866	7/39	Pantyffynon	4/40	Neath
1875	4/39	Severn Tunnel	10/39	Pill
1889	8/39	Pill	11/39	Cardiff
1891	7/39	Pill	6/40	Cardiff
1907	8/39	Whitland	12/39	Sandy
1910	9/39	Whitland	11/39	Whitland
1919	10/38	Worcester	5/40	Worcester
1968	6/39	Shrewsbury	1/40	Wellington
1979	4/39	Cardigan	11/39	Whitland
1989	7/39	Gloucester	12/39	Gloucester
2004	10/39	Birkenhead	12/39	Wellington
2014	5/39	Swindon	1/40	Cirencester
2019	5/39	Worcester	12/39	Sandy
2195	3/39	Weymouth	12/39	Swindon
2315	7/39	Portmadoc	10/39	Welshpool
2340	6/39	Wells	2/40	Carmarthen
2354	5/39	Southall	11/39	Carmarthen
2362	6/39	Bristol SPM	1/40	Llanidloes
2380	8/39	Worcester	11/39	Brecon
2640	8/39	Neyland	11/39	Severn Tunnel
2648	3/39	Severn Tunnel	1/40	Pontypool
2649	9/39	Cardiff	12/39	Pontypool
2652	6/39	Duffryn Yard	10/39	Newbury
2657	7/39	Swansea E Dock	11/39	Ebbw Junction
3170	8/39	Cardiff	11/39	Ebbw Junction
3335	7/39	Cardiff	11/39	Exeter
3378	2/39	Bath Road	10/39	Andover Jct
3500	4/39	Southall	2/40	Oxford
3561	5/39	Gloucester	11/39	Gloucester

3571	4/39	Swindon	1/40	Swindon
3577	4/39	Birkenhead	11/39	N'castle Emlyn

In addition the following 2301 Class 0-6-0 which had been withdrawn in 1939 were re-instated and requisitioned for Military use:-

2419 withdrawn 5/39 and sent to France 10/39,
2439 withdrawn 4/39 and sent to France 10/39,
2447 withdrawn 5/39 and sent to France 11/39,
2455 withdrawn 8/39 and sent to France 11/39,
2480 withdrawn 5/39 and sent to France 11/39,
2524 withdrawn 8/39 and sent to France 10/39.

Their last working sheds were shown in the main table.

The following engines were withdrawn and re-instated outside the 1939-40 period:-

162	5/32	Danygraig	6/32	Danygraig
242	1/46	Barry	6/47	Swindon
1749	11/49	Croes Newydd	12/49	Croes Newydd
2723	10/46	Tondu	6/47	Swindon
2731	7/47	Pill	8/47	Swindon
2733	8/47	Tyseley	11/47	Swindon
2735	8/47	Tondu	8/47	Swindon

Nos. 242, 2723/31/3/5 were used for shunting at Swindon Works.

The re-instatement of 4-4-0 No. 3717 City of Truro in 3/1957 was a special case; as No. 3440 it was allocated to Didcot in 4/1958 and worked until 5/1961.

0-6-0T No. 1393, allocated in 1922 to Neath, was shown as on loan to Gwaen-Cae-Gurwen Colliery at the time.

In May 1922 the GWR still had 40 'ROD' 2-8-0 engines on loan from the Government, but none were recorded as working, all being at Swindon Works under repair prior to return to their owner, the last by October 1922.

Two locomotives were recorded as on loan to the GWR in 1922 under the heading 'CR Engines'. Both were stationed at Westbury. No. 10 had been returned to 'the Government' in 3/22 but No. 6 was not returned until 3/23. No mention of Caledonian Railway engines being used by the GWR is known to the author, but it seems unlikely that they could be anything other than that railway's stock (4-4-0T 1 Class).

SPECIAL NOTE ON Nos. 3265 and 3365:

It should be noted that No. 3365, although listed as condemned 1/30, in reality took the identity of No. 3265 that month, so that it was the latter that was withdrawn. The post-1930 details shown against No. 3265 therefore apply to the engine that had been No. 3365 in 1930.

Locomotives Added to Stock August 1922

The following list of engines and their allocation uses the GWR numbers applicable in 1922 but those marked (+) after the number were renumbered subequently. The engines concerned were:-

1st No.	New No.	Date	1st No.	New No.	Date	1st No.	New No.	Date	1st No.	New No.	Date
11	421	-	698	425	7/47	1372	431	7/49	1375	434	4/49
21	422	9/47	888	426	4/47	1373	432	11/48	1668	435	4/50
332	423	-	1084	427	-	1374	433	5/47	1670	436	10/49
504	424	-	1113	428	10/47						

(Nos. 11, 332, 504, 1084 were not actually renumbered).

Engines marked (*) in the allocation list were condemned without being renumbered.

No.	Date	1922 Shed	Last GWR Shed	Wdn GWR	Last BR shed	Wdn BR
11+	12/09	Bassaleg	Pill	-	Pill	1/49
21+	12/09	Machen	Severn Tunnel	-	Severn Tunnel	6/50
332+	12/09	Bassaleg	Severn Tunnel	-	Severn Tunnel	12/49
504+	1/10	Bassaleg	Pill	-	Pill	1/48
698+	3/14	Bassaleg	Pill	-	Cardiff	1/51
888+	3/14	Bassaleg	Pill	-	Llantrisant	3/50
1084+	4/14	Bassaleg	Pill	11/47	-	-
1113+	4/14	Bassaleg	Duffryn Yard	-	Duffryn Yard	8/50
1114	5/06	Neath N&B	Neath N&B	1/30	-	-
1117	5/08	Neath N&B	Neath N&B	8/30	-	-
1277	5/10	Neath N&B	Neath N&B	11/29	-	-
1327	12/97	Neath N&B	Neath N&B	10/29	-	-
1371	7/97	Neath N&B	Landore	10/29	-	-
1372+	8/15	Machen	Ebbw Junction	-	Ebbw Junction	10/53
1373+	9/15	Bassaleg	Ebbw Junction	-	Ebbw Junction	5/53
1374+	9/15	Bassaleg	Radyr	-	Radyr	2/51
1375+	2/21	Bassaleg	Cathays	-	Radyr	9/53
1391*	8/79	Bassaleg	Bassaleg	11/22	-	-
1392	-/71	Neath N&B	Danygraig	8/26	-	-
1400	-/93	Neath N&B	Neath N&B	10/26	-	-
1402*	12/88	Brecon	Brecon	6/23	-	-
1412*	12/88	Bassaleg	Bassaleg	4/23	-	-
1452*	2/89	Bassaleg	Brecon	2/23	-	-
1458*	3/98	Bassaleg	Bassaleg	10/22	-	-
1460	1/89	Bassaleg	Oswestry	10/24	-	-
1563	11/78	Neath N&B	Neath N&B	2/31	-	-
1591*	7/79	Neath N&B	Neath N&B	8/22	-	-
1668+	2/21	Brecon	Ebbw Junction	-	Ebbw Junction	2/54
1670+	2/21	Bassaleg	Ebbw Junction	-	Ebbw Junction	2/54
1674*	7/05	Brecon	Brecon	1/23	-	-
1677	8/05	Brecon	Machynlleth	9/28	-	-
1685	12/86	Bassaleg	Cardiff	10/34	-	-
1692	5/94	Brecon	Brecon	7/25	-	-
1693	3/87	Bassaleg	Ebbw Junction	4/31	-	-
1694	4/87	Machen	Cardiff	12/26	-	-
1715	5/91	Neath N&B	Neath	-	Neath	10/49
1833	5/94	Machen	Brecon	5/24	-	-
1882	12/90	Neath N&B	Neath	12/46	-	-
2161	8/17	Bassaleg	Pill	5/29	-	-

No.	Date	1922 Shed	Last GWR Shed	Wdn GWR	Last BR shed	Wdn BR
2162	5/14	Burry Port	Llanelly	-	Burry Port	3/55
2163	11/09	Burry Port	Burry Port	4/44	-	-
2164	3/12	Burry Port	Neath	2/29	-	-
2165	5/13	Burry Port	Burry Port	-	Burry Port	3/55
2166	10/16	Burry Port	Swansea E Dock	-	Swansea E Dock	5/55
2167	8/19	Burry Port	Burry Port	-	Llanelly	2/53
2168	2/16	Burry Port	Burry Port	-	Burry Port	5/56
2169	10/96	Bassaleg	Bassaleg	10/27	-	-
2170	10/96	Bassaleg	Cardiff Docks	9/28	-	-
2171	9/00	Bassaleg	Brecon	7/32	-	-
2172	9/00	Brecon	Bassaleg	5/28	-	-
2173	10/00	Brecon	Ebbw Junction	5/32	-	-
2174	-/99	Neath N&B	Neath N&B	7/27	-	-
2175	-/99	Neath N&B	Neath	2/33	-	-
2176	-/07	Burry Port	Llanelly	-	Llanelly	3/55
2177	8/85	Brecon	Brecon	2/28	-	-
2178	8/85	Brecon	Brecon	12/25	-	-
2179	1/86	Brecon	Ebbw Junction	10/26	-	-
2180	1/86	Bassaleg	Ebbw Junction	10/26	-	-
2181	3/84	Bassaleg	Bassaleg	7/27	-	-
2182	2/84	Brecon	Ebbw Junction	12/27	-	-
2183	2/84	Brecon	Ebbw Junction	2/32	-	-
2184	3/84	Brecon	Ebbw Junction	7/33	-	-
2185	1/86	Machen	Brecon	7/23	-	-
2186	1/86	Bassaleg	Ebbw Junction	12/34	-	-
2187	4/84	Bassaleg	Brecon	11/23	-	-
2188	4/84	Bassaleg	Ebbw Junction	3/28	-	-
2189	-/72	Neath N&B	Neath N&B	9/31	-	-
2190	-/81	Bassaleg	Didcot	5/34	-	-
2191	-/81	Bassaleg	Ebbw Junction	4/32	-	-
2192	8/00	Burry Port	Neath	-	Neath	4/51
2193	9/01	Burry Port	Llanelly	-	Llanelly	2/52
2194	5/03	Burry Port	Taunton	-	Taunton	2/53
2195	4/05	Burry Port	Swindon	-	Swindon	1/53
2196	9/06	Burry Port	Burry Port	-	Llanelly	1/56
2197	3/09	Burry Port	Llanelly	-	Llanelly	10/52
2198	12/10	Burry Port	Burry Port	-	Llanelly	3/59
2199	-/72	Neath N&B	Swansea E Dock	6/31	-	-

Locomotives Added to Stock January 1923

No.	Date	1922 Shed	Last GWR Shed	Wdn GWR	Last BR shed	Wdn BR
26	2/05	Kidwelly	Neath	12/27	-	-
-	11/78	Kidwelly	-	3/23	-	-
312	8/07	Sandy	Llanelly	12/34	-	-
339	7/13	Sandy	Sandy	8/43	-	-
359	6/17	Sandy	Danygraig	-	Danygraig	2/54
704	3/20	Sandy	Danygraig	12/43	-	-
803	1/11	Sandy	Danygraig	-	Danygraig	3/51
937*	4/12	Sandy	Sandy	6/23	-	-
944	-/02	Sandy	Swansea H'bour	11/28	-	-
969*	-/75	Sandy	-	7/25	-	-

The second engine listed above was sold at the time of take over and was not allotted a GWR number.
No. 704 was on loan to the Ministry of Fuel & Power (Emlyn Colliery) when condemned. No. 969 never worked as a GWR engine.

Locomotives Added to Stock July 1923 and January 1924

The following list of engines and their allocation uses the GWR numbers applicable in 1923-4 but those marked (+) after the number were renumbered subsequently. The engines concerned were:-

1st no.	New No.	Date	1st No.	New No.	Date	1st No.	New No.	Date	1st No.	New No.	Date
696	1150	12/51	929	1141	3/48	943	1142	11/48	1085	1146	2/49
701	1140	6/48	935	1152	6/49	968	1143	2/49	1086	1147	3/49
779	1151	10/50	942	1153	11/49	974	1144	9/48	1098	1145	1/50

No.	Date	1922 Shed	Last GWR Shed	Wdn GWR	Last BR shed	Wdn BR
150	4/05	Swansea H'bour	Swansea H'bour	5/29	-	-
696+	7/13	Swansea H'bour	Swansea E Dock	-	Swansea E Dock	11/52
701+	4/05	Swansea H'bour	Swansea E Dock	-	Swansea E Dock	5/58
779+	9/16	Swansea H'bour	Danygraig	-	Swansea E Dock	8/63
795	-/03	Swansea H'bour	Swansea H'bour	6/29	-	-
886	8/99	Swansea H'bour	Swansea H'bour	2/28	-	-
921	-/06	Swansea H'bour	Swansea H'bour	9/28	-	-
925	-/74	Swansea H'bour	Swansea H'bour	5/29	-	-
926	10/99	Swansea H'bour	Swansea H'bour	2/29	-	-
927	11/07	Swansea H'bour	Swansea H'bour	8/28	-	-
928	3/12	Swansea H'bour	Swansea H'bour	3/27	-	-
929+	6/06	Swansea H'bour	Danygraig	-	Danygraig	7/52
930	4/02	Swansea H'bour	Swansea H'bour	2/27	-	-
933	4/04	Swansea H'bour	Swansea H'bour	12/27	-	-
935+	1/12	Swansea H'bour	Swansea E Dock	-	Swansea E Dock	12/61
937	4/15	Swansea H'bour	Swansea H'bour	11/26	-	-
942+	3/05	Swansea H'bour	Danygraig	-	Danygraig	10/55
943+	6/11	Swansea H'bour	Danygraig	-	Clee Hill	11/59
968+	6/08	Swansea H'bour	Danygraig	-	Clee Hill	11/60
974+	7/09	Swansea H'bour	Swansea E Dock	-	Swansea E Dock	1/60
1085+	4/12	Swansea H'bour	Danygraig	-	Swansea E Dock	1/51
1086+	2/13	Swansea H'bour	Danygraig	-	Danygraig	4/51
1098+	10/18	Swansea H'bour	Danygraig	-	Danygraig	7/59

Steam Rail Motors

No.	Date	1922 Shed	Last GWR Shed	Wdn GWR	No.	Date	1922 Shed	Last GWR Shed	Wdn GWR
16	11/05	STOCK	STOCK	12/27	68	6/06	Bristol	Stourbridge	12/27
30	1/05	Croes Newydd	Southall	10/35	69	6/06	Bristol	Southall	6/33
31	12/04	CONDEMNED	-	8/22	70	6/06	Stourbridge	Neath	10/35
33	12/04	CONDEMNED	-	8/22	71	6/06	Taunton	Exeter	10/35
34	1/05	CONDEMNED	-	8/22	72	7/06	Exeter	Southall	1/35
35	12/04	CONDEMNED	-	8/22	73	4/06	Gloucester	Exeter	6/33
36	12/04	CONDEMNED	-	8/22	74	4/06	Oswestry	Bristol SPM	6/33
37	2/05	Neath	Exeter	10/35	75	5/06	Bristol	Southall	6/34
38	3/05	Aberdare	Bristol SPM	12/27	76	4/06	Garnant	Bristol SPM	1/35
39	4/05	STOCK	STOCK	6/33	77	4/06	Taunton	Bristol SPM	10/35
40	3/05	Neath	Stourbridge	6/33	78	6/06	Croes Newydd	Croes Newydd	6/34
41	1/05	Chalford	Reading	12/27	79	6/06	Worcester	Croes Newydd	6/34
43	2/05	Gloucester	Gloucester	11/22	80	6/06	Croes Newydd	Exeter	10/35
44	2/05	Gloucester	Gloucester	11/22	81	5/07	Southall	Bristol SPM	11/34
45	2/05	Gloucester	Storubridge	12/27	82	5/07	Llanelly	Bristol SPM	6/33
46	2/05	Stourbridge	Stourbridge	11/22	83	5/07	Garnant	Bristol SPM	6/33
47	2/05	Southall	Southall	11/22	84	12/07	Oswestry	Southall	3/30
50	3/05	Aberdare	Aberdare	11/22	85	12/07	Oswestry	Gloucester	12/27
51	4/05	CONDEMNED	-	8/22	86	12/07	Exeter	Neath	6/33
52	2/05	Swindon	Swindon	10/22	87	12/07	Gloucester	Bristol SPM	12/27
53	9/05	Yeovil	Exeter	6/33	88	1/08	Southall	Stourbridge	10/35
54	9/05	Bristol	Southall	1/30	89	12/07	Yeovil	Taunton	12/27
55	10/05	Bristol	Stourbridge	10/35	90	12/07	Bristol	Southall	12/27
56	10/05	Frome	Birkenhead	1/30	91	1/08	Bath	Southall	1/35
57	9/05	Frome	Neath	12/27	92	1/08	Stourbridge	Bristol SPM	10/35
58	9/05	Exeter	Stourbridge	6/33	93	2/08	Croes Newydd	Stourbridge	11/34
61	3/06	Gloucester	Garnant	12/27	94	2/08	Trowbridge	Reading	4/30
62	4/06	Reading	Exeter	11/29	95	2/08	Yatton	Birkenhead	3/30
63	4/06	Chippenham	Birkenhead	12/27	96	2/08	Southall	Exeter	1/34
64	4/06	Reading	Stourbridge	1/35	97	2/08	Taunton	Southall	4/35
65	4/06	Gloucester	Neath	10/35	98	2/08	Taunton	Southall	6/35
66	5/06	Southall	Stourbridge	3/35	99	2/08	Taunton	Bristol SPM	12/27
67	6/06	Bristol	Bristol SPM	12/27					

Nos. 31/3-6 and 51 were in hand for conversion as driving trailers in the summer of 1922.

New Locomotives June 1922 - October 1956

The locomotives put into GWR and British Railways service between June 1922 and October 1956 were taken into stock under the following Lot numbers:-

Nos.	Type	Class	Date	Lot No.	See Note
2	0-6-0DE	-	1936	302	2
3	0-4-2ST	-	1878	-	3
4	0-4-2ST	-	1921	-	4
5	0-6-0T	-	1877	342	5
6	0-6-0T	-	1875	342	5
7, 8	2-6-2T	-	1923	227	7
12	0-4-0T	-	1926	248	12
13	0-4-0T	-	1926	247	13
111	4-6-0	4073	1924	-	111
1000-6	4-6-0	1000	1945	354	
1007-19	4-6-0	1000	1946	354	
1020-2	4-6-0	1000	1946	358	
1023-9	4-6-0	1000	1947	358	
1101-6	0-4-0T	1101	1926	246	1101
1331	0-6-0ST	1331	1877	245	1331
1366-71	0-6-0PT	1366	1934	286	
1500-9	0-6-0PT	1500	1949	373	
1600-19	0-6-0PT	1600	1949	381	
1620-9	0-6-0PT	1600	1950	381	
1630-49	0-6-0PT	1600	1951	389	
1650-4	0-6-0PT	1600	1954	417	
1655-69	0-6-0PT	1600	1955	417	
2200	0-6-0	2251	1938	312	
2201-10	0-6-0	2251	1939	322	
2211-30	0-6-0	2251	1940	337	
2231-40	0-6-0	2251	1944	347	
2241-50	0-6-0	2251	1945	347	
2251-70	0-6-0	2251	1930	261	
2271-80	0-6-0	2251	1934	283	
2281-90	0-6-0	2251	1936	298	
2291-9	0-6-0	2251	1938	312	
2884-99	2-8-0	2884	1938	321	
2925	4-6-0	4900	1924	-	2925
3020-49	2-8-0	ROD	-	241	3020
3050-99	2-8-0	ROD	-	240	3020
3100	2-6-2T	3100	1938	319	3100
3101-4	2-6-2T	3100	1939	319	3100
3200-7	4-4-0	3200	1936	315	3200
3208-16	4-4-0	3200	1937	315	3200
3217-9	4-4-0	3200	1938	315	3200
3220-2	4-4-0	3200	1938	331	3200
3223-8	4-4-0	3200	1939	331	3200
3200-9	0-6-0	2251	1946	360	
3210-7	0-6-0	2251	1947	360	
3218/9	0-6-0	2251	1948	360	
3400	0-6-0PT	8400	1955	387	9490
3401-9	0-6-0PT	8400	1956	387	9490
3600/1	0-6-0PT	5700	1938	314	
3602-34	0-6-0PT	5700	1939	314	

Nos.	Type	Class	Date	Lot No.	See Note
3635-54	0-6-0PT	5700	1939	325	
3655-84	0-6-0PT	5700	1940	325	
3685/6	0-6-0PT	5700	1940	330	
3687-99	0-6-0PT	5700	1941	330	
3700-14	0-6-0PT	5700	1936	299	
3715-34	0-6-0PT	5700	1937	299	
3735-64	0-6-0PT	5700	1937	306	
3765-84	0-6-0PT	5700	1938	306	
3785-99	0-6-0PT	5700	1938	314	
3800-2	2-8-0	2884	1938	321	
3803	2-8-0	2884	1939	321	
3804-13	2-8-0	2884	1939	328	
3814-23	2-8-0	2884	1940	328	
3824-31	2-8-0	2884	1940	334	
3832/3	2-8-0	2884	1941	334	
3834	2-8-0	2884	1941	341	
3835-43	2-8-0	2884	1942	341	
3844-66	2-8-0	2884	1942	346	
4000	4-6-0	4073	1929	-	4000
4009/16	4-6-0	4073	1925	-	4000
4032/7	4-6-0	4073	1926	-	4000
4063-6	4-6-0	4000	1922	217	4063
4067-72	4-6-0	4000	1923	217	4063
4073/4	4-6-0	4073	1923	224	
4075-82	4-6-0	4073	1924	224	
4083-92	4-6-0	4073	1925	232	
4093-9	4-6-0	4073	1926	234	
4100-9	2-6-2T	5100	1935	292	
4110-9	2-6-2T	5100	1936	292	
4120/1	2-6-2T	5100	1937	313	
4122-9	2-6-2T	5100	1938	313	
4130-9	2-6-2T	5100	1939	323	
4140-9	2-6-2T	5100	1946	335	
4150-9	2-6-2T	5100	1947	361	
4160-9	2-6-2T	5100	1948	369	
4170-9	2-6-2T	5100	1949	369	
4200	2-8-0T	4200	1923	220	
4296-9	2-8-0T	4200	1922	220	
4555-74	2-6-2T	4500	1924	226	
4575-99	2-6-2T	4575	1927	242	
4600-4	0-6-0PT	5700	1941	330	
4605-34	0-6-0PT	5700	1942	330	
4635-41	0-6-0PT	5700	1942	336	
4642-60	0-6-0PT	5700	1943	336	
4661-7	0-6-0PT	5700	1943	352	
4668-89	0-6-0PT	5700	1944	352	
4690-9	0-6-0PT	5700	1945	352	
4706-8	2-8-0	4700	1923	221	
4800-9	0-4-2T	4800	1932	279	
4810-29	0-4-2T	4800	1933	279	
4830-9	0-4-2T	4800	1934	288	
4840-59	0-4-2T	4800	1935	288	
4860-74	0-4-2T	4800	1936	287	
4901-5	4-6-0	4900	1928	254	
4906-70	4-6-0	4900	1929	254	
4971-80	4-6-0	4900	1930	254	
4981-99	4-6-0	4900	1931	268	

Nos.	Type	Class	Date	Lot No.	See Note
5000-2	4-6-0	4073	1926	234	
5003-12	4-6-0	4073	1927	234	
5013-22	4-6-0	4073	1932	280	
5023-32	4-6-0	4073	1934	295	
5033-42	4-6-0	4073	1935	296	
5043-57	4-6-0	4073	1936	303	
5058-67	4-6-0	4073	1937	303	
5068-77	4-6-0	4073	1938	310	
5078-82	4-6-0	4073	1939	310	
5083/4	4-6-0	4073	1937	317	4063
5085	4-6-0	4073	1939	317	4063
5086	4-6-0	4073	1937	317	4063
5087	4-6-0	4073	1940	317	4063
5088-90	4-6-0	4073	1939	317	4063
5091/2	4-6-0	4073	1938	317	4063
5093-7	4-6-0	4073	1939	324	
5098/9	4-6-0	4073	1946	357	
5101-10	2-6-2T	5100	1929	257	
5150-9	2-6-2T	5100	1930	257	
5160-72	2-6-2T	5100	1930	259	
5173-89	2-6-2T	5100	1931	259	
5190-9	2-6-2T	5100	1934	284	
5200-2	2-8-0T	4200	1922	220	
5203/4	2-8-0T	4200	1923	220	
5205-14	2-8-0T	5205	1923	223	
5215-44	2-8-0T	5205	1924	225	
5245-54	2-8-0T	5205	1925	233	
5255/6	2-8-0T	5205	1925	233	5255
5257-74	2-8-0T	5205	1926	233	5255
5275-94	2-8-0T	5205	1930	266	5275
5255-64	2-8-0T	5205	1940	329	
5400	0-6-0PT	5400	1930	277	5400
5400	0-6-0PT	5400	1932	-	5400
5401-6	0-6-0PT	5400	1931	277	
5407-19	0-6-0PT	5400	1932	277	
5420-4	0-6-0PT	5400	1935	301	
5500-4	2-6-2T	4575	1927	242	
5505-24	2-6-2T	4575	1927	249	
5525-44	2-6-2T	4575	1928	251	
5545-64	2-6-2T	4575	1928	253	
5565-74	2-6-2T	4575	1929	253	
5600-3	0-6-2T	5600	1924	228	
5604-49	0-6-2T	5600	1925	228	
5650-84	0-6-2T	5600	1926	235	
5685-99	0-6-2T	5600	1927	235	
5700-49	0-6-0PT	5700	1929	256	
5750-79	0-6-0PT	5700	1929	258	
5780-4	0-6-0PT	5700	1929	260	
5785-99	0-6-0PT	5700	1930	260	
5800-19	0-4-2T	5800	1933	279	
5900	4-6-0	4900	1931	268	
5901-20	4-6-0	4900	1931	275	
5921-40	4-6-0	4900	1933	281	
5941-50	4-6-0	4900	1935	290	
5951-7	4-6-0	4900	1935	297	
5958-65	4-6-0	4900	1936	297	
5966-75	4-6-0	4900	1937	304	

Nos.	Type	Class	Date	Lot No.	See Note
5976-85	4-6-0	4900	1938	311	
5986-94	4-6-0	4900	1939	327	
5995	4-6-0	4900	1940	327	
5996-9	4-6-0	4900	1940	333	
6000-5	4-6-0	6000	1927	243	
6006/7	4-6-0	6000	1928	243	
6007	4-6-0	6000	1936	309	6007
6008-19	4-6-0	6000	1928	243	
6020-9	4-6-0	6000	1930	267	
6100-29	2-6-2T	6100	1931	269	
6130-46	2-6-2T	6100	1932	278	
6147-59	2-6-2T	6100	1933	278	
6160-9	2-6-2T	6100	1935	291	
6342-61	2-6-0	4300	1923	216	6342
6362-9	2-6-0	4300	1925	230	6362
6400-9	0-6-0PT	6400	1932	277	
6410-9	0-6-0PT	6400	1934	294	
6420-4	0-6-0PT	6400	1935	294	
6425-9	0-6-0PT	6400	1935	300	
6430-9	0-6-0PT	6400	1937	305	
6600-18	0-6-2T	5600	1927	244	
6619-29	0-6-2T	5600	1928	244	
6630-49	0-6-2T	5600	1928	252	
6650-99	0-6-2T	5600	1928	255	6650
6700-24	0-6-0PT	6700	1930	262	6700
6725-49	0-6-0PT	6700	1930	265	6725
6750-9	0-6-0PT	6700	1947	362	
6760-5	0-6-0PT	6700	1948	374	
6766-9	0-6-0PT	6700	1949	374	
6770-9	0-6-0PT	6700	1950	379	
6800-19	4-6-0	6800	1936	308	
6820-59	4-6-0	6800	1937	308	
6860-79	4-6-0	6800	1938	308	
6900-5	4-6-0	4900	1940	333	
6906-10	4-6-0	4900	1940	338	
6911-5	4-6-0	4900	1941	338	
6916-35	4-6-0	4900	1941	340	
6936-50	4-6-0	4900	1942	340	
6951-8	4-6-0	4900	1943	340	
6959-70	4-6-0	6959	1944	350	
6971-80	4-6-0	6959	1947	366	
6981-90	4-6-0	6959	1948	366	
6991-5	4-6-0	6959	1948	368	
6996-9	4-6-0	6959	1949	368	
7000-7	4-6-0	4073	1946	357	
7008-17	4-6-0	4073	1948	367	
7018-27	4-6-0	4073	1949	367	
7028-37	4-6-0	4073	1950	375	
7200-19	2-8-2T	7200	1934	266	5275
7220-34	2-8-2T	7200	1935	233	5255
7235-9	2-8-2T	7200	1936	233	5255
7240-2	2-8-2T	7200	1937	318	7240
7243-9	2-8-2T	7200	1938	318	7240
7250-3	2-8-2T	7200	1939	318	7240
7320/1	2-6-0	4300	1925	230	
7400-16	0-6-0PT	7400	1936	307	
7417-29	0-6-0PT	7400	1937	307	

Nos.	Type	Class	Date	Lot No.	See Note
7430-9	0-6-0PT	7400	1948	371	
7440-9	0-6-0PT	7400	1950	380	
7700-24	0-6-0PT	5700	1930	263	7700
7725-36	0-6-0PT	5700	1929	264	7725
7737-49	0-6-0PT	5700	1930	264	7725
7750-61	0-6-0PT	5700	1930	274	7750
7762-74	0-6-0PT	5700	1931	274	7750
7775-99	0-6-0PT	5700	1930	271	7775
7800-11	4-6-0	7800	1938	316	
7812-9	4-6-0	7800	1939	316	
7820-9	4-6-0	7800	1950	377	
7900-6	4-6-0	6959	1949	368	
7907-19	4-6-0	6959	1950	368	
7920-9	4-6-0	6959	1950	376	
8100-3	2-6-2T	8100	1938	320	8100
8104-9	2-6-2T	8100	1939	320	8100
8400-7	0-6-0PT	9400	1949	384	8400
8408-24	0-6-0PT	9400	1950	384	8400
8425-9	0-6-0PT	9400	1951	384	8400
8430-9	0-6-0PT	9400	1953	384	8400
8440-9	0-6-0PT	9400	1954	384	8400
8450-4	0-6-0PT	9400	1949	385	8450
8455-66	0-6-0PT	9400	1950	385	8450
8467-75	0-6-0PT	9400	1951	385	8450
8476-9	0-6-0PT	9400	1952	385	8450
8480-99	0-6-0PT	9400	1952	386	8480
8700	0-6-0PT	5700	1931	273	8700
8700	0-6-0PT	5700	1934	282	
8701-24	0-6-0PT	5700	1931	273	8701
8725-8	0-6-0PT	5700	1930	272	8725
8729-49	0-6-0PT	5700	1931	272	8725
8750-69	0-6-0PT	5700	1933	282	
8770-98	0-6-0PT	5700	1934	282	
8799	0-6-0PT	5700	1934	285	
9300-19	2-6-0	9300	1932	276	
9400-9	0-6-0PT	9400	1947	365	
9410-33	0-6-0PT	9400	1950	382	9410
9434-59	0-6-0PT	9400	1951	382	9410
9460-2	0-6-0PT	9400	1951	383	9460
9463-88	0-6-0PT	9400	1952	383	9460
9489	0-6-0PT	9400	1953	383	9460
9490-7	0-6-0PT	9400	1954	387	9490
9498/9	0-6-0PT	9400	1955	387	9490
9600-21	0-6-0PT	5700	1945	352	
9622-32	0-6-0PT	5700	1945	355	
9633-41	0-6-0PT	5700	1946	355	
9642-51	0-6-0PT	5700	1946	356	
9652-61	0-6-0PT	5700	1946	362	
9662-72	0-6-0PT	5700	1948	370	
9673-82	0-6-0PT	5700	1949	378	
9701-10	0-6-0PT	9700	1933	282	
9711-29	0-6-0PT	5700	1934	285	
9730-59	0-6-0PT	5700	1935	285	
9760-9	0-6-0PT	5700	1935	293	
9770-84	0-6-0PT	5700	1936	293	
9785-99	0-6-0PT	5700	1936	299	

All the above locomotives were built at Swindon, except where noted below.

NOTES:

2 Built by R&W Hawthorn, Leslie (P. No. 3853); was to have been renumbered 500 in 1946, but became BR No. 15100.

3 Narrow gauge (2' 3") taken into stock from the Corris Railway in 8/1930. New in 1878, Falcon Engine Co. (P. No. 323).

4 Narrow gauge (2' 3") taken into stock from the Corris Railway in 8/1930. New in 1921, Kerr, Stuart (order No. 420, P. No. 4047).

5 Acquired under Lot No. 342 from the Weston, Clevedon & Portishead Railway in 7/1940. New in 6/1877 to London, Brighton & South Coast Railway (built at Brighton) and had become WCPR No. 2 PORTISHEAD.

6 Acquired under Lot No. 342 from the Weston, Clevedon & Portishead Railway in 7/1940. New in 12/1875 to London, Brighton & South Coast Railway (built at Brighton) and had become WCPR No. 4.

7 Narrow gauge (1' 11½").

12 Taken into stock 10/1926; returned to makers 12/1926.

13 Taken into stock 12/1926.

111 'Rebuild' of 4-6-2 No. 111 (no Lot issued).

1101 Built by AE (P. Nos. 1987-92).

1331 Originally GWR No. 1387 (built by FW 1877, P. No 340); transferred to departmental stock in 1902 and then overhauled in 1926 under Lot 245. Renumbered 1331 and added to stock in 11/1927.

2925 Rebuilt from 2900 Class (no Lot issued) and renumbered 4900 in 12/1927.

3020 80 locomotives taken into stock as Nos. 3020-99 in 6/1925; all taken out of stock in 10/1925 and sorted into two batches. Nos. 3020-49 were then taken into stock as Lot No. 241 and Nos. 3050-99 were then taken into stock as Lot No. 240. See pp. 163/4 for numbering details, dates and builders.

3100 Reconstructed from Nos. 3173/56/81/55/79 respectively.

3200 The 4-4-0 engines were renumbered 9000-28 in 1946 so that the new 2251 Class 0-6-0 could be numbered 3200-19.

4000 'Rebuilds' of 4000 Class Nos. 4000/9/16/32/7 respectively.

4063 'Renewed' as 4073 Class Nos. 5083-92 in 1937-40; Nos. 4063-72 were allotted Swindon Works progressive numbers 2917-26.

5255 Altered to 2-8-2T Nos. 7220-39.

5275 Altered to 2-8-2T Nos. 7200-19 (these engines never worked as 2-8-0T).

5400 First No. 5400 was actually 2021 Class No. 2062 altered; second No. 5400 was built as a replacement, following No. 5419 (neither the first was 'taken off' stock nor the second 'taken into' stock, so officially only one existed).

5700 Built by NBL (order No. L852 for Nos. 5700-24 and L853 for Nos. 5725-49, P. Nos. 23818-67).

6007 Replacement of first No. 6007 destroyed in an accident.

6342 Swindon Works P. Nos. 2887-2906 allotted to these engines.

6362 Nos. 6362-4 were marked in GWR allocation book as intended to be Nos. 7322-4.

6650 Built by AW (P. Nos. 938-87).

6700 Built by WGB (P. Nos. 2381-2405).

6725 Built by YE (order No. E277, P. Nos. 2249-73).

7240 Alteration of 2-8-0T Nos. 4239/20/02/04/16/05/34/44/49/09/19/40/10/45 respectively.

7700 Built by KS (P. Nos. 4435-59).

7725 Built by NBL (order No. L863, P. Nos. 23921-45).
7750 Built by NBL (order No. L872, P. Nos. 24038-62).
7775 Built by AW (P. Nos. 1131-55).
8100 Reconstructed from Nos. 5100/23/18/45/24/26/20/16/33/15 respectively.
8400 Built by WGB (P. No. 2910-20/22/24/23/21/25-59).
8450 Built by YE (order Nos. E336-E365, P. Nos. 2433-65/67-71/66/72).
8480 Ordered from HC (P. Nos. 1827-31/3-7/9-43/5-9) but built by RSH (order No. E182, P. Nos. 7450-69) as sub-contractors.
8700 Renumbered 9700 1/1934 on alteration to condensing locomotive; built by BP (order No. 148, P. Nos. 6680).
8701 Built by BP (order No. 148, P. Nos. 6681-6704).
8725 Built by WGB (P. Nos. 2422-46).
9410 Built by RSH (P. Nos. 7547-96).
9460 Built by RSH (P. Nos. 7611-40).
9490 Ordered from HE (order No. 3720, P. Nos. 3720-39) but built by YE (order Nos. C437-C446, P. Nos. 2544-53 for Nos. 9490-9 and order Nos. C468-C477, P. Nos. 2575-84 for Nos. 3400-9) as sub-contractors.

GWR ordered non-steam locomotives supplied to British Railways:-

Nos.	Type	Date	Lot No.	Note
15101-6	0-6-0DE	1948	364	Were to have been GWR Nos. 502-7
15107	0-6-0DE	1949	363	Was to have been GWR No. 501
18000	A1A+A1A	1950	372	Gas Turbine locomotive; built by Brown-Boveri (Switzerland)
18100	Co-Co	1952	388	Gas Turbine locomotive; built by Metropolitan-Vickers (Manchester

New GWR engines added to stock between June 1922 and Nationalisation (and various others acquired) are detailed in the list that follows (engines of GWR design put into service by British Railways follow in a separate list). Those marked (+) after the number were renumbered as shown below (the 'Old No. column indicates reversion to the '1st No.' at the date given).

1st No.	New No.	Date	Old No.	1st No.	New No.	Date	Old No.	1st No.	New No.	Date	Old No.
2	15100	3/48	-	4809	1409	11/46	-	4859	1459	11/46	-
2888	4850	12/46	9/48	4810	1410	10/46	-	4860	1460	11/46	-
2925	4900	12/28	-	4811	1411	10/46	-	4861	1461	11/46	-
3200	9000	7/46	-	4812	1412	10/46	-	4862	1462	11/46	-
3201	9001	7/46	-	4813	1413	10/46	-	4863	1463	11/46	-
3202	9002	7/46	-	4814	1414	10/46	-	4864	1464	11/46	-
3203	9003	7/46	-	4815	1415	11/46	-	4865	1465	11/46	-
3204	9004	7/46	-	4816	1416	11/46	-	4866	1466	11/46	-
3205	9005	7/46	-	4817	1417	10/46	-	4867	1467	11/46	-
3206	9006	7/46	-	4818	1418	10/46	-	4868	1468	11/46	-
3207	9007	7/46	-	4819	1419	10/46	-	4869	1469	11/46	-
3208	9008	7/46	-	4820	1420	11/46	-	4870	1470	11/46	-
3209	9009	7/46	-	4821	1421	11/46	-	4871	1471	11/46	-
3210	9010	8/46	-	4822	1422	10/46	-	4872	1472	11/46	-
3211	9011	8/46	-	4823	1423	10/46	-	4873	1473	11/46	-
3212	9012	8/46	-	4824	1424	10/46	-	4874	1474	11/46	-
3213	9013	8/46	-	4825	1425	11/46	-	4907	3903	5/47	4/50
3214	9014	8/46	-	4826	1426	11/46	-	4948	3902	5/47	9/48
3215	9015	8/46	-	4827	1427	11/46	-	4968	3900	5/47	3/49
3216	9016	8/46	-	4828	1428	11/46	-	4971	3901	5/47	4/49
3217	9017	8/46	-	4829	1429	11/46	-	4972	3904	5/47	10/48
3218	9018	8/46	-	4830	1430	11/46	-	5955	3950	10/46	10/48
3219	9019	9/46	-	4831	1431	11/46	-	5976	3951	4/47	11/48
3220	9020	8/46	-	4832	1432	10/46	-	5986	3954	5/47	2/50
3221	9021	9/46	-	4833	1433	11/46	-	6949	3955	5/47	4/49
3222	9022	8/46	-	4834	1434	11/46	-	6953	3953	4/47	9/48
3223	9023	9/46	-	4835	1435	10/46	-	6957	3952	4/47	3/50
3224	9024	9/46	-	4836	1436	11/46	-	7013	4082	2/52	-
3225	9025	9/46	-	4837	1437	12/46	-	8700	9700	1/34	-
3226	9026	8/46	-	4838	1438	10/46	-	9300	7322	3/57	-
3227	9027	9/46	-	4839	1439	11/46	-	9301	7323	10/56	-
3228	9028	8/46	-	4840	1440	10/46	-	9302	7324	2/57	-
3813	4855	7/47	6/49	4841	1441	10/46	-	9303	7325	6/58	-
3818	4852	12/46	9/48	4842	1442	11/46	-	9304	7326	6/58	-
3820	4856	7/47	6/49	4843	1443	10/46	-	9305	7327	1/59	-
3831	4857	8/47	5/49	4844	1444	11/46	-	9306	7328	5/58	-
3837	4854	6/47	8/49	4845	1445	11/46	-	9307	7329	12/56	-
3839	4853	5/47	11/49	4846	1446	11/46	-	9308	7330	6/57	-
3865	4851	12/46	4/49	4847	1447	11/46	-	9309	7331	5/59	-
4009	100	1/36	-	4848	1448	11/46	-	9310	7332	10/58	-
4082	7013	2/52	-	4849	1449	11/46	-	9311	7333	6/57	-
4800	1400	11/46	-	4850	1450	11/46	-	9312	7334	1/59	-
4801	1401	10/46	-	4851	1451	11/46	-	9313	7335	8/58	-
4802	1402	11/46	-	4852	1452	11/46	-	9314	7336	6/58	-
4803	1403	10/46	-	4853	1453	11/46	-	9315	7337	5/59	-
4804	1404	10/46	-	4854	1454	11/46	-	9316	7338	3/58	-
4805	1405	10/46	-	4855	1455	11/46	-	9317	7339	10/56	-
4806	1406	11/46	-	4856	1456	11/46	-	9318	7340	12/57	-
4807	1407	10/46	-	4857	1457	12/46	-	9319	7341	6/57	-
4808	1408	10/46	-	4858	1458	11/46	-				

2-8-0 ROD Class Nos. 3020-99; this list shows the second numbers of these engines after the sorting out that took place in 1925 and the 'Date New' represents their return to traffic. The first numbers and sheds of 1925 are given as an appendix.

No.	Date	First Shed	Last GWR Shed	Wdn GWR	Last BR Shed	Wdn BR
2+	-/35	Paddington	Paddington	-	Swindon	4/65
3	-/78	Corris	Corris	-	Corris	10/48
4	-/21	Corris	Corris	-	Corris	10/48
5	6/77	Bristol SPM	Bristol SPM	-	Newton Abbot	3/54
6	12/75	Bristol SPM	Bristol SPM	-	Bristol SPM	1/48
7	7/23	Aber'wyth VoR	Aber'wyth VoR	-	(STILL IN SERVICE)	
8	7/23	Aber'wyth VoR	Aber'wyth VoR	-	(STILL IN SERVICE)	
12	10/26	Park Royal	Park Royal	12/26	-	-
13	12/26	Paddington	Swindon	5/46	-	-
111	9/24	Paddington	Paddington	-	Laira	7/53
1000	8/45	Paddington	Paddington	-	Swindon	7/64
1001	9/45	Newton Abbot	Newton Abbot	-	Neyland	5/63
1002	9/45	Bath Road	Bath Road	-	Shrewsbury	9/63
1003	10/45	Paddington	Paddington	-	Laira	10/62
1004	10/45	Laira	Laira	-	Penzance	9/62
1005	11/45	Bath Road	Bath Road	-	Bristol SPM	6/63
1006	11/45	Laira	Laira	-	Swindon	9/63
1007	12/45	Bath Road	Bath Road	-	Didcot	10/62
1008	12/45	Paddington	Paddington	-	Swindon	10/63
1009	12/45	Laira	Laira	-	Bristol SPM	2/63
1010	1/46	Paddington	Paddington	-	Swindon	7/64
1011	1/46	Bath Road	Bath Road	-	Swindon	11/64
1012	2/46	Paddington	Paddington	-	Swindon	4/64
1013	2/46	Bath Road	Bath Road	-	Swindon	7/64
1014	2/46	Bath Road	Bath Road	-	Swindon	4/64
1015	3/46	Paddington	Paddington	-	Didcot	11/62
1016	3/46	Stafford Road	Stafford Road	-	Shrewsbury	9/63
1017	3/46	Stafford Road	Stafford Road	-	Shrewsbury	12/62
1018	3/46	Newton Abbot	Newton Abbot	-	Didcot	9/62
1019	4/46	Newton Abbot	Penzance	-	Shrewsbury	2/63
1020	12/46	Laira	Exeter	-	Swindon	2/64
1021	12/46	Paddington	Paddington	-	Bristol SPM	11/63
1022	12/46	Laira	Penzance	-	Shrewsbury	10/62
1023	1/47	Laira	Truro	-	Shrewsbury	3/63
1024	1/47	Stafford Road	Stafford Road	-	Swindon	4/64
1025	1/47	Stafford Road	Stafford Road	-	Shrewsbury	2/63
1026	1/47	Paddington	Paddington	-	Shrewsbury	9/62
1027	3/47	Westbury	Westbury	-	Swindon	10/63
1028	3/47	Bath Road	Bath Road	-	Swindon	12/63
1029	4/47	Stafford Road	Stafford Road	-	Swindon	12/62
1101	6/26	Swansea H'bour	Danygraig	-	Danygraig	11/59
1102	7/26	Danygraig	Danygraig	-	Danygraig	1/60
1103	7/26	Danygraig	Danygraig	-	Danygraig	1/60
1104	8/26	Swansea H'bour	Danygraig	-	Danygraig	1/60
1105	8/26	Danygraig	Danygraig	-	Danygraig	1/60
1106	8/26	Danygraig	Danygraig	-	Danygraig	1/60
1331	-/77	Weymouth	Oswestry	-	Oswestry	1/50
1366	2/34	Swindon	Swindon	-	Bridgwater	1/61
1367	2/34	Swindon	Swindon	-	Wadebridge	10/64
1368	2/34	Swindon	Weymouth	-	Wadebridge	10/64
1369	2/34	Swindon	Weymouth	-	Wadebridge	11/64
1370	2/34	Swindon	Swindon	-	Weymouth	1/60
1371	2/34	Llanelly	Weymouth	-	Swindon	11/60
2200	6/38	Machynlleth	Aberystwyth	-	Westbury	9/62
2201	7/39	Oswestry	Oswestry	-	Reading	9/64

No.	Date	First Shed	Last GWR Shed	Wdn GWR	Last BR Shed	Wdn BR
2202	8/39	Didcot	Didcot	-	Machynlleth	10/60
2203	8/39	Bristol SPM	Tyseley	-	Bristol SPM	6/60
2204	8/39	Oswestry	Portmadoc	-	Templecombe	12/63
2205	8/39	Worcester	Worcester	-	Bristol SPM	6/59
2206	8/39	Bristol SPM	Stratford	-	Reading	12/61
2207	8/39	Worcester	Worcester	-	Lydney	1/61
2208	8/39	Reading	Reading	-	Neyland	6/59
2209	9/39	Tyseley	Tyseley	-	Ebbw Junction	8/62
2210	9/39	Oswestry	Oswestry	-	Banbury	6/65
2211	5/40	Taunton	Taunton	-	Leamington	11/64
2212	5/40	Taunton	Taunton	-	Reading	12/62
2213	5/40	Taunton	Minehead	-	Bristol SPM	11/60
2214	5/40	Taunton	Taunton	-	Exmouth Jct	5/65
2215	5/40	Barnstaple	Taunton	-	Templecombe	10/61
2216	6/40	Carmarthen	Carmarthen	-	Carmarthen	1/62
2217	6/40	Carmarthen	Carmarthen	-	Templecombe	11/64
2218	6/40	Bristol SPM	Ebbw Junction	-	Templecombe	11/64
2219	6/40	Pwllheli	Machynlleth	-	Taunton	3/64
2220	6/40	Bristol SPM	Bristol SPM	-	Neyland	12/61
2221	8/40	Winchester	Didcot	-	Banbury	11/64
2222	8/40	Winchester	Newbury	-	Worcester	5/65
2223	8/40	Oswestry	Aberystwyth	-	Templecombe	5/62
2224	9/40	Swindon	Swindon	-	Carmarthen	9/63
2225	9/40	Bristol SPM	Bristol SPM	-	Brecon	6/59
2226	9/40	Didcot	Didcot	-	Landore	8/59
2227	10/40	Didcot	Croes Newydd	-	Ebbw Junction	1/61
2228	10/40	Chester	Shrewsbury	-	Neyland	6/59
2229	10/40	Shrewsbury	Shrewsbury	-	Barrow Road	9/62
2230	10/40	Evesham	Exeter	-	Didcot	1/62
2231	10/44	Stafford Road	Shrewsbury	-	Severn Tunnel	2/65
2232	10/44	Stafford Road	Stafford Road	-	Worcester	9/64
2233	10/44	Shrewsbury	Shrewsbury	-	Machynlleth	11/61
2234	10/44	Shrewsbury	Shrewsbury	-	Worcester	5/62
2235	10/44	Shrewsbury	Shrewsbury	-	Taunton	8/59
2236	10/44	Carmarthen	Carmarthen	-	Machynlleth	5/65
2237	11/44	Worcester	Worcester	-	Machynlleth	6/59
2238	11/44	Tyseley	Tyseley	-	Tyseley	6/59
2239	11/44	Swindon	Ebbw Junction	-	Aberystwyth	5/62
2240	12/44	Oxford	Didcot	-	Ebbw Junction	6/62
2241	3/45	Worcester	Worcester	-	Hereford	2/64
2242	4/45	Worcester	Worcester	-	Gloucester	5/65
2243	4/45	Hereford	Hereford	-	Ebbw Junction	1/63
2244	4/45	Oswestry	Oswestry	-	Worcester	5/65
2245	4/45	Reading	Reading	-	Gloucester	5/63
2246	8/45	Stourbridge	Stourbridge	-	Worcester	12/63
2247	8/45	Worcester	Worcester	-	Templecombe	2/64
2248	8/45	Gloucester	Gloucester	-	Reading	9/64
2249	8/45	Oxford	Oxford	-	Reading	9/64
2250	9/45	Swindon	Swindon	-	Swindon	8/62
2251	3/30	Bristol SPM	Bristol SPM	-	Templecombe	12/63
2252	3/30	Bristol SPM	Winchester	-	Didcot	12/59
2253	3/30	Reading	Bristol SPM	-	Worcester	3/65
2254	3/30	Didcot	Carmarthen	-	Gloucester	1/59
2255	3/30	Swindon	Llanidloes	-	Machynlleth	5/62
2256	3/30	Stafford Road	Banbury	-	Hereford	9/62
2257	4/30	Stafford Road	Tyseley	-	Reading	9/64
2258	4/30	Gloucester	Wells	-	Swindon	12/58
2259	4/30	Didcot	Croes Newydd	-	Neyland	4/59
2260	4/30	Bristol SPM	Aberystwyth	-	Aberystwyth	11/61
2261	4/30	Bristol SPM	Taunton	-	Reading	9/64

No.	Date	First Shed	Last GWR Shed	Wdn GWR	Last BR Shed	Wdn BR
2262	4/30	Oxley	Chester	-	Reading	11/59
2263	4/30	Worcester	Worcester	-	Whitland	1/59
2264	4/30	Reading	Reading	-	Machynlleth	6/60
2265	5/30	Exeter	Bristol SPM	-	Bristol SPM	6/60
2266	5/30	Taunton	Barnstaple	-	Hereford	10/59
2267	6/30	Barnstaple	Taunton	-	Leamington	11/61
2268	6/30	Barnstaple	Taunton	-	Machynlleth	5/65
2269	7/30	Bristol SPM	Bristol SPM	-	Bristol SPM	1/59
2270	4/30	Oxley	Stourbridge	-	Banbury	9/59
2271	8/34	Bristol SPM	Carmarthen	-	Machynlleth	9/62
2272	8/34	Carmarthen	Carmarthen	-	Carmarthen	8/59
2273	8/34	Carmarthen	Landore	-	Westbury	12/63
2274	8/34	Worcester	Worcester	-	Carmarthen	6/60
2275	8/34	Stourbridge	Barnstaple	-	Machynlleth	2/60
2276	9/34	Reading	Paddington	-	Machynlleth	9/62
2277	9/34	Bristol SPM	Worcester	-	Exmouth Jct	12/63
2278	9/34	Worcester	Worcester	-	Goodwick	9/59
2279	9/34	Leamington	Stourbridge	-	Tyseley	1/59
2280	9/34	Didcot	Ebbw Junction	-	Brecon	9/59
2281	1/36	Leamington	Stourbridge	-	Machynlleth	11/59
2282	1/36	Didcot	Paddington	-	Goodwick	5/60
2283	1/36	Worcester	Aberystwyth	-	Severn Tunnel	12/63
2284	1/36	Neath	Carmarthen	-	Landore	6/59
2285	1/36	Didcot	Southall	-	Machynlleth	11/59
2286	1/36	Worcester	Hereford	-	Hereford	9/64
2287	1/36	Oxley	Croes Newydd	-	Gloucester	5/65
2288	1/36	Whitland	Whitland	-	Gloucester	12/61
2289	1/36	Banbury	Didcot	-	Banbury	5/64
2290	1/36	Gloucester	Worcester	-	Carmarthen	6/59
2291	2/38	Carmarthen	Carmarthen	-	Worcester	9/64
2292	1/38	Chester	Tyseley	-	Severn Tunnel	6/62
2293	2/38	Andover Jct	Bristol SPM	-	Swindon	7/59
2294	2/38	Gloucester	Worcester	-	Machynlleth	9/62
2295	2/38	Banbury	Banbury	-	Hereford	7/62
2296	6/38	Portmadoc	Tyseley	-	Worcester	10/59
2297	6/38	Aberystwyth	Stratford	-	Banbury	6/60
2298	6/38	Pwllheli	Aberystwyth	-	Ebbw Junction	12/63
2299	6/38	Reading	Reading	-	Reading	9/59
2884	3/38	Severn Tunnel	Severn Tunnel	-	Pontypool	4/64
2885	3/38	Pontypool	Banbury	-	Severn Tunnel	1/64
2886	3/38	Carmarthen	Chester	-	Westbury	6/64
2887	3/38	Severn Tunnel	Severn Tunnel	-	Taunton	6/64
2888+	3/38	Banbury	Llanelly	-	Banbury	2/63
2889	3/38	Cardiff	Cardiff	-	Reading	4/63
2890	4/38	Pantyffynon	Cardiff	-	Ebbw Junction	4/65
2891	4/38	Cardiff	Cardiff	-	Cardiff E Dock	10/64
2892	4/38	Pontypool	Severn Tunnel	-	Severn Tunnel	5/63
2893	4/38	Aberdare	Pontypool	-	Didcot	11/64
2894	11/38	Swindon	Ebbw Junction	-	Ebbw Junction	8/63
2895	11/38	Oxley	Birkenhead	-	Ebbw Junction	4/65
2896	11/38	Ebbw Junction	Ebbw Junction	-	Pontypool	6/64
2897	11/38	Pontypool	Shrewsbury	-	Pontypool	3/63
2898	11/38	Oxley	Banbury	-	Didcot	10/64
2899	12/38	Oxley	Banbury	-	Southall	3/65
2925+	12/24	Laira	Paddington	-	Paddington	4/59
3020	4/26	Paddington	Stafford Road	-	Banbury	6/54
3021	5/26	Oxley	Worcester	-	Worcester	7/48
3022	4/26	Oxley	Bristol SPM	-	Bristol SPM	6/56
3023	5/26	Paddington	Pontypool	-	Tyseley	10/55
3024	7/26	Oxley	Oxley	-	Carmarthen	10/58

No.	Date	First Shed	Last GWR Shed	Wdn GWR	Last BR Shed	Wdn BR
3025	6/26	Paddington	Reading	-	Pontypool	7/54
3026	8/26	Reading	Croes Newydd	-	Pontypool	12/54
3027	9/26	Worcester	Worcester	-	Worcester	5/48
3028	9/26	Gloucester	Croes Newydd	-	Oxley	8/56
3029	9/26	Gloucester	Worcester	-	Oxley	5/56
3030	9/26	Oxley	Worcester	-	Worcester	7/48
3031	9/26	Newton Abbot	Oxley	-	Oxley	5/56
3032	10/26	Newton Abbot	Westbury	-	Bristol SPM	10/55
3033	10/26	Exeter	Oxley	-	Shrewsbury	5/53
3034	12/26	Carmarthen	Bristol SPM	-	Bristol SPM	2/53
3035	12/26	Cardiff	Westbury	-	Westbury	8/48
3036	1/27	Cardiff	Aberdare	-	Carmarthen	3/58
3037	1/27	Cardiff	Pontypool	-	Pontypool	8/48
3038	1/27	Pontypool	Cardiff	-	Pontypool	7/56
3039	1/27	Laira	Oxley	-	Oxley	8/48
3040	1/27	Laira	Pontypool	-	Pontypool	6/56
3041	2/27	Swindon	Bristol SPM	-	Carmarthen	3/58
3042	3/27	Llanelly	Cardiff	-	Pontypool	7/56
3043	3/27	Swindon	Stafford Road	-	Oxley	9/56
3044	3/27	Neath	Cardiff	-	Tyseley	10/56
3045	3/27	Bristol SPM	Bristol SPM	8/47	-	-
3046	4/27	Bristol SPM	Bristol SPM	-	Bristol SPM	8/48
3047	4/27	Paddington	Reading	-	Oxley	6/53
3048	4/27	Gloucester	Worcester	-	Bristol SPM	5/56
3049	5/27	Laira	Tyseley	-	Tyseley	11/48
3050	1/26	Bristol SPM	Bristol SPM	3/29	-	-
3051	1/26	Bristol SPM	Bristol SPM	5/29	-	-
3052	2/26	Bristol SPM	Bristol SPM	4/29	-	-
3053	2/26	Bristol SPM	Bristol SPM	2/29	-	-
3054	3/26	Swindon	Bristol SPM	9/28	-	-
3055	3/26	Bristol SPM	Bristol SPM	4/29	-	-
3056	3/26	Swindon	Westbury	4/29	-	-
3057	3/26	Paddington	Paddington	7/29	-	-
3058	4/26	Paddington	Paddington	11/28	-	-
3059	5/26	Paddington	Paddington	7/29	-	-
3060	6/26	Paddington	Paddington	5/29	-	-
3061	6/26	Paddington	Paddington	4/29	-	-
3062	7/26	Paddington	Paddington	7/29	-	-
3063	7/26	Paddington	Paddington	6/29	-	-
3064	7/26	Oxley	Oxley	12/28	-	-
3065	7/26	Paddington	Paddington	3/29	-	-
3066	7/26	Oxley	Oxley	12/28	-	-
3067	8/26	Oxley	Oxley	10/27	-	-
3068	9/26	Oxley	Oxley	10/28	-	-
3069	9/26	Oxley	Oxley	3/29	-	-
3070	9/26	Oxley	Oxley	9/28	-	-
3071	10/26	Newton Abbot	Stourbridge	10/28	-	-
3072	10/26	Worcester	Worcester	8/29	-	-
3073	10/26	Tyseley	Stafford Road	9/28	-	-
3074	10/26	Bristol SPM	Cardiff	1/29	-	-
3075	12/26	Cardiff	Cardiff	6/29	-	-
3076	12/26	Stourbridge	Aberdare	4/29	-	-
3077	1/27	Cardiff	Cardiff	4/29	-	-
3078	1/27	Llanelly	Llanelly	3/29	-	-
3079	10/26	Pontypool	Pontypool	1/29	-	-
3080	12/26	Llanelly	Llanelly	8/29	-	-
3081	1/27	Llanelly	Llanelly	8/29	-	-
3082	1/27	Pontypool	Pontypool	3/30	-	-
3083	1/27	Cardiff	Cardiff	8/29	-	-
3084	1/27	Paddington	Severn Tunnel	7/30	-	-

No.	Date	First Shed	Last GWR Shed	Wdn GWR	Last BR Shed	Wdn BR
3085	1/27	Swindon	Carmarthen	10/29	-	-
3086	1/27	Llanelly	Llanelly	7/30	-	-
3087	2/27	Llanelly	Llanelly	9/29	-	-
3088	2/27	Carmarthen	Carmarthen	4/29	-	-
3089	3/27	Bristol SPM	Exeter	10/29	-	-
3090	2/27	Llanelly	Llanelly	2/30	-	-
3091	3/27	Neyland	Neath	3/29	-	-
3092	3/27	Bristol SPM	Bristol SPM	12/28	-	-
3093	3/27	Neyland	Ebbw Junction	12/31	-	-
3094	2/27	Banbury	Newton Abbot	6/28	-	-
3095	3/27	Bristol SPM	Bristol SPM	2/30	-	-
3096	3/27	Bristol SPM	Bristol SPM	3/29	-	-
3097	4/27	Neath	Aberdare	7/30	-	-
3098	4/27	Neath	Neath	12/28	-	-
3099	2/27	Banbury	Banbury	12/29	-	-
3100	12/38	Gloucester	Tondu	-	Tondu	5/57
3101	2/39	Bristol SPM	Tyseley	-	Tyseley	9/57
3102	5/39	Bristol SPM	Stafford Road	-	Stafford Road	10/58
3103	8/39	Leamington	Ebbw Junction	-	Ebbw Junction	1/60
3104	10/39	Llanelly	Oxley	-	Stafford Road	6/57
3200+	5/36	Oswestry	Machynlleth	-	Machynlleth	3/55
3201+	4/36	Oswestry	Oswestry	-	Oswestry	4/54
3202+	6/36	Aberystwyth	Aberystwyth	-	Aberystwyth	5/54
3203+	7/36	Oswestry	Oswestry	-	Machynlleth	10/55
3204+	8/36	Machynlleth	Machynlleth	-	Shrewsbury	6/60
3205+	9/36	Machynlleth	Machynlleth	-	Oswestry	7/59
3206+	11/36	Didcot	Didcot	-	Didcot	8/48
3207+	12/36	Stratford	Tyseley	-	Tyseley	7/48
3208+	2/37	Crewe	Tyseley	-	Machynlleth	7/57
3209+	2/37	Cheltenham	Machynlleth	-	Machynlleth	7/57
3210+	4/37	Aberystwyth	Tyseley	-	Oswestry	7/57
3211+	3/37	Oswestry	Swindon	-	Swindon	7/57
3212+	5/37	Oswestry	Machynlleth	-	Machynlleth	7/57
3213+	7/37	Aberystwyth	Aberystwyth	-	Machynlleth	12/58
3214+	8/37	Portmadoc	Pwllheli	-	Croes Newydd	10/60
3215+	10/37	Didcot	Didcot	-	Machynlleth	6/60
3216+	12/37	Oswestry	Oswestry	-	Machynlleth	7/57
3217+	3/38	Aberystwyth	Aberystwyth	-	Machynlleth	10/60
3218+	4/38	Tyseley	Swindon	-	Croes Newydd	6/60
3219+	6/38	Gloucester	Tyseley	-	Tyseley	11/48
3220+	11/38	Oswestry	Oswestry	-	Machynlleth	7/57
3221+	11/38	Aberystwyth	Aberystwyth	-	Machynlleth	12/58
3222+	12/38	Oswestry	Oswestry	-	Aberystwyth	8/57
3223+	2/39	Andover Jct	Swindon	-	Swindon	7/57
3224+	2/39	Shrewsbury	Shrewsbury	-	Machynlleth	9/57
3225+	4/39	Aberystwyth	Aberystwyth	-	Aberystwyth	8/57
3226+	6/39	Oswestry	Oswestry	-	Oswestry	8/57
3227+	6/39	Aberystwyth	Machynlleth	-	Oswestry	8/57
3228+	11/39	Oswestry	Machynlleth	-	Croes Newydd	9/57
3200	9/46	Portmadoc	Portmadoc	-	Templecombe	1/65
3201	9/46	Machynlleth	Machynlleth	-	Templecombe	5/65
3202	10/46	Oswestry	Oswestry	-	Oswestry	6/60
3203	10/46	Shrewsbury	Croes Newydd	-	Gloucester	12/63
3204	10/46	Gloucester	Gloucester	-	Hereford	3/63
3205	10/46	Gloucester	Gloucester	-	Templecombe	5/65
3206	10/46	Croes Newydd	Croes Newydd	-	Templecombe	12/63
3207	10/46	Pwllheli	Portmadoc	-	Worcester	12/62
3208	10/46	Oswestry	Oswestry	-	Machynlleth	5/65
3209	11/46	Hereford	Hereford	-	Machynlleth	6/64
3210	12/47	Didcot	Didcot	-	Templecombe	11/64

No.	Date	First Shed	Last GWR Shed	Wdn GWR	Last BR Shed	Wdn BR
3211	12/47	Didcot	Didcot	-	Didcot	9/62
3212	12/47	Didcot	Didcot	-	Severn Tunnel	10/63
3213	12/47	Gloucester	Gloucester	-	Reading	12/63
3214	12/47	Worcester	Worcester	-	Neyland	10/63
3215	12/47	Bristol SPM	Bristol SPM	-	Templecombe	1/63
3216	12/47	Banbury	Banbury	-	Templecombe	12/63
3217	12/47	Shrewsbury	Shrewsbury	-	Leamington	11/64
3600	12/38	Paddington	Paddington	-	Neath	12/63
3601	12/38	Hereford	Hereford	-	Stourbridge	10/64
3602	1/39	Stafford Road	Shrewsbury	-	Shrewsbury	2/62
3603	1/39	Exeter	Exeter	-	Neath	7/64
3604	1/39	Bristol SPM	Bristol SPM	-	Swansea E Dock	12/63
3605	1/39	Ebbw Junction	Aberdare	-	Oxley	9/66
3606	1/39	Exeter	Exeter	-	Barrow Road	12/62
3607	2/39	Worcester	Worcester	-	Tyseley	10/66
3608	2/39	Oxford	Oxford	-	Southall	6/65
3609	2/39	Worcester	Gloucester	-	Lydney	8/60
3610	2/39	Aberdare	Aberdare	-	Aberdare	2/65
3611	2/39	Neath	Neath	-	Neath	8/62
3612	3/39	Aberbeeg	Aberbeeg	-	Llantrisant	10/64
3613	3/39	Tyseley	Tyseley	-	Neath	5/64
3614	3/39	Bristol SPM	Bristol SPM	-	Westbury	3/62
3615	3/39	Stafford Road	Shrewsbury	-	Worcester	10/65
3616	3/36	Cardiff	Aberbeeg	-	Gloucester	9/65
3617	4/39	Llantrisant	Llantrisant	-	Llantrisant	9/64
3618	4/39	Paddington	Paddington	-	Paddington	6/64
3619	4/39	Paddington	Paddington	-	Tyseley	9/66
3620	5/39	Southall	Southall	-	Southall	6/65
3621	5/39	Neath	Neath N&B	-	Ebbw Junction	11/64
3622	5/39	Didcot	Didcot	-	Paddington	9/64
3623	5/39	Bristol SPM	Bristol SPM	-	Bristol SPM	9/62
3624	5/39	Tyseley	Tyseley	-	Neyland	5/62
3625	5/39	Tyseley	Tyseley	-	Tyseley	7/66
3626	6/39	Birkenhead	Birkenhead	-	Duffryn Yard	8/63
3627	6/39	Aberbeeg	Tondu	-	Aberdare	2/63
3628	6/39	Pontypool	Branches Fork	-	Pontypool	1/63
3629	6/39	Laira	Laira	-	Westbury	1/63
3630	6/39	Banbury	Banbury	-	Croes Newydd	9/62
3631	7/39	Leamington	Leamington	-	Oxley	7/65
3632	7/39	Bristol SPM	Swindon	-	Barrow Road	11/62
3633	7/39	Swansea E Dock	Swansea E Dock	-	Gloucester	10/63
3634	7/39	Ebbw Junction	Ebbw Junction	-	Aberdare	7/64
3635	9/39	Paddington	Paddington	-	Tyseley	4/65
3636	9/39	Tondu	Ebbw Junction	-	Ebbw Junction	2/62
3637	9/39	Whitland	Goodwick	-	Goodwick	10/62
3638	9/39	Brecon	Brecon	-	Brecon	1/61
3639	10/39	Laira	Laira	-	Neyland	1/63
3640	10/39	Tondu	Aberbeeg	-	Pontypool	5/62
3641	10/39	Swansea E Dock	Swansea E Dock	-	Swansea E Dock	8/62
3642	10/39	Paddington	Llanelly	-	Neath	4/65
3643	10/39	Bristol SPM	Bristol SPM	-	Gloucester	10/65
3644	10/39	Paddington	Paddington	-	Radyr	6/65
3645	11/39	Swindon	Swindon	-	Swindon	5/62
3646	11/39	Paddington	Paddington	-	Paddington	6/64
3647	11/39	Ebbw Junction	Ebbw Junction	-	Neath	6/65
3648	11/39	Paddington	Paddington	-	Tondu	12/63
3649	11/39	Stourbridge	Stourbridge	-	Stourbridge	1/61
3650	12/39	Tyseley	Tyseley	-	Neath	9/63
3651	12/39	Pontypool	Pontypool	-	Pontypool	4/63
3652	12/39	Paddington	Slough	-	Neath	10/63

No.	Date	First Shed	Last GWR Shed	Wdn GWR	Last BR Shed	Wdn BR
3653	12/39	Tyseley	Tyseley	-	Oxford	10/63
3654	12/39	Milford Haven	Neyland	-	Cardiff E Dock	8/65
3655	1/40	Aberdare	Aberdare	-	Aberdare	12/62
3656	1/40	Llantrisant	Llantrisant	-	Llantrisant	2/62
3657	1/40	Tyseley	Tyseley	-	Whitland	5/61
3658	1/40	Duffryn Yard	Paddington	-	Stourbridge	9/65
3659	1/40	Paddington	Paddington	-	Templecombe	10/65
3660	2/40	Tyseley	Tyseley	-	Tyseley	4/63
3661	2/40	Danygraig	Llanelly	-	Pontypool	4/65
3662	2/40	Ebbw Junction	Ebbw Junction	-	Ebbw Junction	8/65
3663	2/40	Reading	Henley	-	Barry	12/62
3664	2/40	Tyseley	Tyseley	-	Aberbeeg	5/64
3665	3/40	Chester	Chester	-	Southall	1/64
3666	3/40	Swindon	Swindon	-	Paddington	12/62
3667	3/40	Stourbridge	Stourbridge	-	Ebbw Junction	5/61
3668	3/40	Aberbeeg	Tondu	-	Barry	7/63
3669	3/40	Neath	Paddington	-	Westbury	9/65
3670	4/40	Cardiff	Aberbeeg	-	Cardiff	5/61
3671	4/40	Yeovil	Yeovil	-	Llanelly	7/65
3672	4/40	Paddington	Paddington	-	Swanse E Dock	4/64
3673	4/40	Tyseley	Tyseley	-	Tyseley	5/64
3674	4/40	Aberbeeg	Tondu	-	Pill	12/62
3675	6/40	Laira	Laira	-	Gloucester	12/65
3676	6/40	Bristol SPM	Bristol SPM	-	Aberbeeg	5/61
3677	6/40	Didcot	Slough	-	Oxford	12/65
3678	7/40	Landore	Landore	-	Llanelly	12/63
3679	7/40	Swansea E Dock	Swansea E Dock	-	Exmouth Jct	3/63
3680	7/40	Pontypool	Aberbeeg	-	Llantrisant	5/64
3681	7/40	Slough	Slough	-	Bath LMS	3/66
3682	7/40	Swindon	Swindon	-	Worcester	12/65
3683	7/40	Tondu	Aberbeeg	-	Hereford	10/64
3684	7/40	Swindon	Chippenham	-	Swindon	5/62
3685	12/40	Paddington	Paddington	-	Pontypool	4/64
3686	12/40	Taunton	Laira	-	Severn Tunnel	7/65
3687	1/41	Oxford	Oxford	-	Neath	6/65
3688	1/41	Paddington	Paddington	-	Duffryn Yard	9/62
3689	1/41	Paddington	Tyseley	-	Barry	5/64
3690	2/41	Aberbeeg	Pontypool	-	Neath	6/65
3691	2/41	Aberbeeg	Llantrisant	-	Ebbw Junction	4/65
3692	2/41	Pontypool	Pontypool	-	Neath	5/64
3693	3/41	Tyseley	Tyseley	-	Gloucester	7/64
3694	3/41	Banbury	Banbury	-	Ebbw Junction	9/62
3695	8/41	Aberbeeg	Tondu	-	Abercynon	7/64
3696	8/41	Westbury	Westbury	-	Barrow Road	11/65
3697	9/41	Reading	Reading	-	Didcot	5/62
3698	9/41	Llanelly	Llanelly	-	Llanelly	4/64
3699	9/41	Aberbeeg	Tondu	-	Severn Tunnel	2/65
3700	9/36	Aberbeeg	Ebbw Junction	-	Severn Tunnel	2/65
3701	9/36	Neath N&B	Landore	-	Llanelly	5/64
3702	10/36	Ebbw Junction	Shrewsbury	-	Barrow Road	4/64
3703	10/36	Llantrisant	Llantrisant	-	Pontypool	5/62
3704	10/36	Southall	Southall	-	Southall	1/61
3705	10/36	Penzance	Laira	-	Ebbw Junction	3/65
3706	10/36	Brecon	Brecon	-	Ebbw Junction	11/63
3707	10/36	Tyseley	Cardiff E Dock	-	Merthyr	9/64
3708	10/36	Brecon	Cardiff	-	Ebbw Junction	7/65
3709	10/36	Ebbw Junction	Didcot	-	Croes Newydd	8/66
3710	10/36	Paddington	Paddington	-	Barry	5/63
3711	11/36	Pontypool	Pontypool	-	Paddington	5/63
3712	11/36	Tondu	Ebbw Junction	-	Whitland	12/63

No.	Date	First Shed	Last GWR Shed	Wdn GWR	Last BR Shed	Wdn BR
3713	11/36	Landore	Landore	-	Llanelly	2/62
3714	11/36	Ebbw Junction	Ebbw Junction	-	Ebbw Junction	12/63
3715	1/37	Reading	Reading	-	Paddington	3/65
3716	1/37	Ebbw Junction	Cardiff	-	Aberdare	12/63
3717	1/37	Pontypool	Pontypool	-	Radyr	6/65
3718	2/37	Duffryn Yard	Duffryn Yard	-	Duffryn Yard	5/62
3719	2/37	Llanelly	Llanelly	-	Llanely	1/63
3720	2/37	Bristol SPM	Bristol SPM	-	Westbury	12/63
3721	2/37	Oxford	Didcot	-	Swansea E Dock	4/64
3722	2/37	Oxford	Oxford	-	Ebbw Junction	5/62
3723	2/37	Oxford	Paddington	-	Reading	4/62
3724	2/37	Swindon	Swindon	-	Swindon	5/61
3725	6/37	Hereford	Ledbury	-	Worcester	1/65
3726	6/37	Pontypool	Ebbw Junction	-	Ebbw Junction	1/62
3727	6/37	Henley	Southall	-	Barry	4/64
3728	7/37	Worcester	Ledbury	-	Radyr	4/65
3729	7/37	Ebbw Junction	Aberbeeg	-	Hereford	3/63
3730	7/37	Pontypool	Branches Fork	-	Abercynon	11/64
3731	7/37	Bristol SPM	Frome	-	Neath	5/64
3732	8/37	Stafford Road	Wellington	-	Wellington	5/62
3733	8/37	Bath	Yeovil	-	Swansea E Dock	12/63
3734	8/37	Paddington	Paddington	-	Abercynon	4/64
3735	8/37	Frome	Frome	-	Westbury	9/65
3736	9/37	Westbury	Reading	-	Taunton	3/63
3737	9/37	Swindon	Swindon	-	Gloucester	9/64
3738	9/37	Paddington	Paddington	-	Cardiff E Dock	8/65
3739	9/37	Swindon	Swindon	-	Westbury	10/64
3740	9/37	Tyseley	Stourbridge	-	Lydney	1/59
3741	9/37	Oxford	Oxford	-	Neath	9/62
3742	9/37	Birkenhead	Birkenhead	-	Bath LMS	11/64
3743	9/37	Tyseley	Tyseley	-	Neath	5/62
3744	10/37	Wellington	Oxley	-	Oxley	8/66
3745	10/37	Shrewsbury	Oxley	-	Gloucester	12/64
3746	10/37	Chippenham	Bristol SPM	-	Westbury	7/64
3747	10/37	Aberdare	Aberdare	-	Ebbw Junction	2/65
3748	10/37	Swindon	Chippenham	-	Cardiff E Dock	5/64
3749	10/37	Chester	Crewe	-	Croes Newydd	11/65
3750	11/37	Southall	Southall	-	Paddington	9/62
3751	11/37	Banbury	Banbury	-	Oxford	9/65
3752	11/37	Llanelly	Llanelly	-	Didcot	7/64
3753	11/37	Aberdare	Aberdare	-	Aberdare	1/65
3754	11/37	Paddington	Paddington	-	Shrewsbury	11/65
3755	11/37	Aberbeeg	Cardiff	-	Cardiff E Dock	10/62
3756	11/37	Stafford Road	Stafford Road	-	Tondu	3/63
3757	12/37	Neath	Neath N&B	-	Neath	5/64
3758	12/37	Westbury	Frome	-	Bath LMS	3/66
3759	12/37	Bristol SPM	Bristol SPM	-	Gloucester	12/65
3760	12/37	Tyseley	Stafford Road	-	Croes Newydd	9/62
3761	12/37	Llanelly	Llanelly	-	Llanelly	5/64
3762	12/37	Chester	Chester	-	Duffryn Yard	12/63
3763	12/37	Bristol SPM	Bristol SPM	-	Southall	6/65
3764	12/37	Swindon	Bristol SPM	-	Ebbw Junction	1/63
3765	3/38	Bristol SPM	Bristol SPM	-	Barrow Road	7/63
3766	4/38	Paddington	Paddington	-	Aberbeeg	8/63
3767	4/38	Neath N&B	Brecon	-	Ebbw Junction	10/65
3768	4/38	Landore	Landore	-	Neath	5/64
3769	4/38	Slough	Slough	-	Shrewsbury	10/62
3770	5/38	Reading	Reading	-	Tyseley	4/65
3771	5/38	Llanelly	Pantyffynon	-	Llanelly	1/63
3772	5/38	Tondu	Tondu	-	Newport	6/65

No.	Date	First Shed	Last GWR Shed	Wdn GWR	Last BR Shed	Wdn BR
3773	5/38	Bristol SPM	Bristol SPM	-	Cardiff E Dock	10/62
3774	5/38	Neath	Neath N&B	-	Neath	5/62
3775	7/38	Stafford Road	Wellington	-	Gloucester	12/65
3776	7/38	Aberbeeg	Aberbeeg	-	Oxley	4/66
3777	7/38	Ledbury	Llanelly	-	Llanelly	12/63
3778	7/38	Stafford Road	Stafford Road	-	Oxley	3/64
3779	7/38	Tondu	Pontypool	-	Pontypool	12/63
3780	8/38	Westbury	Swindon	-	Swindon	12/62
3781	8/38	Paddington	Danygraig	-	Llanelly	8/63
3782	8/38	Shrewsbury	Shrewsbury	-	Oxley	10/66
3783	8/38	Basingstoke	Reading	-	Abercynon	9/62
3784	8/38	Bristol SPM	Bristol SPM	-	Radyr	6/65
3785	9/38	Danygraig	Landore	-	Swansea E Dock	7/62
3786	9/38	Stourbridge	Chester	-	Stourbridge	3/63
3787	9/38	Laira	Laira	-	Westbury	1/63
3788	9/38	Shrewsbury	Shrewsbury	-	Oxley	11/65
3789	10/38	Kidderminster	Hereford	-	Croes Newydd	10/65
3790	10/38	Penzance	Laira	-	Cardiff E Dock	6/65
3791	10/38	Duffryn Yard	Duffryn Yard	-	Treherbert	12/63
3792	10/38	Oxley	Oxley	-	Oxley	11/65
3793	10/38	Stafford Road	Oxley	-	Stafford Road	9/58
3794	10/38	Exeter	Exeter	-	Oxford	12/64
3795	11/38	Yeovil	Bristol SPM	-	Barrow Road	5/63
3796	11/38	Ebbw Junction	Ebbw Junction	-	Severn Tunnel	3/65
3797	12/38	Landore	Landore	-	Neath	11/64
3798	12/38	Oxford	Slough	-	Barrow Road	10/64
3799	12/38	Southall	Southall	-	Southall	1/61
3800	12/38	Pontypool	Ebbw Junction	-	Ebbw Junction	8/64
3801	12/38	Ebbw Junction	Ebbw Junction	-	Severn Tunnel	8/64
3802	12/38	Leamington	Banbury	-	Barrow Road	8/65
3803	1/39	Tyseley	Banbury	-	Severn Tunnel	7/63
3804	8/39	Ebbw Junction	Ebbw Junction	-	Cardiff E Dock	7/64
3805	8/39	Ebbw Junction	Ebbw Junction	-	Ebbw Junction	9/64
3806	8/39	Llanelly	Stourbridge	-	Banbury	12/63
3807	8/39	Cardiff	Ebbw Junction	-	Ebbw Junction	2/65
3808	9/39	Llanelly	Stourbridge	-	Ebbw Junction	7/65
3809	9/39	Llanelly	Cardiff	-	Croes Newydd	10/64
3810	9/39	Cardiff	Ebbw Junction	-	Southall	11/64
3811	9/39	Llanelly	Laira	-	Severn Tunnel	9/63
3812	9/39	Stourbridge	Cardiff	-	Southall	6/65
3813+	9/39	Neath	Laira	-	Croes Newydd	7/65
3814	3/40	Ebbw Junction	Aberdare	-	Didcot	12/64
3815	3/40	Pontypool	Severn Tunnel	-	Croes Newydd	5/64
3816	3/40	Ebbw Junction	Ebbw Junction	-	Severn Tunnel	7/65
3817	3/40	Severn Tunnel	Cardiff	-	Croes Newydd	8/65
3818+	4/40	Cardiff	Severn Tunnel	-	Southall	5/65
3819	4/40	Tyseley	Banbury	-	Didcot	12/64
3820+	4/40	Leamington	Paddington	-	Southall	7/65
3821	4/40	Oxley	Banbury	-	Oswestry	10/64
3822	4/40	Llanelly	Pontypool	-	Cardiff E Dock	1/64
3823	5/40	Neath	Cardiff	-	Severn Tunnel	7/65
3824	9/40	Pontypool	Cardiff	-	Ebbw Junction	6/64
3825	9/40	Pontypool	Banbury	-	Oswestry	9/64
3826	9/40	Pontypool	Pontypool	-	Severn Tunnel	1/65
3827	10/40	Tyseley	Banbury	-	Ebbw Junction	11/62
3828	11/40	Cardiff	Pontypool	-	Banbury	10/64
3829	11/40	Chester	Banbury	-	Croes Newydd	3/64
3830	11/40	Ebbw Junction	Ebbw Junction	-	Ebbw Junction	6/65
3831+	12/40	Severn Tunnel	Ebbw Junction	-	Croes Newydd	9/63
3832	1/41	Paddington	Paddington	-	Neath	4/64

No.	Date	First Shed	Last GWR Shed	Wdn GWR	Last BR Shed	Wdn BR
3833	1/41	Ebbw Junction	Ebbw Junction	-	Ebbw Junction	7/63
3834	12/41	Bristol SPM	Exeter	-	Taunton	4/64
3835	1/42	Oxford	Oxford	-	Didcot	1/65
3836	1/42	Oxford	Oxford	-	Barrow Road	11/65
3837+	1/42	Oxford	Paddington	-	Ebbw Junction	7/65
3838	1/42	Oxford	Oxford	-	Southall	11/64
3839+	1/42	Oxford	Paddington	-	Aberdare	12/63
3840	1/42	Reading	Reading	-	Ebbw Junction	7/65
3841	2/42	Reading	Reading	-	Pontypool	3/64
3842	2/42	Reading	Westbury	-	Severn Tunnel	7/65
3843	2/42	Reading	Reading	-	Severn Tunnel	10/63
3844	4/42	Reading	Reading	-	Barrow Road	10/65
3845	4/42	Reading	Reading	-	Banbury	6/64
3846	5/42	Reading	Reading	-	Croes Newydd	11/63
3847	5/42	Oxford	Oxford	-	Neath	3/64
3848	5/42	Oxford	Oxford	-	Southall	7/65
3849	6/42	Reading	Westbury	-	Croes Newydd	5/65
3850	6/42	Bristol SPM	Westbury	-	Croes Newydd	8/65
3851	6/42	Paddington	Paddington	-	Southall	7/65
3852	6/42	Paddington	Paddington	-	Banbury	4/64
3853	6/42	Paddington	Paddington	-	Severn Tunnel	12/63
3854	9/42	Southall	Southall	-	Southall	6/65
3855	10/42	Southall	Southalll	-	Croes Newydd	8/65
3856	10/42	Southall	Southall	-	Severn Tunnel	10/64
3857	10/42	Southall	Southall	-	Banbury	3/64
3858	10/42	Southall	Southall	-	Reading	9/63
3859	10/42	Southall	Southall	-	Southall	5/65
3860	10/42	Southall	Southall	-	Neath	8/64
3861	11/42	Banbury	Banbury	-	Severn Tunnel	7/65
3862	11/42	Oxford	Pontypool	-	Croes Newydd	2/65
3863	11/42	Bristol SPM	Westbury	-	Barrow Road	10/65
3864	11/42	Swindon	Laira	-	Ebbw Junction	7/65
3865+	11/42	Oxford	Severn Tunnel	-	Croes Newydd	3/65
3866	12/42	Cardiff	Oxford	-	Southall	7/65
4000	11/29	Newton Abbot	Stafford Road	-	Landore	5/57
4009+	4/25	Laira	Paddington	-	Paddington	3/50
4016	10/25	Stafford Road	Newton Abbot	-	Paddington	9/51
4032	4/26	Laira	Laira	-	Taunton	9/51
4037	6/26	Stafford Road	Paddington	-	Exeter	9/62
4063	11/22	Newton Abbot	Landore	3/37	-	-
4064	12/22	Paddington	Bath Road	2/37	-	-
4065	12/22	Exeter	Stafford Road	3/39	-	-
4066	12/22	Wolverhampton	Gloucester	12/37	-	-
4067	1/23	Wolverhampton	Weymouth	9/40	-	-
4068	1/23	Wolverhampton	Shrewsbury	11/38	-	-
4069	1/23	Wolverhampton	Bath Road	4/39	-	-
4070	2/23	Wolverhampton	Gloucester	1/39	-	-
4071	2/23	Paddington	Bath Road	9/38	-	-
4072	2/23	Paddington	Taunton	3/38	-	-
4073	8/23	Paddington	Paddington	-	Cardiff	5/60
4074	12/23	Paddington	Landore	-	Paddington	5/63
4075	1/24	Paddington	Paddington	-	Paddington	11/61
4076	2/24	Paddington	Paddington	-	Llanelly	2/63
4077	2/24	Paddington	Newton Abbot	-	Bristol SPM	8/62
4078	2/24	Paddington	Landore	-	Llanelly	7/62
4079	2/24	Paddington	Hereford	-	Bristol SPM	5/64
4080	3/24	Paddington	Bath Road	-	Southall	8/64
4081	3/24	Paddington	Landore	-	Carmarthen	1/63
4082+	4/24	Paddington	Gloucester	-	Tyseley (as 7013)	2/65

No.	Date	First Shed	Last GWR Shed	Wdn GWR	Last BR Shed	Wdn BR
4083	5/25	Paddington	Cardiff	-	Cardiff	12/61
4084	5/25	Laira	Bath Road	-	Cardiff	10/60
4085	5/25	Laira	Reading	-	Paddington	5/62
4086	6/25	Laira	Worcester	-	Oxford	4/62
4087	6/25	Laira	Laira	-	Bristol SPM	10/63
4088	7/25	Laira	Laira	-	Bristol SPM	5/64
4089	7/25	Paddington	Bath Road	-	Reading	9/64
4090	7/25	Paddington	Laira	-	Cardiff E Dock	6/63
4091	7/25	Paddington	Paddington	-	Paddington	1/59
4092	8/25	Paddington	Worcester	-	Oxford	12/61
4093	5/26	Paddington	Bath Road	-	Gloucester	9/64
4094	5/26	Paddington	Cardiff	-	Carmarthen	3/62
4095	6/26	Laira	Landore	-	Reading	12/62
4096	6/26	Laira	Bath Road	-	Llanelly	1/63
4097	6/26	Paddington	Penzance	-	Landore	5/60
4098	7/26	Paddington	Newton Abbot	-	Paddington	12/63
4099	8/26	Paddington	Newton Abbot	-	Llanelly	9/62
4100	8/35	Worcester	Worcester	-	Gloucester	10/65
4101	8/35	Stafford Road	Tyseley	-	Banbury	6/64
4102	8/35	Stafford Road	Leamington	-	Bristol SPM	9/62
4103	8/35	Tyseley	Stafford Road	-	Barrow Road	9/64
4104	8/35	Chester	Stourbridge	-	Gloucester	5/64
4105	9/35	Birkenhead	Stafford Road	-	Banbury	2/64
4106	9/35	Tyseley	Tyseley	-	Gloucester	9/62
4107	9/35	Tyseley	Tyseley	-	Severn Tunnel	6/65
4108	9/35	Truro	Stafford Road	-	Neath	10/64
4109	9/35	Truro	Newton Abbot	-	Gloucester	4/64
4110	10/36	Severn Tunnel	Stafford Road	-	Severn Tunnel	6/65
4111	10/36	Tyseley	Tyseley	-	Tyseley	9/65
4112	10/36	Leamington	Leamington	-	Banbury	8/62
4113	10/36	Swindon	Taunton	-	Worcester	11/65
4114	10/36	Worcester	Worcester	-	Kidderminster	11/63
4115	10/36	Stafford Road	Stafford Road	-	Severn Tunnel	6/65
4116	11/36	Stafford Road	Tyseley	-	Cheltenham	9/62
4117	11/36	Newton Abbot	Taunton	-	Taunton	9/61
4118	11/36	Tyseley	Shrewsbury	-	Leamington	9/62
4119	11/36	Severn Tunnel	Severn Tunnel	-	Severn Tunnel	9/63
4120	12/37	Shrewsbury	Birkenhead	-	Leamington	11/64
4121	12/37	Leamington	Birkenhead	-	Severn Tunnel	6/65
4122	1/38	Leamington	Birkenhead	-	Aberdare	6/64
4123	1/38	Tyseley	Birkenhead	-	Gloucester	11/61
4124	1/38	Chester	Birkenhead	-	Worcester	8/64
4125	5/38	Stourbridge	Birkenhead	-	Tyseley	6/65
4126	5/38	Leamington	Birkenhead	-	Tyseley	1/62
4127	5/38	Stafford Road	Birkenhead	-	Severn Tunnel	1/63
4128	5/38	Birkenhead	Birkenhead	-	Severn Tunnel	6/64
4129	5/38	Birkenhead	Birkenhead	-	Kidderminster	9/62
4130	10/39	Severn Tunnel	Severn Tunnel	-	Severn Tunnel	7/64
4131	10/39	Severn Tunnel	Pontypool	-	Barrow Road	9/64
4132	10/39	Neath	Whitland	-	Radyr	7/64
4133	10/39	Swansea E Dock	Newton Abbot	-	Leamington	10/64
4134	10/39	Llanelly	Landore	-	Neath	5/63
4135	10/39	Pontypool	Pontypool	-	Cardiff E Dock	6/64
4136	11/39	Swansea E Dock	Taunton	-	Aberdare	6/64
4137	11/39	Severn Tunnel	Severn Tunnel	-	Severn Tunnel	10/64
4138	11/39	Cardiff	Pontypool	-	Pontypool	7/58
4139	12/39	Worcester	Worcester	-	Gloucester	7/58
4140	8/46	Gloucester	Gloucester	-	Stourbridge	11/63
4141	8/46	Cheltenham	Cheltenham	-	Gloucester	3/63
4142	8/46	Bath Road	Bath Road	-	Neath	12/63

No.	Date	First Shed	Last GWR Shed	Wdn GWR	Last BR Shed	Wdn BR
4143	9/46	Bath Road	Bath Road	–	Radyr	6/64
4144	9/46	Tondu	Severn Tunnel	–	Severn Tunnel	6/65
4145	9/46	Ebbw Junction	Tondu	–	Severn Tunnel	12/62
4146	9/46	Stourbridge	Stourbridge	–	Severn Tunnel	8/62
4147	9/46	Tyseley	Tyseley	–	Tyseley	10/65
4148	10/46	Severn Tunnel	Severn Tunnel	–	Oxley	9/65
4149	10/46	Stourbridge	Stourbridge	–	Banbury	2/63
4150	6/47	Stourbridge	Stourbridge	–	Severn Tunnel	6/65
4151	6/47	Bath Road	Bath Road	–	Banbury	4/65
4152	6/47	Bath Road	Bath Road	–	Severn Tunnel	9/62
4153	6/47	Kidderminster	Kidderminster	–	Stourbridge	11/64
4154	7/47	Wellington	Crewe	–	Oxley	10/65
4155	8/47	Bath Road	Bath Road	–	Tyseley	9/65
4156	8/47	Severn Tunnel	Severn Tunnel	–	Severn Tunnel	6/65
4157	8/47	Tyseley	Tyseley	–	Severn Tunnel	6/65
4158	8/47	Severn Tunnel	Severn Tunnel	–	Tyseley	6/65
4159	9/47	Chester	Chester	–	Severn Tunnel	6/64
4200	2/23	Ebbw Junction	Severn Tunnel	–	Ebbw Junction	3/59
4296	8/22	Barry	Swansea E Dock	–	Llanelly	12/63
4297	8/22	Barry	Aberdare	–	Ebbw Junction	2/65
4298	10/22	Barry	St. Blazey	–	Llanelly	6/63
4299	10/22	Barry	Danygraig	–	Ebbw Junction	12/63
4555	9/24	Tyseley	Machynlleth	–	Laira	12/63
4556	9/24	Tyseley	Whitland	–	Whitland	6/60
4557	9/24	Laira	Tondu	–	Whitland	9/62
4558	9/24	Laira	Worcester	–	Whitland	7/62
4559	9/24	Tyseley	St. Blazey	–	St. Blazey	10/60
4560	9/24	Tyseley	Pwllheli	–	Oswestry	8/59
4561	10/24	Tyseley	Truro	–	Laira	5/62
4562	10/24	Tyseley	Bridport	–	Weymouth	3/60
4563	10/24	Laira	Yatton	–	Penzance	10/61
4564	10/24	Newton Abbot	Cheltenham	–	Gloucester	9/64
4565	10/24	Moreton'stead	Moorswater	–	St. Blazey	10/61
4566	10/24	Newton Abbot	St. Ives	–	Laira	4/62
4567	10/24	Newton Abbot	Cheltenham	–	Laira	9/62
4568	10/24	Newton Abbot	Bodmin	–	Penzance	2/59
4569	10/24	Ebbw Junction	Truro	–	Swindon	8/64
4570	11/24	Ebbw Junction	St. Blazey	–	Laira	1/63
4571	11/24	Ebbw Junction	Machynlleth	–	Penzance	3/61
4572	11/24	Bristol	Yeovil	–	Westbury	12/58
4573	11/24	Westbury	Westbury	–	Whitland	8/61
4574	11/24	Westbury	Penzance	–	Exmouth Jct	3/63
4575	2/27	Laira	Machynlleth	–	Machynlleth	8/60
4576	2/27	Newton Abbot	Pembroke Dock	–	Whitland	9/58
4577	2/27	Exeter	Bath Road	–	Penzance	11/59
4578	2/27	Newton Abbot	Cheltenham	–	Oswestry	8/58
4579	2/27	Tyseley	Whitland	–	Whitland	9/58
4580	2/27	Tyseley	Bath Road	–	Oswestry	6/58
4581	2/27	St. Blazey	Truro	–	Tondu	4/58
4582	2/27	Newquay	Newton Abbot	–	Bath Road	4/58
4583	2/27	Taunton	Laira	–	Laira	1/58
4584	2/27	Tyseley	Kidderminster	–	St. Blazey	2/59
4585	3/27	Newton Abbot	Andover Jct	–	St. Blazey	10/59
4586	3/27	Tyseley	Kidderminster	–	Cheltenham	4/56
4587	3/27	Ebbw Junction	Newton Abbot	–	St. Blazey	8/60
4588	3/27	Tyseley	Truro	–	Laira	7/62
4589	3/27	Newton Abbot	Truro	–	Laira	9/60
4590	3/27	Taunton	Swindon	–	Laira	10/58
4591	3/27	Taunton	Laira	–	Swindon	8/64
4592	3/27	Paddington	Swindon	–	Laira	1/60

No.	Date	First Shed	Last GWR Shed	Wdn GWR	Last BR Shed	Wdn BR
4593	3/27	Paddington	Ebbw Junction	-	Gloucester	9/64
4594	4/27	Paddington	Kidderminster	-	Whitland	11/60
4595	4/27	Paddington	Bath Road	-	Oswestry	12/58
4596	4/27	Paddington	Worcester	-	Worcester	9/57
4597	4/27	Paddington	Aberbeeg	-	Bath Road	1/58
4598	4/27	Laira	Bodmin	-	Truro	12/56
4599	4/27	Paddington	Ebbw Junction	-	Machynlleth	3/59
4600	9/41	Hereford	Hereford	-	Pontypool	7/64
4601	9/41	Didcot	Didcot	-	Nine Elms	11/62
4602	10/41	Ludlow	Shrewsbury	-	Oxley	11/64
4603	10/41	Bristol SPM	Bristol SPM	-	Pontypool	4/64
4604	10/41	Southall	Southall	-	Llanelly	7/65
4605	1/42	Tyseley	Tyseley	-	Wellington	10/62
4606	1/42	Paddington	Paddington	-	Paddington	2/65
4607	1/42	Bristol SPM	Bristol SPM	-	Westbury	9/65
4608	1/42	Southall	Southall	-	Southall	9/64
4609	1/42	Paddington	Paddington	-	Southall	5/65
4610	2/42	Southall	Southall	-	Exmouth Jct	10/64
4611	2/42	Tondu	Pontypool	-	Southall	6/65
4612	2/42	Bristol SPM	Bristol SPM	-	Cardiff E Dock	8/65
4613	2/42	Worcester	Worcester	-	Worcester	12/64
4614	2/42	Worcester	Worcester	-	Gloucester	7/64
4615	4/42	Paddington	Paddington	-	Paddington	10/64
4616	4/42	Cardiff	Cathays	-	Pontypool	10/64
4617	4/42	Paddington	Watlington	-	Shrewsbury	10/63
4618	4/42	Aberdare	Cardiff E Dock	-	Barry	10/63
4619	5/42	Bristol SPM	Bristol SPM	-	Barrow Road	9/64
4620	5/42	Tyseley	Tyseley	-	Ebbw Junction	7/65
4621	5/42	Neath	Neath N&B	-	Severn Tunnel	7/65
4622	6/42	Cardiff	Cardiff	-	Southall	6/64
4623	6/42	Shrewsbury	Shrewsbury	-	Cardiff E Dock	6/65
4624	6/42	Bristol SPM	Bristol SPM	-	Gloucester	9/64
4625	9/42	Kidderminster	Kidderminster	-	Oxley	5/62
4626	9/42	Bristol SPM	Bristol SPM	-	Salisbury SR	3/64
4627	9/42	Gloucester	Gloucester	-	Ebbw Junction	10/64
4628	10/42	Gloucester	Gloucester	-	Worcester	5/64
4629	10/42	Worcester	Worcester	-	Gloucester	9/63
4630	11/42	Cardiff	Cardiff E Dock	-	Barrow Road	10/65
4631	11/42	Banbury	Banbury	-	Templecombe	6/65
4632	11/42	Cardiff E Dock	Merthyr	-	Merthyr	1/62
4633	11/42	Cardiff	Cardiff	-	Cardiff E Dock	4/63
4634	11/42	Tondu	Tondu	-	Templecombe	9/64
4635	12/42	Cardiff E Dock	Merthyr	-	Tyseley	7/66
4636	12/42	Westbury	Frome	-	Westbury	9/65
4637	12/42	Cardiff	Cardiff	-	Radyr	5/64
4638	12/42	Stourbridge	Stourbridge	-	Southall	6/65
4639	12/42	Pontypool	Pontypool	-	Cardiff E Dock	6/65
4640	12/42	Duffryn Yard	Duffryn Yard	-	Duffryn Yard	12/63
4641	12/42	Gloucester	Worcester	-	Kidderminster	5/62
4642	1/43	Paddington	Paddington	-	Pontypool	1/64
4643	1/43	Pontypool	Tondu	-	Ebbw Junction	4/65
4644	1/43	Paddington	Paddington	-	Goodwick	10/63
4645	2/43	Oxford	Oxford	-	Croes Newydd	11/65
4646	2/43	Banbury	Banbury	-	Tyseley	11/66
4647	3/43	Bristol SPM	Bristol SPM	-	Westbury	10/62
4648	3/43	Tyseley	Tyseley	-	Tyseley	9/64
4649	3/43	Reading	Basingstoke	-	Oxford	9/64
4650	4/43	Slough	Slough	-	Ebbw Junction	7/65
4651	4/43	Swindon	Chippenham	-	Duffryn Yard	9/63
4652	4/43	Branches Fork	Cardiff	-	Tondu	4/64

No.	Date	First Shed	Last GWR Shed	Wdn GWR	Last BR Shed	Wdn BR
4653	4/43	Laira	Laira	-	Neath	11/64
4654	4/43	Milford Haven	Milford Haven	-	Whitland	12/63
4655	5/43	Laira	Bristol SPM	-	Exmouth Jct	6/65
4656	5/43	Laira	Laira	-	Yeovil Town	12/62
4657	5/43	St. Blazey	Hereford	-	Neath	7/64
4658	6/43	Truro	Laira	-	Swansea E Dock	5/64
4659	6/43	Gloucester	Gloucester	-	Hereford	5/64
4660	6/43	Bristol SPM	Bridport	-	Neath	7/64
4661	10/43	Didcot	Basingstoke	-	Westbury	12/63
4662	10/43	Pill	Pill	-	Ebbw Junction	9/65
4663	11/43	Southall	Southall	-	Cardiff E Dock	6/65
4664	11/43	Worcester	Worcester	-	Worcester	7/65
4665	12/43	Paddington	Paddington	-	Tyseley	6/65
4666	12/43	Paddington	Paddington	-	Exmouth Jct	6/65
4667	12/43	Paddington	Paddington	-	Barry	4/64
4668	1/44	Pontypool	Pontypool	-	Llanelly	7/65
4669	1/44	Pontypool	Tondu	-	Neath	6/65
4670	1/44	Duffryn Yard	Basingstoke	-	Paddington	10/64
4671	2/44	Pill	Ebbw Junction	-	Severn Tunnel	11/65
4672	3/44	Shrewsbury	Shrewsbury	-	Feltham	7/63
4673	3/44	Southall	Southall	-	Pontypool	6/65
4674	3/44	Cardiff	Llantrisant	-	Radyr	11/64
4675	3/44	Tondu	Tondu	-	Llanelly	6/65
4676	4/44	Slough	Oxford	-	Llanelly	10/65
4677	4/44	Ebbw Junction	Cardiff	-	Neath	3/63
4678	5/44	Hereford	Hereford	-	Hereford	7/64
4679	5/44	Laira	Laira	-	Radyr	5/65
4680	6/44	Paddington	Paddington	-	Worcester	12/65
4681	11/44	Duffryn Yard	Duffryn Yard	-	Barrow Road	12/63
4682	11/44	Ebbw Junction	Aberbeeg	-	Neath	9/63
4683	11/44	Tyseley	Tyseley	-	Croes Newydd	10/65
4684	12/44	Duffryn Yard	Duffryn Yard	-	Barrow Road	7/65
4685	12/44	Cardiff	Aberbeeg	-	Aberbeeg	5/62
4686	12/44	Cardiff	Aberbeeg	-	Nine Elms	8/59
4687	12/44	Stourbridge	Stourbridge	-	Oxley	11/64
4688	12/44	Reading	Reading	-	Aberdare	12/63
4689	12/44	Yeovil	Yeovil	-	Gloucester	12/65
4690	1/45	Didcot	Reading	-	Merthyr	10/63
4691	1/45	Paddington	Paddington	-	Templecombe	9/64
4692	1/45	Reading	Reading	-	Exmouth Jct	9/64
4693	1/45	Laira	Laira	-	Neath	4/64
4694	1/45	Danygraig	Danygraig	-	Exmouth Jct	6/65
4695	1/45	Southall	Southall	-	Neath	6/64
4696	2/45	Stourbridge	Stourbridge	-	Tyseley	11/66
4697	2/45	Swindon	Swindon	-	Westbury	7/65
4698	2/45	Didcot	Paddington	-	Gloucester	10/65
4699	2/45	Didcot	Paddington	-	Swansea E Dock	6/64
4706	3/23	Paddington	Exeter	-	Paddington	2/64
4707	4/23	Laira	Paddington	-	Paddington	5/64
4708	4/23	Paddington	Oxley	-	Paddington	10/62
4800+	8/32	Gloucester	Swindon	-	Swindon	6/57
4801+	8/32	Cheltenham	Croes Newydd	-	Cheltenham	11/58
4802+	8/32	Gloucester	Cheltenham	-	Bath Road	10/56
4803+	9/32	Chalford	Weymouth	-	Weymouth	11/57
4804+	9/32	Gloucester	Ross-on-Wye	-	Lydney	2/56
4805+	9/32	Exeter	Exeter	-	Exeter	9/58
4806+	9/32	Tiverton Jct	Gloucester	-	Southall	3/58
4807+	9/32	Exeter	Reading	-	Reading	6/60
4808+	9/32	Exeter	Worcester	-	Laira	3/58
4809+	10/32	Exeter	Lydney	-	Gloucester LMS	10/63

No.	Date	First Shed	Last GWR Shed	Wdn GWR	Last BR Shed	Wdn BR
4810+	3/33	Stourbridge	Stourbridge	-	Radyr	6/61
4811+	3/33	Birkenhead	Croes Newydd	-	Slough	10/56
4812+	3/33	Oswestry	Oswestry	-	Bath Road	6/60
4813+	4/33	Gloucester	Gloucester	-	Gloucester	3/56
4814+	4/33	Banbury	Stourbridge	-	Stourbridge	4/57
4815+	4/33	Oswestry	Yatton	-	Southall	2/57
4816+	4/33	Kidderminster	Croes Newydd	-	Croes Newydd	10/56
4817+	4/33	Cheltenham	Oswestry	-	Swindon	2/59
4818+	4/33	Gloucester	Worcester	-	Weymouth	10/58
4819+	4/33	Exeter	Goodwick	-	St. Blazey	4/61
4820+	11/33	Pontypool	Cathays	-	Gloucester	11/64
4821+	11/33	Pontypool	Ebbw Junction	-	Gloucester LMS	12/63
4822+	11/33	Pontypool	Pontypool	-	Swindon	6/57
4823+	11/33	Pontypool	Goodwick	-	Oswestry	1/59
4824+	11/33	Goodwick	Gloucester	-	Gloucester	12/63
4825+	11/33	Southall	Cathays	-	Oxford	2/56
4826+	11/33	Southall	Aylesbury	-	Gloucester LMS	4/62
4827+	11/33	Reading	Newton Abbot	-	Cheltenham	6/60
4828+	11/33	Reading	Croes Newydd	-	Gloucester	6/59
4829+	11/33	Swindon	Ashburton	-	Newton Abbot	3/59
4830+	7/34	Aylesbury	Bath Road	-	Lydney	9/58
4831+	7/34	Ebbw Junction	Goodwick	-	Lydney	4/61
4832+	7/34	Tiverton Jct	Oswestry	-	Oswestry	7/63
4833+	7/34	Swindon	Chippenham	-	Cheltenham	1/61
4834+	7/34	Penmaenpool	Machynlleth	-	Exeter	7/62
4835+	8/34	Stourbridge	Tiverton Jct	-	Oxford	1/62
4836+	8/34	Swindon	Swindon	-	Southall	10/58
4837+	8/34	Aberayron	Slough	-	Oxford	2/59
4838+	8/34	Reading	Stourbridge	-	Oswestry	11/62
4839+	8/34	Neath	Newton Abbot	-	Newton Abbot	8/57
4840+	3/35	Exeter	Exeter	-	Gloucester	12/63
4841+	4/35	Gloucester	Chalford	-	Gloucester	6/60
4842+	4/35	Staines	Marlow	-	Exmouth Jct	5/65
4843+	4/35	Southall	Staines	-	Southall	6/57
4844+	4/35	Reading	Reading	-	Gloucester	10/64
4845+	4/35	Kidderminster	Ross-on-Wye	-	Gloucester	9/64
4846+	4/35	Cheltenham	Swindon	-	Southall	9/58
4847+	4/35	Southall	Reading	-	Hereford	3/64
4848+	4/35	Southall	Abingdon	-	Slough	6/60
4849+	4/35	Exeter	Tiverton Jct	-	Oswestry	6/60
4850+	7/35	Oxford	Oxford	-	Exmouth Jct	5/65
4851+	7/35	Exeter	Exter	-	Gloucester	7/64
4852+	7/35	Goodwick	Goodwick	-	Newton Abbot	6/60
4853+	7/35	Stourbridge	Chippenham	-	Gloucester	11/64
4854+	7/35	Bristol SPM	Weymouth	-	Gloucester	12/60
4855+	7/35	Yatton	Leominster	-	Gloucester	5/64
4856+	7/35	Croes Newydd	Lydney	-	Gloucester	2/59
4857+	8/35	Stourbridge	Croes Newydd	-	Oswestry	2/59
4858+	8/35	Stourbridge	Banbury	-	Hereford	11/64
4859+	8/35	Oswestry	Oswestry	-	Weymouth	9/58
4860+	2/36	Goodwick	Leominster	-	Leominster	2/56
4861+	2/36	Merthyr	Abercynon	-	Worcester	5/58
4862+	2/36	Reading	Staines	-	Exeter	9/62
4863+	2/36	Gloucester	Yatton	-	Gloucester	4/61
4864+	2/36	Gloucester	Gloucester	-	Swindon	6/60
4865+	2/36	Ashburton	Penmaenpool	-	Croes Newydd	9/58
4866+	2/36	Newton Abbot	Newton Abbot	-	Taunton	12/63
4867+	2/36	Landore	Weymouth	-	Cheltenham	4/59
4868+	2/36	Newton Abbot	Exeter	-	St. Blazey	3/62
4869+	2/36	Exter	Exeter	-	Exeter	9/58

No.	Date	First Shed	Last GWR Shed	Wdn GWR	Last BR Shed	Wdn BR
4870+	4/36	Newton Abbot	Newton Abbot	-	Exeter	10/62
4871+	4/36	Llantrisant	Llantrisant	-	Exeter	10/63
4872+	4/36	Danygraig	Carmarthen	-	Gloucester	11/64
4873+	4/36	Chippenham	Banbury	-	Gloucester	8/62
4874+	4/36	Brecon	Aberayron	-	Gloucester	9/64
4901	12/28	Penzance	Cardiff	-	Oxley	9/60
4902	12/28	Penzance	Oxford	-	Taunton	9/63
4903	12/28	Penzance	Oxford	-	Worcester	10/64
4904	12/28	Penzance	Oxley	-	Taunton	12/63
4905	12/28	Truro	Swindon	-	Oxford	11/63
4906	1/29	Truro	Truro	-	Oxley	9/62
4907+	1/29	Laira	Paddington	-	Hereford	8/63
4908	1/29	Laira	Llanelly	-	Oxford	10/63
4909	1/29	Laira	Banbury	-	Swindon	9/62
4910	1/29	Laira	Carmarthen	-	Oxford	12/63
4911	2/29	Laira	Truro	6/41	-	-
4912	2/29	Laira	Pontypool	-	Oxley	8/62
4913	2/29	Laira	Cardiff	-	Hereford	9/62
4914	2/29	Laira	Reading	-	Bristol SPM	12/63
4915	2/29	Gloucester	Carmarthen	-	Reading	2/63
4916	2/29	Gloucester	Oxley	-	Swindon	8/64
4917	3/29	Paddington	Tyseley	-	Westbury	9/62
4918	3/29	Paddington	Chester	-	Cardiff E Dock	6/63
4919	3/29	Paddington	Shrewsbury	-	Worcester	10/64
4920	3/29	Paddington	Reading	-	Barrow Road	12/65
4921	4/29	Paddington	Oxford	-	Reading	9/62
4922	4/29	Paddington	Carmarthen	-	Oxford	7/63
4923	4/29	Bristol SPM	Oxley	-	Banbury	5/64
4924	5/29	Paddington	Tyseley	-	Swindon	10/63
4925	5/29	Paddington	Swindon	-	Southall	8/62
4926	5/29	Paddington	Westbury	-	Pontypool	9/61
4927	5/29	Paddington	Westbury	-	Llanelly	9/63
4928	5/29	Stafford Road	Oxford	-	Duffryn Yard	12/63
4929	5/29	Stafford Road	Truro	-	Gloucester	3/65
4930	5/29	Stafford Road	Tyseley	-	Swindon	12/63
4931	5/29	Stafford Road	Reading	-	Cardiff	7/62
4932	6/29	Stafford Road	Pontypool	-	Severn Tunnel	11/64
4933	6/29	Bath Road	Pontypool	-	Shrewsbury	8/64
4934	6/29	Bath Road	Leamington	-	Taunton	9/62
4935	6/29	Chester	Paddington	-	Paddington	3/63
4936	6/29	Chester	Truro	-	Cardiff E Dock	1/64
4937	6/29	Stafford Road	Neyland	-	Pontypool	9/62
4938	6/29	Bath Road	Oxford	-	Llanelly	11/62
4939	7/29	Truro	Tyseley	-	Oxford	2/63
4940	7/29	Taunton	St. Blazey	-	Taunton	11/59
4941	7/29	Newton Abbot	Ebbw Junction	-	Westbury	10/62
4942	7/29	Newton Abbot	Bath Road	-	Didcot	12/63
4943	7/29	Paddington	Paddington	-	Pontypool	12/63
4944	7/29	Paddington	Oxley	-	Southall	9/62
4945	8/29	Basingstoke	Swindon	-	Southall	11/61
4946	8/29	Bath Road	Penzance	-	Shrewsbury	6/63
4947	8/29	Gloucester	Penzance	-	Bristol SPM	9/62
4948+	8/29	Bath Road	Laira	-	Swindon	9/62
4949	8/29	Stafford Road	Penzance	-	Barrow Road	9/64
4950	8/29	Stafford Road	Stafford Road	-	Didcot	5/64
4951	7/29	Paddington	Paddington	-	Oxford	6/64
4952	8/29	Bath Road	Cardiff	-	Cardiff	9/62
4953	8/29	Bath Road	Cardiff	-	Cardiff E Dock	5/63
4954	8/29	Stafford Road	Taunton	-	Oxley	11/64
4955	8/29	Bath Road	Oxley	-	Pontypool	10/63

No.	Date	First Shed	Last GWR Shed	Wdn GWR	Last BR Shed	Wdn BR
4956	9/29	Paddington	Swindon	-	Westbury	7/63
4957	9/29	Oxley	Neyland	-	Reading	3/62
4958	9/29	Oxley	Paddington	-	Gloucester	9/64
4959	9/29	Westbury	Tyseley	-	Didcot	12/64
4960	9/29	Westbury	Stafford Road	-	Bristol SPM	9/62
4961	11/29	Goodwick	Paddington	-	Reading	11/62
4962	11/29	Cardiff	Paddington	-	Oxford	10/65
4963	11/29	Carmarthen	Westbury	-	Worcester	6/62
4964	11/29	Goodwick	Oxley	-	Pontypool	10/63
4965	11/29	Oxley	Bristol SPM	-	Didcot	3/62
4966	11/29	Basingstoke	Laira	-	Oxford	11/63
4967	12/29	Tyseley	Tyseley	-	Neath	9/62
4968+	12/29	Paddington	Bristol SPM	-	Bristol SPM	7/62
4969	12/29	Paddington	Bristol SPM	-	Southall	9/62
4970	12/29	Paddington	Penzance	-	Duffryn Yard	7/63
4971+	1/30	Paddington	Laira	-	Cardiff	8/62
4972+	1/30	Paddington	Swindon	-	Westbury	2/64
4973	1/30	Penzance	Oxford	-	Cardiff	7/62
4974	1/30	Penzance	Cardiff	-	Gloucester	4/62
4975	1/30	Penzance	Cardiff	-	Oxford	9/63
4976	1/30	Penzance	Chester	-	Oxford	5/64
4977	1/30	Penzance	Gloucester	-	Cardiff	5/62
4978	2/30	Laira	Paddington	-	Severn Tunnel	9/64
4979	2/30	Laira	Cardiff	-	Oxford	12/63
4980	2/30	Penzance	Worcester	-	Swindon	7/63
4981	12/30	Oxley	Carmarthen	-	Oxford	10/63
4982	1/31	Weymouth	Goodwick	-	Laira	5/62
4983	1/31	Goodwick	Newton Abbot	-	Bristol SPM	12/63
4984	1/31	Worcester	Carmarthen	-	Cardiff	9/62
4985	1/31	Paddington	Paddington	-	Neath	9/64
4986	1/31	Penzance	Bristol SPM	-	Southall	5/62
4987	1/31	Carmarthen	Oxley	-	Southall	4/62
4988	1/31	Paddington	Weymouth	-	Oxford	2/64
4989	2/31	Penzance	Reading	-	Severn Tunnel	11/64
4990	2/31	Penzance	Bristol SPM	-	Hereford	4/62
4991	2/31	Bristol SPM	Oxley	-	Swindon	12/63
4992	2/31	Penzance	Tyseley	-	Barrow Road	4/65
4993	2/31	Penzance	Tyseley	-	Severn Tunnel	2/65
4994	2/31	Penzance	Reading	-	Didcot	3/63
4995	2/31	Laira	Reading	-	Southall	6/62
4996	3/31	Paddington	Oxley	-	Taunton	9/63
4997	3/31	Oxford	Neyland	-	Oxley	10/61
4998	3/31	Paddington	Paddington	-	Banbury	10/63
4999	3/31	Weymouth	Tyseley	-	Bristol SPM	9/62
5000	9/26	Paddington	Paddington	-	Oxley	10/64
5001	9/26	Paddington	Cardiff	-	Paddington	2/63
5002	9/26	Paddington	Landore	-	Southall	9/64
5003	5/27	Paddington	Taunton	-	Newton Abbot	8/62
5004	6/27	Laira	Paddington	-	Neath	4/62
5005	6/27	Paddington	Cardiff	-	Swindon	2/60
5006	6/27	Paddington	Landore	-	Carmarthen	4/62
5007	6/27	Laira	Cardiff	-	Gloucester	9/62
5008	6/27	Laira	Paddington	-	Paddington	9/62
5009	6/27	Laira	Laira	-	Swindon	10/60
5010	7/27	Paddington	Cardiff	-	Reading	10/59
5011	7/27	Newton Abbot	Newton Abbot	-	Paddington	9/62
5012	7/27	Newton Abbot	Exeter	-	Oxford	4/62
5013	6/32	Newton Abbot	Landore	-	Neath	7/62
5014	6/32	Laira	Paddington	-	Tyseley	2/65
5015	6/32	Cardiff	Stafford Road	-	Cardiff E Dock	4/63

No.	Date	First Shed	Last GWR Shed	Wdn GWR	Last BR Shed	Wdn BR
5016	7/32	Paddington	Landore	-	Llanelly	9/62
5017	7/32	Taunton	Worcester	-	Gloucester	9/62
5018	7/32	Paddington	Stafford Road	-	Reading	3/64
5019	7/32	Laira	Bath Road	-	Stafford Road	9/62
5020	7/32	Laira	Cardiff	-	Llanelly	11/62
5021	8/32	Laira	Shrewsbury	-	Cardiff	9/62
5022	8/32	Paddington	Paddington	-	Stafford Road	6/63
5023	4/34	Paddington	Paddington	-	Swindon	2/63
5024	4/34	Newton Abbot	Bath Road	-	Newton Abbot	5/62
5025	4/34	Paddington	Bath Road	-	Hereford	11/63
5026	4/34	Newton Abbot	Laira	-	Oxley	11/64
5027	4/34	Paddington	Paddington	-	Llanelly	11/62
5028	5/34	Laira	Laira	-	Laira	5/60
5029	5/34	Paddington	Paddington	-	Cardiff E Dock	12/63
5030	5/34	Exeter	Cardiff	-	Carmarthen	9/62
5031	5/34	Stafford Road	Stafford Road	-	Stafford Road	10/63
5032	5/34	Shrewsbury	Shrewsbury	-	Paddington	9/62
5033	5/35	Stafford Road	Chester	-	Oxford	9/62
5034	5/35	Newton Abbot	Newton Abbot	-	Paddington	9/62
5035	5/35	Cardiff	Paddington	-	Swindon	5/62
5036	5/35	Cardiff	Paddington	-	Paddington	9/62
5037	5/35	Paddington	Paddington	-	Bristol SPM	3/64
5038	6/35	Paddington	Paddington	-	Reading	9/63
5039	6/35	Paddington	Paddington	-	Reading	6/64
5040	6/35	Paddington	Paddington	-	Bristol SPM	10/63
5041	7/35	Laira	Laira	-	Paddington	12/63
5042	7/35	Paddington	Gloucester	-	Gloucester	6/65
5043	3/36	Paddington	Paddington	-	Cardiff E Dock	12/63
5044	3/36	Paddington	Paddington	-	Cardiff	4/62
5045	3/36	Paddington	Paddington	-	Stafford Road	9/62
5046	4/36	Cardiff	Cardiff	-	Stafford Road	9/62
5047	4/36	Landore	Newton Abbot	-	Stafford Road	9/62
5048	4/36	Bath Road	Bath Road	-	Llanelly	8/62
5049	4/36	Worcester	Cardiff	-	Bristol SPM	3/63
5050	5/36	Worcester	Laira	-	Bristol SPM	8/63
5051	5/36	Landore	Landore	-	Llanelly	5/63
5052	5/36	Cardiff	Cardiff	-	Bristol SPM	9/62
5053	5/36	Shrewsbury	Stafford Road	-	Cardiff	7/62
5054	6/36	Paddington	Cardiff	-	Gloucester	10/64
5055	6/36	Paddington	Paddington	-	Gloucester	9/64
5056	6/36	Paddington	Paddington	-	Oxley	11/64
5057	6/36	Newton Abbot	Laira	-	Paddington	3/64
5058	5/37	Newton Abbot	Newton Abbot	-	Gloucester	3/63
5059	5/37	Exeter	Exeter	-	Shrewsbury	6/62
5060	6/37	Stafford Road	Laira	-	Paddington	4/63
5061	6/37	Shrewsbury	Shrewsbury	-	Cardiff	9/62
5062	6/37	Stafford Road	Newton Abbot	-	Llanelly	8/62
5063	6/37	Worcester	Worcester	-	Oxley	2/65
5064	6/37	Newton Abbot	Shrewsbury	-	Gloucester	9/62
5065	7/37	Exeter	Paddington	-	Paddington	1/63
5066	7/37	Paddington	Paddington	-	Paddington	9/62
5067	7/37	Paddington	Swindon	-	Reading	7/62
5068	6/38	Bath Road	Swindon	-	Oxford	9/62
5069	6/38	Paddington	Paddington	-	Laira	2/62
5070	6/38	Paddington	Stafford Road	-	Paddington	3/64
5071	6/38	Newton Abbot	Newton Abbot	-	Bristol SPM	10/63
5072	6/38	Newton Abbot	Landore	-	Stafford Road	10/62
5073	7/38	Laira	Shrewsbury	-	Cardiff E Dock	2/64
5074	7/38	Paddington	Bath Road	-	Bristol SPM	5/64
5075	8/38	Stafford Road	Stafford Road	-	Bristol SPM	9/62

No.	Date	First Shed	Last GWR Shed	Wdn GWR	Last BR Shed	Wdn BR
5076	8/38	Exeter	Bath Road	-	Southall	9/64
5077	8/38	Paddington	Taunton	-	Llanelly	7/62
5078	5/39	Laira	Newton Abbot	-	Neath	11/62
5079	5/39	Paddington	Newton Abbot	-	Newton Abbot	5/60
5080	5/39	Paddington	Cardiff	-	Llanelly	4/63
5081	5/39	Stafford Road	Paddington	-	Cardiff E Dock	10/63
5082	6/39	Bath Road	Bath Road	-	Paddington	7/62
5083	6/37	Landore	Bath Road	-	Worcester	1/59
5084	4/37	Weymouth	Bath Road	-	Paddington	7/62
5085	7/39	Paddington	Paddington	-	Bristol SPM	2/64
5086	12/37	Stafford Road	Shrewsbury	-	Worcester	11/58
5087	11/40	Paddington	Paddington	-	Llanelly	8/63
5088	2/39	Shrewsbury	Stafford Road	-	Stafford Road	9/62
5089	10/39	Landore	Landore	-	Oxley	11/64
5090	4/39	Laira	Laira	-	Paddington	5/62
5091	12/38	Bath Road	Bath Road	-	Tyseley	10/64
5092	4/38	Gloucester	Worcester	-	Cardiff E Dock	7/63
5093	6/39	Paddington	Landore	-	Paddington	9/63
5094	6/39	Newton Abbot	Newton Abbot	-	Bristol SPM	9/62
5095	6/39	Laira	Laira	-	Shrewsbury	8/62
5096	6/39	Bath Road	Weston-S-Mare	-	Worcester	6/64
5097	7/39	Shrewsbury	Shrewsbury	-	Cardiff E Dock	3/63
5098	5/46	Stafford Road	Exeter	-	Reading	6/64
5099	5/46	Paddington	Paddington	-	Gloucester	2/63
5101	11/29	Stourbridge	Stourbridge	-	Leamington	6/63
5102	11/29	Tyseley	Tyseley	-	Neath	3/60
5103	11/29	Tyseley	Tyseley	-	Pontypool	12/60
5104	11/29	Tyseley	Leamington	-	Bristol SPM	11/60
5105	11/29	Stourbridge	Stourbridge	-	Gloucester	10/58
5106	11/29	Stafford Road	Tyseley	-	Laira	3/60
5107	12/29	Bristol SPM	Stourbridge	-	Stourbridge	6/57
5108	12/29	Tyseley	Newton Abbot	-	Newton Abbot	7/58
5109	12/29	Stafford Road	Leamington	-	Tyseley	6/57
5110	12/29	Bristol SPM	Kidderminster	-	Severn Tunnel	12/60
5150	2/30	Chester	Newton Abbot	-	Newton Abbot	8/60
5151	2/30	Tyseley	Stafford Road	-	Kidderminster	8/62
5152	2/30	Stourbridge	Tyseley	-	Worcester	11/63
5153	3/30	Chester	Newton Abbot	-	Leamington	11/64
5154	2/30	Stafford Road	Shrewsbury	-	Hereford	8/63
5155	3/30	Stafford Road	Stourbridge	-	Stourbridge	1/60
5156	3/30	Stafford Road	Tyseley	-	Hereford	9/58
5157	3/30	Stourbridge	Newton Abbot	-	Cheltenham	7/58
5158	3/30	Bath Road	St. Blazey	-	Newton Abbot	4/61
5159	3/30	Pembroke Dock	Treherbert	-	Treherbert	4/56
5160	10/30	Leamington	Stourbridge	-	Hereford	11/58
5161	11/30	Banbury	Leamington	-	Leamington	4/57
5162	11/30	Tyseley	Tyseley	-	Gloucester	7/58
5163	11/30	Severn Tunnel	Leamington	-	Tyseley	11/59
5164	11/30	Tyseley	Tyseley	-	Pontypool	4/63
5165	11/30	Tyseley	Stourbridge	-	Gloucester	2/58
5166	11/30	Stourbridge	Tyseley	-	Severn Tunnel	5/61
5167	12/30	Stafford Road	Stourbridge	-	Banbury	1/62
5168	12/30	Reading	Shrewsbury	-	Newton Abbot	8/58
5169	12/30	Oxford	Bath Road	-	Severn Tunnel	9/60
5170	12/30	Stratford	Stourbridge	-	Banbury	12/59
5171	12/30	Tyseley	Tyseley	-	Tondu	7/58
5172	12/30	Stourbridge	Taunton	-	Taunton	10/58
5173	12/30	Stafford Road	Worcester	-	Gloucester	8/62
5174	12/30	Tyseley	Chester	-	Tyseley	11/61
5175	1/31	Stourbridge	Tyseley	-	Exeter	4/61

No.	Date	First Shed	Last GWR Shed	Wdn GWR	Last BR Shed	Wdn BR
5176	1/31	Tyseley	Chester	-	Stourbridge	1/61
5177	1/31	Tyseley	Tyseley	-	Leamington	5/61
5178	1/31	Tyseley	Wellington	-	Newton Abbot	3/60
5179	2/31	Paddington	Chester	-	Worcester	7/60
5180	2/31	Stourbridge	Stourbridge	-	Westbury	7/62
5181	3/31	Tyseley	Chester	-	Severn Tunnel	8/62
5182	3/31	Stourbridge	Tyseley	-	Gloucester	5/62
5183	3/31	Stourbridge	Severn Tunnel	-	Stafford Road	5/62
5184	3/31	Leamington	Chester	-	Gloucester	10/64
5185	3/31	Chester	Leamington	-	Taunton	3/60
5186	3/31	Leamington	Chester	-	Bath Road	8/59
5187	4/31	Bath Road	Leamington	-	Stourbridge	5/62
5188	4/31	Stafford Road	Tyseley	-	Severn Tunnel	7/62
5189	4/31	Stourbridge	Stourbridge	-	Stourbridge	8/59
5190	10/34	Tyseley	Tyseley	-	Exeter	9/62
5191	10/34	Stafford Road	Stourbridge	-	Severn Tunnel	6/64
5192	10/34	Leamington	Leamington	-	Stourbridge	6/63
5193	10/34	Stourbridge	Stourbridge	-	Severn Tunnel	6/62
5194	10/34	Tyseley	Leamington	-	Leamington	4/61
5195	10/34	Banbury	Barry	-	Newton Abbot	6/61
5196	10/34	Stourbridge	Stourbridge	-	Newton Abbot	12/59
5197	10/34	Stourbridge	Stourbridge	-	Newton Abbot	7/60
5198	11/34	Stafford Road	Tyseley	-	Taunton	6/61
5199	11/34	Tyseley	Tyseley	-	Gloucester	3/63
5200	12/22	Swansea E Dock	Cardiff	-	Ebbw Junction	3/65
5201	12/22	Ebbw Junction	Ebbw Junction	-	Llanelly	6/63
5202	12/22	Ebbw Junction	Tondu	-	Cardiff E Dock	6/65
5203	2/23	Ebbw Junction	Llanelly	-	Tondu	12/63
5204	2/23	Llanelly	Llanelly	-	Neath	11/62
5205	6/23	Ebbw Junction	Severn Tunnel	-	Worcester	12/63
5206	6/23	Ebbw Junction	Ebbw Junction	-	Pontypool	5/65
5207	7/23	Ebbw Junction	Carmarthen	-	Cardiff	1/61
5208	8/23	Ebbw Junction	Ebbw Junction	-	Cardiff E Dock	6/65
5209	8/23	Ebbw Junction	Llanelly	-	Ebbw Junction	7/65
5210	9/23	Neath	Swansea E Dock	-	Aberdare	9/64
5211	9/23	Ebbw Junction	Landore	-	Aberdare	5/64
5212	10/23	Ebbw Junction	Llanelly	-	Severn Tunnel	5/62
5213	10/23	Llanelly	Llanelly	-	Neath	10/64
5214	11/23	Landore	Swansea E Dock	-	Aberbeeg	9/64
5215	3/24	Duffryn Yard	Llanelly	-	Llanelly	7/64
5216	4/24	Duffryn Yard	Duffryn Yard	-	Neath	10/63
5217	4/24	Llanelly	Ebbw Junction	-	Cardiff E Dock	5/63
5218	4/24	Ebbw Junction	Ebbw Junction	-	Aberbeeg	9/64
5219	4/24	Neath	Landore	-	Llanelly	12/62
5220	5/24	Neath	Pantyffynon	-	Cardiff E Dock	12/63
5221	5/24	Ebbw Junction	Swansea E Dock	-	Neath	10/63
5222	5/24	Landore	Ebbw Junction	-	Cardiff E Dock	5/64
5223	5/24	Duffryn Yard	Llanelly	-	Ebbw Junction	3/65
5224	5/24	Ebbw Junction	Ebbw Junction	-	Cardiff E Dock	4/63
5225	5/24	Ebbw Junction	Neath	-	Cardiff E Dock	8/63
5226	5/24	Duffryn Yard	Llanelly	-	Severn Tunnel	3/65
5227	6/24	Ebbw Junction	Swansea E Dock	-	Cardiff E Dock	2/63
5228	6/24	Duffryn Yard	Llanelly	-	Llantrisant	4/64
5229	6/24	Llantrisant	Ebbw Junction	-	Cardiff E Dock	5/63
5230	6/24	Llanelly	Llanelly	-	Aberdare	7/64
5231	6/24	Aberdare	Carmarthen	-	Cardiff E Dock	4/64
5232	7/24	Llanelly	Swansea E Dock	-	Duffryn Yard	1/63
5233	7/24	Aberdare	Ebbw Junction	-	Cardiff E Dock	8/63
5234	7/24	Westbury	Ebbw Junction	-	Cardiff E Dock	6/63
5235	7/24	Neath	Pill	-	Severn Tunnel	9/65

No.	Date	First Shed	Last GWR Shed	Wdn GWR	Last BR Shed	Wdn BR
5236	8/24	Aberdare	Cardiff	-	Ebbw Junction	10/63
5237	8/24	Swansea E Dock	Aberdare	-	Aberdare	9/64
5238	8/24	Aberdare	Ebbw Junction	-	Neath	1/64
5239	8/24	Neath	Neath	-	Neath	4/63
5240	8/24	Aberdare	Pantyffynon	-	Aberdare	2/64
5241	8/24	Aberdare	Bristol SPM	-	Ebbw Junction	6/65
5242	8/24	Westbury	Neath	-	Aberdare	12/64
5243	8/24	Laira	Ebbw Junction	-	Severn Tunnel	11/64
5244	9/24	Truro	Pill	-	Ebbw Junction	5/64
5245	11/25	Neath	Aberdare	-	Neath	10/64
5246	12/25	Landore	Swansea E Dock	-	Duffryn Yard	12/63
5247	12/25	Llanelly	Llanelly	-	Llanelly	3/63
5248	12/25	Ebbw Junction	Pantyffynon	-	Llantrisant	10/63
5249	12/25	Ebbw Junction	Cardiff	-	Aberdare	10/63
5250	12/25	Llanelly	Duffryn Yard	-	Ebbw Junction	12/63
5251	12/25	Duffryn Yard	Ebbw Junction	-	Ebbw Junction	1/64
5252	12/25	Ebbw Junction	Pill	-	Ebbw Junction	5/65
5253	12/25	Neath	Swansea E Dock	-	Severn Tunnel	4/63
5254	12/25	Llanelly	Neath	-	Neath	7/64
5255	12/25	Ebbw Junction	Aberdare	-	-	-
5256	12/25	Ebbw Junction	Aberdare	-	-	-
5257	1/26	Cardiff	Cardiff	-	-	-
5258	1/26	Neath	Duffryn Yard	-	-	-
5259	1/26	Ebbw Junction	Ebbw Junction	-	-	-
5260	3/26	Ebbw Junction	Pill	-	-	-
5261	3/26	Dock Street	Ebbw Junction	-	-	-
5262	3/26	Swansea E Dock	Swansea E Dock	-	-	-
5263	3/26	Glyn Neath	Neath	-	-	-
5264	3/26	Cardiff	Cathays (stored from 8/30 until converted)			
5265	3/26	Cardiff	Llanelly	-	-	-
5266	3/26	Ebbw Junction	Ebbw Junction	-	-	-
5267	3/26	Ebbw Junction	Ebbw Junction	-	-	-
5268	3/26	Duffryn Yard	Neath	-	-	-
5269	3/26	Duffryn Yard	Swansea E Dock	-	-	-
5270	4/26	Landore	Llanelly	-	-	-
5271	4/26	Aberdare	Aberdare	-	-	-
5272	4/26	Landore	Swansea E Dock	-	-	-
5273	4/26	Swansea E Dock	Cardiff	-	-	-
5274	4/26	Ebbw Junction	Duffryn Yard	-	-	-
5275	7/30	STOCK	STOCK	-	-	-
5276	8/30	STOCK	STOCK	-	-	-
5277	8/30	STOCK	STOCK	-	-	-
5278	8/30	STOCK	STOCK	-	-	-
5279	8/30	STOCK	STOCK	-	-	-
5280	8/30	STOCK	STOCK	-	-	-
5281	8/30	STOCK	STOCK	-	-	-
5282	8/30	STOCK	STOCK	-	-	-
5283	9/30	STOCK	STOCK	-	-	-
5284	9/30	STOCK	STOCK	-	-	-
5285	9/30	STOCK	STOCK	-	-	-
5286	9/30	STOCK	STOCK	-	-	-
5287	9/30	STOCK	STOCK	-	-	-
5288	9/30	STOCK	STOCK	-	-	-
5289	10/30	STOCK	STOCK	-	-	-
5290	10/30	STOCK	STOCK	-	-	-
5291	10/30	STOCK	STOCK	-	-	-
5292	10/30	STOCK	STOCK	-	-	-
5293	10/30	STOCK	STOCK	-	-	-
5294	10/30	STOCK	STOCK	-	-	-
5255	1/40	Ebbw Junction	Ebbw Junction	-	Ebbw Junction	5/63

No.	Date	First Shed	Last GWR Shed	Wdn GWR	Last BR Shed	Wdn BR
5256	2/40	Ebbw Junction	Ebbw Junction	-	Ebbw Junction	3/65
5257	2/40	Duffryn Yard	Duffryn Yard	-	Neath	10/64
5258	2/40	Aberdare	Aberdare	-	Aberdare	12/62
5259	2/40	Aberdare	Aberbeeg	-	Ebbw Junction	3/64
5260	2/40	Duffryn Yard	Duffryn Yard	-	Aberbeeg	3/63
5261	2/40	Llanelly	Llanelly	-	Cardiff E Dock	3/65
5262	3/40	Cardiff	Severn Tunnel	-	Llanelly	8/63
5263	3/40	Aberdare	Aberdare	-	Aberdare	12/63
5264	3/40	Ebbw Junction	Ebbw Junction	-	Neath	9/64
5400	8/30	Cathays	Cardiff	6/32	-	-
5400	6/32	Southall	Landore	-	Oswestry	4/59
5401	11/31	Westbury	Southall	-	Oswestry	2/57
5402	11/31	Frome	Westbury	-	Westbury	9/58
5403	12/31	Frome	Frome	-	Westbury	8/57
5404	12/31	Banbury	Banbury	-	Banbury	12/57
5405	12/31	Banbury	Southall	-	Oswestry	10/57
5406	12/31	Gloucester	Frome	-	Westbury	9/57
5407	1/32	Stafford Road	Banbury	-	Banbury	6/60
5408	1/32	Worcester	Landore	-	Lydney	12/56
5409	1/32	Southall	Southall	-	Banbury	6/59
5410	1/32	Southall	Southall	-	Westbury	10/63
5411	5/32	Southall	Southall	-	Taunton	6/58
5412	5/32	Southall	Laira	-	Exeter	4/62
5413	5/32	Southall	Oxford	-	Gloucester	10/57
5414	5/32	Southall	Southall	-	Westbury	10/59
5415	5/32	Southall	Southall	-	Southall	7/57
5416	6/32	Southall	Southall	-	Yeovil Town	8/63
5417	6/32	Southall	Southall	-	Banbury	1/61
5418	6/32	Southall	Southall	-	Cheltenham	6/60
5419	6/32	Southall	Westbury	-	Westbury	2/58
5420	11/35	Southall	Southall	-	Lydney	10/63
5421	11/35	Southall	Marlow	-	Oswestry	9/62
5422	11/35	Southall	Westbury	-	Oswestry	6/60
5423	11/35	Westbury	Westbury	-	Westbury	6/59
5424	12/35	Stafford Road	Banbury	-	Banbury	4/59
5500	4/27	Paddington	Truro	-	Truro	10/59
5501	5/27	Laira	Taunton	-	Taunton	7/58
5502	5/27	Newton Abbot	Moorswater	-	St. Blazey	7/58
5503	5/27	Newton Abbot	Taunton	-	Taunton	5/61
5504	5/27	Truro	Taunton	-	Westbury	10/60
5505	10/27	Kidderminster	Newton Abbot	-	Truro	5/27
5506	10/27	Kidderminster	Wells	-	Laira	5/58
5507	10/27	Kidderminster	Pwllheli	-	Machynlleth	8/58
5508	10/27	Kidderminster	Westbury	-	Southall	12/64
5509	10/27	Bath Road	Westbury	-	Truro	12/61
5510	11/27	Bath Road	Swindon	-	Laira	10/60
5511	11/27	Bath Road	Bath Road	-	Laira	12/61
5512	11/27	Bath Road	Bath Road	-	Bath Road	2/57
5513	11/27	Bath Road	Pembroke Dock	-	Whitland	6/57
5514	11/27	Bath Road	Bath Road	-	Truro	11/60
5515	11/27	Stourbridge	Cheltenham	-	Laira	11/61
5516	11/27	Tyseley	Ebbw Junction	-	Taunton	8/61
5517	12/27	Stourbridge	Portmadoc	-	Machynlleth	12/58
5518	11/27	Tyseley	Kidderminster	-	Gloucester	5/64
5519	12/27	Stourbridge	St. Blazey	-	Laira	6/60
5520	12/27	Stafford Road	Aberdare	-	Whitland	9/62
5521	12/27	Newton Abbot	Taunton	-	Laira	4/62
5522	12/27	Newton Abbot	Taunton	-	Taunton	3/59
5523	12/27	Bath Road	Bath Road	-	St. Blazey	6/60
5524	12/27	Stourbridge	Pwllheli	-	Exeter	6/60

No.	Date	First Shed	Last GWR Shed	Wdn GWR	Last BR Shed	Wdn BR
5525	5/28	St. Blazey	Exeter	-	Newton Abbot	9/62
5526	5/28	St. Blazey	Truro	-	Westbury	6/62
5527	5/28	Bath Road	Bath Road	-	Whitland	6/60
5528	5/28	Bath Road	Bath Road	-	Swindon	11/59
5529	5/28	Bath Road	Yeovil	-	Bath Road	8/60
5530	6/28	Stourbridge	Newton Abbot	-	Bath Road	1/60
5531	6/28	Tyseley	St. Blazey	-	Southall	12/64
5532	6/28	Stourbridge	Pontypool	-	Laira	7/62
5533	6/28	Stafford Road	Taunton	-	Truro	12/59
5534	6/28	Bath Road	Swindon	-	St. Blazey	9/60
5535	6/28	Cheltenham	Bath Road	-	Bath Road	6/57
5536	7/28	Westbury	Bath Road	-	Swindon	12/60
5537	7/28	Cheltenham	Truro	-	Penzance	8/62
5538	7/28	Cheltenham	Cheltenham	-	Newton Abbot	10/61
5539	7/28	Cheltenham	Bath Road	-	St. Blazey	4/62
5540	7/28	Gloucester	Laira	-	Oswestry	8/60
5541	8/28	Bath Road	Machynlleth	-	Laira	7/62
5542	7/28	Gloucester	Taunton	-	Westbury	12/61
5543	7/28	Cheltenham	Taunton	-	Taunton	7/60
5544	8/28	Stourbridge	Worcester	-	Laira	9/62
5545	10/28	Ebbw Junction	Ebbw Junction	-	Gloucester	11/64
5546	10/28	Tyseley	Bath Road	-	Truro	9/60
5547	10/28	Stafford Road	Bath Road	-	Swindon	2/62
5548	10/28	Tyseley	Bath Road	-	Yeovil Town	6/63
5549	10/28	Stafford Road	Whitland	-	Whitland	1/62
5550	10/28	Ebbw Junction	Ebbw Junction	-	Whitland	9/62
5551	10/28	Tyseley	Kingsbridge	-	St. Blazey	1/60
5552	11/28	Tyseley	Newton Abbot	-	Truro	10/60
5553	11/28	Bath Road	Bath Road	-	St. Blazey	11/61
5554	11/28	Bath Road	Westbury	-	Whitland	8/63
5555	11/28	Bath Road	Bath Road	-	Exeter	7/63
5556	11/28	Frome	Tondu	-	Machynlleth	12/59
5557	11/28	Bath Road	Newton Abbot	-	St. Blazey	10/60
5558	11/28	Swindon	Bath Road	-	Newton Abbot	10/60
5559	11/28	Swindon	Bath Road	-	Truro	1/60
5560	12/28	Bath Road	Aberystwyth	-	Laira	4/62
5561	12/28	Bath Road	Bath Road	-	Whitland	7/60
5562	12/28	Bath Road	Truro	-	Penzance	9/62
5563	12/28	Bath Road	Swindon	-	Yeovil	9/64
5564	12/28	Bath Road	Bath Road	-	Southall	12/64
5565	1/29	Swindon	Yeovil	-	Machynlleth	9/60
5566	1/29	Frome	Swindon	-	Swindon	1/59
5567	1/29	Bath Road	Laira	-	Truro	1/60
5568	1/29	Stratford	Whitland	-	Exmouth Jct	1/63
5569	1/29	Stratford	Laira	-	Southall	12/64
5570	2/29	Evesham	Aberystwyth	-	Southall	12/63
5571	2/29	Exeter	Taunton	-	Southall	10/64
5572	2/29	Exeter	Bath Road	-	Laira	4/62
5573	2/29	Taunton	Kidderminster	-	Southall	1/64
5574	2/29	Worcester	Cheltenham	-	Tondu	12/58
5600	12/24	Cathays	Ferndale	-	Radyr	7/62
5601	12/24	Cardiff Docks	Treherbert	-	Rhymney	1/65
5602	12/24	Cardiff Docks	Ebbw Junction	-	Radyr	9/64
5603	12/24	Cardiff Docks	Ebbw Junction	-	Rhymney	9/64
5604	1/25	Cardiff Docks	Landore	-	Llanelly	11/62
5605	1/25	Barry	Cathays	-	Croes Newydd	5/66
5606	2/25	Trehafod	Oxley	-	Tyseley	11/65
5607	2/25	Trehafod	Treherbert	-	Treherbert	12/63
5608	2/25	Abercynon	Treherbert	-	Treherbert	8/63
5609	2/25	Barry	Barry	-	Merthyr	10/64

No.	Date	First Shed	Last GWR Shed	Wdn GWR	Last BR Shed	Wdn BR
5610	2/25	Barry	Ferndale	-	Rhymney	12/63
5611	2/25	Cathays	Treherbert	-	Treherbert	1/63
5612	2/25	Cathays	Duffryn Yard	-	Llanelly	4/63
5613	2/25	Cathays	Treherbert	-	Radyr	5/65
5614	2/25	Cathays	Cathays	-	Abercynon	8/63
5615	3/25	Cathays	Ferndale	-	Rhymney	7/63
5616	3/25	Cathays	Oxford	-	Swansea E Dock	9/63
5617	3/25	Cathays	Cathays	-	Abercynon	9/62
5618	3/25	Ebbw Junction	Abercynon	-	Rhymney	3/65
5619	3/25	Chester	Abercynon	-	Barry	6/64
5620	6/25	Cathays	Severn Tunnel	-	Aberdare	7/63
5621	6/25	Cathays	Barry	-	Rhymney	6/65
5622	6/25	Cathays	Barry	-	Rhymney	6/63
5623	6/25	Cathays	Cathays	-	Radyr	2/64
5624	6/25	Cathays	Cathays	-	Radyr	6/64
5625	8/25	Cathays	Severn Tunnel	-	Radyr	10/63
5626	8/25	Cathays	Severn Tunnel	-	Severn Tunnel	12/63
5627	8/25	Cathays	Barry	-	Abercynon	4/63
5628	8/25	Cathays	Cardiff	-	Swansea E Dock	2/63
5629	8/25	Cathays	Duffryn Yard	-	Abercynon	5/64
5630	8/25	Cathays	Abercynon	-	Abercynon	12/62
5631	8/25	Cathays	Landore	-	Swansea E Dock	9/62
5632	9/25	Cathays	Barry	-	Abercynon	9/64
5633	9/25	Cathays	Tondu	-	Radyr	5/65
5634	9/25	Cathays	Leamington	-	Whitland	7/64
5635	9/25	Cathays	Rhymney	-	Radyr	5/64
5636	9/25	Cathays	Treherbert	-	Cae Harris	5/62
5637	9/25	Cathays	Abercynon	-	Barry	6/64
5638	9/25	Cathays	Ebbw Junction	-	Pontypool	2/64
5639	9/25	Cathays	Duffryn Yard	-	Barry	5/62
5640	10/25	Cathays	Radyr	-	Leamington	6/63
5641	10/25	Coke Ovens	Abercynon	-	Abercynon	9/64
5642	10/25	Cathays	Shrewsbury	-	Aberdare	9/62
5643	10/25	Cathays	Abercynon	-	Barry	7/63
5644	11/25	Cathays	Abercynon	-	Abercynon	6/63
5645	11/25	Cathays	Severn Tunnel	-	Rhymney	6/63
5646	11/25	Cathays	Cathays	-	Treherbert	9/62
5647	11/25	Cathays	Cathays	-	Aberdare	3/64
5648	11/25	Cathays	Barry	-	Radyr	9/64
5649	11/25	Cathays	Pontypool	-	Aberdare	3/63
5650	1/26	Cathays	Treherbert	-	Rhymney	6/63
5651	1/26	Cathays	Cathays	-	Rhymney	12/64
5652	1/26	Cathays	Cae Harris	-	Cae Harris	9/62
5653	1/26	Cathays	Cae Harris	-	Barry	1/63
5654	1/26	Cathays	Merthyr	-	Radyr	12/63
5655	1/26	Cathays	Radyr	-	Rhymney	6/65
5656	2/26	Cathays	Duffryn Yard	-	Swansea E Dock	2/64
5657	2/26	Cathays	Oxley	-	Abercynon	7/62
5658	2/26	Cathays	Cathays	-	Tyseley	11/65
5659	2/26	Cathays	Cathays	-	Croes Newydd	11/65
5660	2/26	Cathays	Rhymney	-	Radyr	10/64
5661	2/26	Cathays	Cathays	-	Merthyr	7/62
5662	2/26	Abercynon	Barry	-	Rhymney	11/64
5663	2/26	Cathays	Treherbert	-	Rhymney	8/62
5664	2/26	Cathays	Barry	-	Barry	8/62
5665	6/26	Cathays	Barry	-	Pontypool	6/65
5666	6/26	Cathays	Cae Harris	-	Rhymney	7/63
5667	6/26	Cathays	Barry	-	Croes Newydd	7/65
5668	6/26	Cathays	Ferndale	-	Barry	9/64
5669	6/26	Cathays	Barry	-	Radyr	9/64

No.	Date	First Shed	Last GWR Shed	Wdn GWR	Last BR Shed	Wdn BR
5670	8/26	Ebbw Junction	Oxley	-	Rhymney	10/64
5671	8/26	Ebbw Junction	Cae Harris	-	Cae Harris	1/64
5672	8/26	Ebbw Junction	Cathays	-	Radyr	9/63
5673	8/26	Ebbw Junction	Shrewsbury	-	Radyr	3/65
5674	8/26	Treherbert	Cae Harris	-	Treherbert	4/64
5675	9/26	Coke Ovens	Pantyffynon	-	Llanelly	12/64
5676	9/26	Cathays	Treherbert	-	Croes Newydd	11/65
5677	9/26	Cardiff	Merthyr	-	Croes Newydd	11/65
5678	9/26	Cardiff	Rhymney	-	Treherbert	1/64
5679	9/26	Cardiff	Cardiff	-	Wellington	7/63
5680	9/26	Barry	Treherbert	-	Aberdare	12/63
5681	10/26	Barry	Cathays	-	Rhymney	5/65
5682	10/26	Barry	Abercynon	-	Abercynon	5/62
5683	10/26	Cathays	Rhymney	-	Cardiff E Dock	3/64
5684	10/26	Neath	Oxley	-	Tyseley	7/65
5685	1/27	Cardiff	Cardiff	-	Abercynon	2/64
5686	1/27	Cathays	Abercynon	-	Rhymney	3/65
5687	1/27	Barry	Cathays	-	Rhymney	12/63
5688	1/27	Cathays	Treherbert	-	Rhymney	6/65
5689	1/27	Cathays	Westbury	-	Radyr	5/65
5690	1/27	Cathays	Chester	-	Tondu	8/63
5691	1/27	Cathays	Treherbert	-	Radyr	6/65
5692	1/27	Cathays	Rhymney	-	Radyr	7/65
5693	1/27	Swindon	Barry	-	Treherbert	1/63
5694	1/27	Cathays	Cae Harris	-	Radyr	11/64
5695	1/27	Cathays	Treherbert	-	Treherbert	12/62
5696	1/27	Cathays	Rhymney	-	Rhymney	5/65
5697	1/27	Cathays	Gloucester	-	Radyr	4/63
5698	1/27	Cathays	Merthyr	-	Aberdare	9/62
5699	1/27	Cathays	Barry	-	Radyr	11/64
5700	1/29	Banbury	Tyseley	-	Oswestry	3/56
5701	1/29	Tyseley	Tyseley	-	Westbury	1/58
5702	1/29	Duffryn Yard	Llanelly	-	Llanelly	5/60
5703	1/29	Neath	Neath	-	Pantyffynon	2/59
5704	1/29	Glyn Neath	Swansea E Dock	-	Danygraig	5/60
5705	1/29	Llanelly	Llanelly	-	Llanelly	8/59
5706	2/29	Ebbw Junction	Severn Tunnel	-	Aberbeeg	9/61
5707	2/29	Cardiff	Tondu	-	Tondu	3/59
5708	2/29	Tondu	Llantrisant	-	Llantrisant	4/59
5709	2/29	Cardiff	Ebbw Junction	-	Ebbw Junction	3/60
5710	2/29	Cardiff	Didcot	-	Cathays	8/57
5711	2/29	Merthyr	Merthyr	-	Merthyr	1/58
5712	2/29	Banbury	Tyseley	-	Wellington	10/57
5713	2/29	Bristol SPM	Duffryn Yard	-	Goodwick	1/60
5714	3/29	Bristol SPM	Severn Tunnel	-	Pill	1/58
5715	3/29	Paddington	Slough	-	Slough	8/58
5716	3/29	Bristol SPM	Goodwick	-	Goodwick	3/58
5717	3/29	Paddington	Paddington	-	Paddington	5/60
5718	3/29	Bristol SPM	Frome	-	Westbury	7/58
5719	3/29	Oxley	Stourbridge	-	Chester	11/58
5720	3/29	Danygraig	Neath	-	Neath	1/62
5721	3/29	Merthyr	Merthyr	-	Taunton	8/59
5722	3/29	Llanelly	Llanelly	-	Llanelly	9/58
5723	3/29	Tyseley	Chester	-	Chester	11/57
5724	3/29	Tyseley	Banbury	-	Cathays	8/57
5725	1/29	Leamington	Chester	-	Chester	9/58
5726	1/29	Tyseley	Stourbridge	-	Oswestry	10/59
5727	2/29	Southall	Southall	-	Cardiff	5/60
5728	2/29	Branches Fork	Pontypool	-	Duffryn Yard	5/62
5729	2/29	Aberdare	Severn Tunnel	-	Aberbeeg	10/57

No.	Date	First Shed	Last GWR Shed	Wdn GWR	Last BR Shed	Wdn BR
5730	2/29	Neath	Danygraig	-	Danygraig	3/58
5731	2/29	Duffryn Yard	Duffryn Yard	-	Aberdare	5/60
5732	2/29	Ebbw Junction	Ebbw Junction	-	Ebbw Junction	1/58
5733	2/29	Ebbw Junction	Aberbeeg	-	Pill	8/58
5734	3/29	Duffryn Yard	Duffryn Yard	-	Pill	5/59
5735	3/29	Neath	Didcot	-	Didcot	11/57
5736	3/29	Oxley	Banbury	-	Pill	1/58
5737	3/29	Paddington	Slough	-	Didcot	10/59
5738	3/29	Oxley	Tyseley	-	Duffryn Yard	9/59
5739	3/29	Tyseley	Wellington	-	Chester	9/58
5740	3/29	Llantrisant	Pill	-	Pill	6/59
5741	3/29	Aberbeeg	Ebbw Junction	-	Pill	6/57
5742	3/29	Tyseley	Tyseley	-	Croes Newydd	9/58
5743	3/29	Swansea E Dock	Swansea E Dock	-	Danygraig	1/59
5744	4/29	Paddington	Didcot	-	Westbury	4/62
5745	4/29	Paddington	Tyseley	-	Wellington	11/59
5746	4/29	Swansea E Dock	Neath N&B	-	Didcot	9/62
5747	4/29	Merthyr	Pill	-	Pill	9/59
5748	4/29	Ebbw Junction	Oxley	-	Neyland	9/60
5749	4/29	Ebbw Junction	Cardiff	-	Cardiff E Dock	7/63
5750	4/29	Neath	Southall	-	Pontypool	5/60
5751	4/29	Paddington	Reading	-	Taunton	4/58
5752	4/29	Paddington	Didcot	-	Didcot	2/57
5753	4/29	Paddington	Southall	-	Southall	9/59
5754	4/29	Paddington	Stourbridge	-	Stourbridge	6/60
5755	4/29	Cardiff E Dock	Southall	-	Slough	7/60
5756	5/29	Ebbw Junction	Tondu	-	Pontypool	9/61
5757	5/29	Paddington	Westbury	-	Westbury	12/60
5758	5/29	Paddington	Wellington	-	Pill	5/62
5759	5/29	Paddington	Landore	-	Pontypool	9/60
5760	5/29	Paddington	Exeter	-	Taunton	10/57
5761	6/29	Paddington	Reading	-	Neath	5/62
5762	6/29	Paddington	Reading	-	Reading	3/56
5763	6/29	Paddington	Reading	-	Lydney	5/60
5764	6/29	Paddington	Paddington	-	Paddington	5/60
5765	6/29	Paddington	Hereford	-	Hereford	3/59
5766	7/29	Paddington	Reading	-	Slough	5/62
5767	7/29	Southall	Yeovil	-	Westbury	9/58
5768	7/29	Ebbw Junction	Pontypool	-	Pill	3/61
5769	7/29	Ebbw Junction	Merthyr	-	Bristol SPM	5/60
5770	9/29	Bristol SPM	Aberdare	-	Duffryn Yard	12/61
5771	9/29	Bristol SPM	Westbury	-	Bristol SPM	3/61
5772	9/29	Paddington	Reading	-	Ebbw Junction	6/58
5773	9/29	Paddington	Duffryn Yard	-	Neath	9/62
5774	9/29	Swindon	Shrewsbury	-	Croes Newydd	10/62
5775	9/29	Neath	Danygraig	-	Pontypool	7/63
5776	9/29	Ebbw Junction	Pill	-	Cardiff	5/60
5777	9/29	Ebbw Junction	Llantrisant	-	Pill	6/58
5778	9/29	Duffryn Yard	Neath N&B	-	Neath	7/62
5779	9/29	Paddington	Truro	-	Bristol SPM	5/62
5780	12/29	Chester	Oxley	-	Taunton	10/61
5781	12/29	Bristol SPM	Westbury	-	Yeovil	10/58
5782	12/29	Llanelly	Llanelly	-	Llanelly	11/58
5783	12/29	Ebbw Junction	Slough	-	Swansea E Dock	3/62
5784	12/29	Bristol SPM	Bristol SPM	-	Weymouth	6/59
5785	1/30	Bath Road	Frome	-	Aberbeeg	12/58
5786	1/30	Aberdare	Aberbeeg	-	Cardiff	4/58
5787	1/30	Severn Tunnel	Aberdare	-	Duffryn Yard	10/63
5788	1/30	Ebbw Junction	Aberbeeg	-	Llantrisant	10/59
5789	1/30	Ebbw Junction	Aberbeeg	-	Pontypool	5/62

No.	Date	First Shed	Last GWR Shed	Wdn GWR	Last BR Shed	Wdn BR
5790	9/30	Stourbridge	Tyseley	-	Tredegar	3/59
5791	9/30	Stourbridge	Chester	-	Kidderminster	4/61
5792	10/30	Tondu	Pontypool	-	Pontypool	10/56
5793	10/30	Gloucester	Gloucester	-	Taunton	4/62
5794	10/30	Stourbridge	Stourbridge	-	Aberbeeg	12/59
5795	10/30	Stourbridge	Stourbridge	-	Stourbridge	4/60
5796	11/30	Llantrisant	Aberdare	-	Didcot	3/59
5797	11/30	Tondu	Tondu	-	Aberdare	9/58
5798	11/30	Paddington	Newton Abbot	-	Taunton	9/62
5799	11/30	Paddington	Southall	-	Southall	7/59
5800	1/33	Swindon	Swindon	-	Swindon	7/58
5801	1/33	Swindon	Brecon	-	Machynlleth	9/58
5802	1/33	Swindon	Swindon	-	Swindon	12/58
5803	1/33	Swindon	Bath Road	-	Penmaenpool	7/57
5804	1/33	Swindon	Swindon	-	Swindon	6/59
5805	1/33	Swindon	Tetbury	-	Swindon	3/58
5806	1/33	Swindon	Llanfyllin	-	Oswestry	6/57
5807	1/33	Swindon	Leominster	-	Leominster	6/57
5808	1/33	Swindon	Kington	-	Oxford	2/57
5809	2/33	Swindon	Bath Road	-	Machynlleth	8/59
5810	8/33	Llanelly	Bala	-	Croes Newydd	1/59
5811	8/33	Wellington	Bala	-	Oxford	5/57
5812	8/33	Tiverton Jct	Taunton	-	Oswestry	6/57
5813	8/33	Ludlow	Bath Road	-	Leamington	11/57
5814	8/33	Worcester	Kington	-	Kington	6/57
5815	8/33	Worcester	Worcester	-	Swindon	4/61
5816	8/33	Llanfyllin	Worcester	-	Oxford	7/57
5817	8/33	Ebbw Junction	Leominster	-	Leominster	6/57
5818	8/33	Ebbw Junction	Pontrilas	-	Oxford	9/59
5819	8/33	Llandovery	N'castle Emlyn	-	Carmarthen	6/57
5900	3/31	Paddington	Westbury	-	Bristol SPM	12/63
5901	5/31	Truro	Reading	-	Reading	6/64
5902	5/31	Truro	Exeter	-	Bristol SPM	11/62
5903	5/31	Chester	Didcot	-	Llanelly	9/63
5904	5/31	Chester	Oxford	-	Bristol SPM	11/63
5905	5/31	Westbury	Goodwick	-	Goodwick	7/63
5906	5/31	Bristol SPM	Ebbw Junction	-	Reading	5/62
5907	5/31	Truro	Tyseley	-	Reading	11/61
5908	6/31	Swindon	Llanelly	-	Bristol SPM	7/63
5909	6/31	Bristol SPM	Stafford Road	-	Cardiff	7/62
5910	6/31	Oxford	Cardiff	-	Oxley	9/62
5911	6/31	Carmarthen	Ebbw Junction	-	Cardiff	9/62
5912	6/31	Oxford	Chester	-	Banbury	12/62
5913	6/31	Landore	Landore	-	Exeter	5/62
5914	7/31	Paddington	Worcester	-	Gloucester	1/64
5915	7/31	Shrewsbury	Penzance	-	Reading	1/60
5916	7/31	Shrewsbury	Oxley	-	Oxley	7/62
5917	7/31	Bath Road	Worcester	-	Southall	9/62
5918	7/31	Bath Road	Oxley	-	Oxford	9/62
5919	7/31	Bath Road	Stafford Road	-	Paddington	8/63
5920	8/31	Leamington	Oxley	-	Westbury	1/62
5921	5/33	Reading	Oxley	-	Westbury	1/62
5922	5/33	Paddington	Paddington	-	Oxford	1/64
5923	5/33	Oxford	Chester	-	Oxford	12/63
5924	5/33	Reading	Westbury	-	Bristol SPM	12/63
5925	5/33	Paddington	Westbury	-	Westbury	10/62
5926	6/33	Paddington	St. Blazey	-	Banbury	9/62
5927	6/33	Paddington	Stafford Road	-	Tyseley	10/64
5928	6/33	Chester	Goodwick	-	Goodwick	5/62
5929	6/33	Westbury	Neyland	-	Pontypool	10/63

No.	Date	First Shed	Last GWR Shed	Wdn GWR	Last BR Shed	Wdn BR
5930	6/33	Oxley	Banbury	-	Worcester	9/62
5931	6/33	Paddington	Paddington	-	Paddington	9/62
5932	6/33	Bath Road	Paddington	-	Barrow Road	10/65
5933	6/33	Bath Road	Reading	-	Oxford	8/65
5934	6/33	Shrewsbury	Swindon	-	Bath Road	5/64
5935	7/33	Stafford Road	Didcot	-	Cardiff	5/62
5936	7/33	Paddington	Paddington	-	Gloucester	1/65
5937	7/33	Newton Abbot	Paddington	-	Cardiff E Dock	11/63
5938	7/33	Oxford	Paddington	-	Cardiff E Dock	5/63
5939	7/33	Laira	Paddington	-	Ebbw Junction	10/64
5940	8/33	Stafford Road	Paddington	-	Bristol SPM	9/62
5941	2/35	Paddington	Paddington	-	Bristol SPM	7/62
5942	2/35	Paddington	Stafford Road	-	Tyseley	12/63
5943	3/35	Westbury	Swindon	-	Swindon	6/63
5944	3/35	Oxley	Stafford Road	-	Gloucester	4/63
5945	3/35	Oxley	Oxley	-	Oxford	4/63
5946	3/35	Cardiff	Cardiff	-	Goodwick	7/62
5947	3/35	Oxley	Oxley	-	Banbury	7/62
5948	3/35	Didcot	Reading	-	Pontypool	8/63
5949	4/35	Weymouth	Bath Road	-	Swindon	5/61
5950	4/35	Paddington	Tyseley	-	Westbury	11/61
5951	12/35	Gloucester	Gloucester	-	Gloucester	4/64
5952	12/35	Penzance	Paddington	-	Cardiff E Dock	6/64
5953	12/35	Penznace	Cardiff	-	Bristol SPM	10/62
5954	12/35	Paddington	Banbury	-	Bristol SPM	10/63
5955+	12/35	Tyseley	Bath Road	-	Barrow Road	4/65
5956	12/35	Worcester	Reading	-	Oxford	3/63
5957	12/35	Oxley	Oxley	-	Oxford	7/64
5958	1/36	Taunton	Cardiff	-	Bristol SPM	3/64
5959	1/36	Reading	Reading	-	Tyseley	9/62
5960	1/36	Oxford	Oxford	-	Oxford	9/62
5961	6/36	Stafford Road	Westbury	-	Ebbw Junction	8/65
5962	7/36	Paddington	Paddington	-	Worcester	11/64
5963	7/36	Landore	Carmarthen	-	Bristol SPM	6/64
5964	7/36	Swindon	Bristol SPM	-	Westbury	9/62
5965	8/36	Hereford	Gloucester	-	Tyseley	7/62
5966	3/37	Oxley	Chester	-	Oxford	9/62
5967	3/37	Chester	Banbury	-	Westbury	6/64
5968	3/37	Westbury	Weymouth	-	Gloucester	9/62
5969	4/37	Bath Road	Weymouth	-	Carmarthen	8/62
5970	4/37	Cardiff	Cardiff	-	Hereford	11/63
5971	4/37	Weymouth	Westbury	-	Oxford	12/65
5972	4/37	Neath	Cardiff	-	Cardiff E Dock	12/63
5973	5/37	Paddington	Reading	-	Reading	9/62
5974	4/37	Westbury	Westbury	-	Severn Tunnel	12/64
5975	5/37	Westbury	Pontypool	-	Barrow Road	7/64
5976+	9/38	Tyseley	Bristol SPM	-	Pontypool	7/64
5977	9/38	Worcester	Cardiff	-	Reading	8/63
5978	9/38	Paddington	Swindon	-	Bristol SPM	10/63
5979	9/38	Bath Road	Oxley	-	Worcester	11/64
5980	9/38	Gloucester	Gloucester	-	Gloucester	9/62
5981	10/38	Stafford Road	Shrewsbury	-	Neath	9/62
5982	10/38	Carmarthen	Taunton	-	Reading	9/62
5983	10/38	Worcester	Worcester	-	Tyseley	4/65
5984	10/38	Weymouth	Bristol SPM	-	Cardiff E Dock	1/65
5985	10/38	Paddington	Westbury	-	Oxford	9/63
5986+	11/39	Reading	Paddington	-	Westbury	9/63
5987	11/39	Paddington	Paddington	-	Oxford	1/64
5988	11/39	Hereford	Gloucester	-	Tyseley	10/65
5989	12/39	Oxley	Oxley	-	Neath	7/62

No.	Date	First Shed	Last GWR Shed	Wdn GWR	Last BR Shed	Wdn BR
5990	12/39	Gloucester	Gloucester	-	Banbury	1/65
5991	12/39	Banbury	Banbury	-	Shrewsbury	7/64
5992	12/39	Banbury	Banbury	-	Ebbw Junction	8/65
5993	12/39	Tyseley	Tyseley	-	Reading	5/63
5994	12/39	Shrewsbury	Shrewsbury	-	Shrewsbury	3/63
5995	1/40	Stafford Road	Stafford Road	-	Oxley	4/63
5996	6/40	Paddington	Paddington	-	Oxley	8/62
5997	6/40	Tyseley	Tyseley	-	Swindon	7/62
5998	6/40	Laira	Laira	-	Pontypool	3/64
5999	6/40	Newton Abbot	Taunton	-	Westbury	9/62
6000	6/27	Paddington	Laira	-	Paddington	12/62
6001	7/27	Paddington	Paddington	-	Stafford Road	9/62
6002	7/27	Laira	Laira	-	Stafford Road	9/62
6003	7/27	Paddington	Paddington	-	Cardiff	6/62
6004	7/27	Laira	Laira	-	Cardiff	6/62
6005	7/27	Paddington	Stafford Road	-	Paddington	11/62
6006	2/28	Laira	Stafford Road	-	Stafford Road	2/62
6007	3/28	Paddington	Paddington	3/36	-	-
6007	3/36	Paddington	Paddington	-	Stafford Road	9/62
6008	3/28	Laira	Stafford Road	-	Stafford Road	6/62
6009	3/28	Paddington	Paddington	-	Paddington	9/62
6010	4/28	Laira	Laira	-	Paddington	6/62
6011	4/28	Paddington	Stafford Road	-	Paddington	12/62
6012	4/28	Newton Abbot	Laira	-	Stafford Road	9/62
6013	5/28	Paddington	Paddington	-	Stafford Road	6/62
6014	5/28	Newton Abbot	Paddington	-	Stafford Road	9/62
6015	6/28	Paddington	Paddington	-	Stafford Road	9/62
6016	6/28	Laira	Laira	-	Stafford Road	9/62
6017	6/28	Paddington	Laira	-	Stafford Road	7/62
6018	6/28	Laira	Newton Abbot	-	Paddington	12/62
6019	7/28	Stafford Road	Laira	-	Paddington	9/62
6020	5/30	Laira	Laira	-	Stafford Road	7/62
6021	6/30	Paddington	Paddington	-	Paddington	9/62
6022	6/30	Laira	Laira	-	Stafford Road	9/62
6023	6/30	Newton Abbot	Newton Abbot	-	Cardiff	6/62
6024	6/30	Laira	Newton Abbot	-	Cardiff	6/62
6025	7/30	Paddington	Paddington	-	Paddington	12/62
6026	7/30	Paddington	Laira	-	Paddington	9/62
6027	7/30	Paddington	Newton Abbot	-	Stafford Road	9/62
6028	7/30	Paddington	Newton Abbot	-	Paddington	11/62
6029	8/30	Paddington	Laira	-	Paddington	7/62
6100	4/31	Paddington	Slough	-	Reading	9/58
6101	5/31	Southall	Slough	-	Tyseley	3/62
6102	5/31	Reading	Southall	-	Reading	8/59
6103	5/31	Paddington	Oxford	-	Reading	12/64
6104	5/31	Slough	Slough	-	Didcot	6/60
6105	5/31	Slough	Slough	-	Paddington	3/60
6106	5/31	Slough	Slough	-	Oxford	12/65
6107	5/31	Southall	Slough	-	Reading	11/64
6108	5/31	Slough	Slough	-	Oxford	8/65
6109	6/31	Reading	Reading	-	Didcot	8/62
6110	8/31	Slough	Southall	-	Oxford	6/65
6111	8/31	Paddington	Aylesbury	-	Oxford	12/65
6112	8/31	Southall	Paddington	-	Southall	9/65
6113	8/31	Paddington	Slough	-	Gloucester	11/65
6114	9/31	Reading	Slough	-	Severn Tunnel	10/64
6115	9/31	Slough	Reading	-	Severn Tunnel	11/64
6116	9/31	Slough	Slough	-	Radyr	6/65
6117	9/31	Slough	Paddington	-	Southall	9/65
6118	9/31	Paddington	Southall	-	Reading	11/63

No.	Date	First Shed	Last GWR Shed	Wdn GWR	Last BR Shed	Wdn BR
6119	9/31	Paddington	Slough	-	Reading	3/63
6120	9/31	Aylesbury	Paddington	-	Didcot	4/62
6121	10/31	Paddington	Paddington	-	Paddington	7/60
6122	10/31	Oxford	Oxford	-	Swindon	9/64
6123	10/31	Slough	Slough	-	Oxford	4/62
6124	10/31	Oxford	Slough	-	Radyr	6/64
6125	10/31	Slough	Southall	-	Southall	1/65
6126	10/31	Oxford	Slough	-	Oxford	12/65
6127	10/31	Paddington	Slough	-	Slough	3/62
6128	10/31	Southall	Southall	-	Gloucester	3/65
6129	11/31	Didcot	Aylesbury	-	Stourbridge	9/65
6130	9/32	Slough	Reading	-	Swindon	7/64
6131	10/32	Reading	Reading	-	Reading	9/64
6132	10/32	Reading	Paddington	-	Oxford	10/65
6133	10/32	Paddington	Slough	-	Southall	12/63
6134	10/32	Paddington	Paddington	-	Oxford	12/65
6135	10/32	Reading	Paddington	-	Oxford	12/65
6136	10/32	Slough	Reading	-	Oxford	12/65
6137	10/32	Paddington	Paddington	-	Stourbridge	11/64
6138	11/32	Reading	Oxford	-	Reading	9/63
6139	11/32	Paddington	Southall	-	Southall	11/64
6140	11/32	Slough	Reading	-	Worcester	7/64
6141	11/32	Paddington	Paddington	-	Gloucester	12/65
6142	11/32	Southall	Paddington	-	Southall	9/64
6143	11/32	Paddington	Slough	-	Oxford	11/65
6144	11/32	Paddington	Paddington	-	Aberdare	6/64
6145	12/32	Paddington	Slough	-	Oxford	12/65
6146	12/32	Paddington	Slough	-	Taunton	9/62
6147	1/33	Southall	Southall	-	Worcester	12/65
6148	1/33	Southall	Southall	-	Taunton	9/64
6149	1/33	Slough	Paddington	-	Neath	6/64
6150	2/33	Slough	Slough	-	Didcot	3/65
6151	2/33	Paddington	Slough	-	Pontypool	11/63
6152	2/33	Slough	Slough	-	Didcot	1/62
6153	2/33	Southall	Slough	-	Didcot	1/62
6154	3/33	Reading	Reading	-	Oxford	6/65
6155	3/33	Slough	Paddington	-	Worcester	10/65
6156	3/33	Slough	Southall	-	Oxford	12/65
6157	3/33	Paddington	Slough	-	Taunton	5/62
6158	3/33	Paddington	Paddington	-	Aberdare	6/64
6159	4/33	Southall	Paddington	-	Didcot	6/65
6160	10/35	Truro	Slough	-	Gloucester	12/65
6161	10/35	Penzance	Slough	-	Oxford	10/65
6162	10/35	Truro	Reading	-	Southall	3/62
6163	10/35	Truro	Reading	-	Southall	10/65
6164	10/35	Penzance	Slough	-	Reading	11/63
6165	10/35	Reading	Southall	-	Worcester	12/65
6166	10/35	Paddington	Paddington	-	Newton Abbot	1/62
6167	10/35	Slough	Slough	-	Southall	10/65
6168	11/35	Aylesbury	Paddington	-	Paddington	3/62
6169	11/35	Southall	Southall	-	Worcester	10/65
6342	3/23	Wolverhampton	Oxley	-	Shrewsbury	9/62
6343	3/23	Taunton	Taunton	-	Taunton	9/60
6344	3/23	Paddington	Carmarthen	-	Gloucester	11/63
6345	4/23	Worcester	Newton Abbot	-	Taunton	9/64
6346	4/23	Reading	Birkenhead	-	Taunton	9/64
6347	4/23	Bristol	Neyland	-	Llanelly	12/63
6348	4/23	Weymouth	Shrewsbury	-	Hereford	2/62
6349	5/23	Laira	Hereford	-	Gloucester	8/64
6350	5/23	Plymouth	Birkenhead	-	Didcot	1/64

No.	Date	First Shed	Last GWR Shed	Wdn GWR	Last BR Shed	Wdn BR
6351	5/23	Neyland	Westbury	-	Bristol SPM	11/60
6352	9/23	Reading	Hereford	-	Cardiff	11/60
6353	9/23	Wolverhampton	Cardiff	-	Westbury	5/63
6354	10/23	Reading	Penzance	-	Gloucester	2/59
6355	10/23	Paddington	Neyland	-	Whitland	3/59
6356	11/23	Reading	St. Blazey	-	Westbury	1/63
6357	11/23	Paddington	Swindon	-	Severn Tunnel	7/64
6358	11/23	Cardiff	Swindon	-	Westbury	10/59
6359	11/23	Bristol SPM	Didcot	-	Hereford	9/59
6360	11/23	Plymouth	Swindon	-	Bristol SPM	11/60
6361	12/23	Newton Abbot	Oxley	-	Aberdare	5/64
6362	4/25	Penzance	Oxley	-	Severn Tunnel	9/62
6363	4/25	Exeter	Reading	-	Taunton	9/64
6364	5/25	Oxley	Taunton	-	Stourbridge	11/64
6365	10/25	Paddington	Westbury	-	Gloucester	10/63
6366	10/25	Basingstoke	Reading	-	Cardiff	9/62
6367	10/25	Westbury	Carmarthen	-	Didcot	11/64
6368	10/25	Bath Road	Westbury	-	Gloucester	12/63
6369	10/25	Shrewsbury	Westbury	-	Severn Tunnel	2/63
6400	2/32	Aberdare	Pontypool	-	Yeovil Town	4/64
6401	2/32	Merthyr	Abercynon	-	Ebbw Junction	6/60
6402	2/32	Cardiff	Cathays	-	Laira	6/59
6403	3/32	Stafford Road	Bath Road	-	Stourbridge	10/63
6404	3/32	Stourbridge	Birkenhead	-	Oswestry	6/59
6405	3/32	Stourbridge	Birkenhead	-	Croes Newydd	6/59
6406	4/32	Laira	Laira	-	Laira	6/60
6407	4/32	Laira	Southall	-	Laira	8/58
6408	4/32	Laira	Merthyr	-	Tondu	2/62
6409	4/32	Laira	Ebbw Junction	-	Ebbw Junction	3/59
6410	11/34	Aberdare	Aberdare	-	Tondu	11/62
6411	11/34	Cathays	Abercynon	-	Radyr	3/61
6412	11/34	Landore	Landore	-	Gloucester	11/64
6413	11/34	Cardiff	Aberdare	-	Laira	11/61
6414	11/34	Laira	Laira	-	Laira	6/59
6415	11/34	Ebbw Junction	Ebbw Junction	-	Lydney	11/61
6416	11/34	Llantrisant	Cathays	-	Merthyr	9/63
6417	12/34	Laira	Laira	-	Aberdare	6/59
6418	12/34	Stourbridge	Stafford Road	-	Stourbridge	11/62
6419	12/34	Laira	Laira	-	Yeovil Town	12/64
6420	8/35	Laira	St. Blazey	-	Laira	11/59
6421	8/35	Laira	Laira	-	Laira	1/63
6422	8/35	Stafford Road	Stafford Road	-	Stourbridge	9/62
6423	9/35	Cathays	Cathays	-	Merthyr	8/58
6424	9/35	Ebbw Junction	Pontypool	-	Stourbridge	9/64
6425	11/35	Kidderminster	Landore	-	Ebbw Junction	1/61
6426	11/35	Ebbw Junction	Ebbw Junction	-	Ebbw Junction	3/61
6427	11/35	Cathays	Merthyr	-	Merthyr	8/58
6428	11/35	Ebbw Junction	Ebbw Junction	-	Stourbridge	3/59
6429	11/35	Ebbw Junction	Pontypool	-	Wellington	3/62
6430	3/37	Kidderminster	Pontypool	-	Yeovil Town	10/64
6431	3/37	Landore	Landore	-	Tondu	1/63
6432	3/37	Pontypool	Pontypool	-	Pontypool	3/59
6433	3/37	Cardiff	Cathays	-	Merthyr	1/63
6434	3/37	Merthyr	Abercynon	-	Stourbridge	9/64
6435	4/37	Cathays	Cathays	-	Yeovil Town	10/64
6436	4/37	Cathays	Cathays	-	Tondu	9/62
6437	4/37	Pontypool	Aberdare	-	Gloucester	7/63
6438	4/37	Ebbw Junction	Abercynon	-	Laira	11/62
6439	4/37	Ebbw Junction	Ebbw Junction	-	Ebbw Junction	5/60
6600	8/27	Neath	Oxley	-	Llantrisant	8/62

No.	Date	First Shed	Last GWR Shed	Wdn GWR	Last BR Shed	Wdn BR
6601	8/27	Landore	Bristol SPM	-	Tondu	11/62
6602	8/27	Pontypool	Barry	-	Llanelly	10/64
6603	9/27	Neath	Radyr	-	Abercynon	3/64
6604	9/27	Llanelly	Landore	-	Croes Newydd	10/65
6605	9/27	Pontypool	Aberdare	-	Radyr	1/64
6606	9/27	Cathays	Shrewsbury	-	Radyr	1/65
6607	9/27	Cathays	Radyr	-	Radyr	3/63
6608	9/27	Cathays	Radyr	-	Rhymney	6/64
6609	9/27	Cathays	Oxley	-	Tyseley	9/63
6610	9/27	Cathays	Oxley	-	Croes Newydd	2/63
6611	9/27	Cathays	Tyseley	-	Croes Newydd	11/65
6612	10/27	Cathays	Ebbw Junction	-	Rhymney	5/65
6613	10/27	Cathays	Neath	-	Llanelly	10/65
6614	10/27	Cathays	Barry	-	Neath	6/65
6615	12/27	Cathays	Cathays	-	Croes Newydd	5/63
6616	12/27	Cathays	Duffryn Yard	-	Neath	9/62
6617	12/27	Cathays	Stourbridge	-	Croes Newydd	9/62
6618	12/27	Cathays	Radyr	-	Tyseley	11/63
6619	1/28	Cathays	Barry	-	Barry	3/63
6620	1/28	Cathays	Barry	-	Duffryn Yard	12/63
6621	1/28	Cathays	Aberbeeg	-	Radyr	12/64
6622	1/28	Cathays	Cardiff	-	Pontypool	12/64
6623	1/28	Cathays	Gloucester	-	Llanelly	6/63
6624	1/28	Radyr	Chester	-	Radyr	6/64
6625	1/28	Bristol SPM	Leamington	-	Tyseley	11/65
6626	1/28	Cathays	Cathays	-	Croes Newydd	11/65
6627	2/28	Cathays	Cathays	-	Aberdare	11/63
6628	2/28	Cathays	Aberdare	-	Neath	5/65
6629	2/28	Cathays	Duffryn Yard	-	Radyr	10/62
6630	8/28	Banbury	Tyseley	-	Cardiff	9/62
6631	8/28	Bristol SPM	Gloucester	-	Oxley	9/63
6632	8/28	Banbury	Leamington	-	Croes Newydd	9/63
6633	8/28	Britsol SPM	Shrewsbury	-	Tyseley	5/65
6634	8/28	Cathays	Pontypool	-	Pontypool	4/64
6635	8/28	Cathays	Cathays	-	Radyr	6/64
6636	8/28	Ebbw Junction	Pontypool	-	Pontypool	6/63
6637	8/28	Cathays	Barry	-	Radyr	6/64
6638	9/28	Ebbw Junction	Oxley	-	Radyr	6/64
6639	9/28	Ebbw Junction	Severn Tunnel	-	Leamington	10/63
6640	9/28	Cathays	Oxley	-	Oxley	9/62
6641	9/28	Penarth Dock	Barry	-	Neath	9/62
6642	9/28	Penarth Dock	Tondu	-	Severn Tunnel	1/63
6643	9/28	Cathays	Barry	-	Llanelly	8/65
6644	9/28	Cathays	Leamington	-	Tyseley	7/65
6645	9/28	Cathays	Oxley	-	Oxley	8/62
6646	9/28	Barry	Stourbridge	-	Tyseley	5/64
6647	9/28	Barry	Barry	-	Radyr	9/62
6648	9/28	Barry	Cathays	-	Radyr	5/65
6649	9/28	Barry	Ebbw Junction	-	Rhymney	2/65
6650	8/28	Cathays	Leamington	-	Radyr	5/65
6651	8/28	Ebbw Junction	Pontypool	-	Croes Newydd	10/65
6652	8/28	Cathays	Aberdare	-	Llanelly	12/63
6653	8/28	Cathays	Barry	-	Llanelly	12/63
6654	8/28	Dock Street	Ebbw Junction	-	Radyr	6/65
6655	8/28	Cathays	Cathays	-	Rhymney	3/65
6656	8/28	Dock Street	Bristol SPM	-	Stourbridge	9/65
6657	8/28	Cathays	Leamington	-	Radyr	6/65
6658	9/28	Cathays	Barry	-	Pontypool	4/65
6659	8/28	Cathays	Cathays	-	Radyr	10/63
6660	9/28	Cathays	Cathays	-	Radyr	6/64

No.	Date	First Shed	Last GWR Shed	Wdn GWR	Last BR Shed	Wdn BR
6661	9/28	Cathays	Cathays	-	Radyr	6/65
6662	9/28	Cathays	Barry	-	Swansea E Dock	4/63
6663	9/28	Cathays	Ebbw Junction	-	Leamington	6/63
6664	9/28	Cathays	Radyr	-	Aberdare	12/63
6665	9/28	Cathays	Stourbridge	-	Tyseley	10/65
6666	9/28	Cathays	Severn Tunnel	-	Severn Tunnel	7/63
6667	9/28	Cathays	Stourbridge	-	Tyseley	11/65
6668	9/28	Cathays	Barry	-	Croes Newydd	12/65
6669	9/28	Cathays	Barry	-	Radyr	12/62
6670	9/28	Cathays	Bristol SPM	-	Barry	10/63
6671	9/28	Cathays	Bristol SPM	-	Banbury	10/65
6672	9/28	Cathays	Ebbw Junction	-	Radyr	7/65
6673	9/28	Cathays	Severn Tunnel	-	Aberdare	3/63
6674	9/28	Cathays	Stourbridge	-	Croes Newydd	7/63
6675	9/28	Barry	Tondu	-	Pontypool	6/63
6676	9/28	Severn Tunnel	Severn Tunnel	-	Tondu	4/63
6677	9/28	Cathays	Stourbridge	-	Pontypool	1/63
6678	9/28	Barry	Stourbridge	-	Stourbridge	11/64
6679	9/28	Cathays	Landore	-	Stourbridge	9/65
6680	10/28	Barry	Landore	-	Radyr	4/64
6681	10/28	Barry	Gloucester	-	Tyseley	10/65
6682	10/28	Barry	Oxford	-	Cardiff E Dock	2/64
6683	10/28	Barry	Landore	-	Croes Newydd	10/65
6684	10/28	Barry	Stourbridge	-	Radyr	11/64
6685	10/28	Barry	Tondu	-	Rhymney	9/64
6686	10/28	Landore	Duffryn Yard	-	Rhymney	4/64
6687	10/28	Barry	Pontypool	-	Aberdare	5/62
6688	10/28	Landore	Duffryn Yard	-	Treherbert	4/64
6689	10/28	Barry	Severn Tunnel	-	Radyr	6/65
6690	10/28	Duffryn Yard	Westbury	-	Radyr	6/64
6691	10/28	Merthyr	Cardiff	-	Llanelly	6/65
6692	10/28	Merthyr	Aberdare	-	Stourbridge	9/65
6693	10/28	Merthyr	Aberdare	-	Pontypool	1/63
6694	10/28	Tyseley	Croes Newydd	-	Croes Newydd	10/63
6695	10/28	Tyseley	Landore	-	Radyr	7/64
6696	10/28	Tyseley	Banbury	-	Barry	12/63
6697	10/28	Oxley	Leamington	-	Croes Newydd	5/66
6698	10/28	Oxley	Croes Newydd	-	Croes Newydd	3/63
6699	10/28	Oxley	Westbury	-	Radyr	12/63
6700	2/30	Cardiff Docks	Cardiff E Dock	-	Gurnos	6/61
6701	2/30	Cardiff Docks	Cardiff E Dock	-	Duffryn Yard	6/59
6702	3/30	Cardiff Docks	Cardiff E Dock	-	Upper Bank	7/60
6703	3/30	Cardiff Docks	Cardiff E Dock	-	Cardiff E Dock	1/58
6704	3/30	Cardiff Docks	Cardiff E Dock	-	Radyr	3/58
6705	4/30	Cardiff Docks	Cardiff E Dock	-	Cardiff E Dock	1/58
6706	4/30	Cardiff Docks	Cardiff E Dock	-	Cardiff E Dock	1/58
6707	4/30	Cardiff Docks	Cardiff E Dock	-	Pill	6/59
6708	5/30	Cardiff Docks	Cardiff E Dock	-	Cardiff E Dock	12/57
6709	5/30	Cardiff Docks	Cardiff E Dock	-	Cardiff E Dock	1/58
6710	6/30	Cardiff Docks	Pill	-	Pill	8/57
6711	6/30	Cardiff Docks	Pill	-	Pill	6/59
6712	6/30	Cardiff Docks	Barry	-	Upper Bank	7/60
6713	6/30	Cardiff Docks	Danygraig	-	Upper Bank	9/57
6714	6/30	Cardiff Docks	Swansea E Dock	-	Swansea E Dock	12/63
6715	7/30	Swansea E Dock	Duffryn Yard	-	Didcot	5/58
6716	7/30	Bristol SPM	Swindon	-	Swindon	2/59
6717	7/30	Duffryn Yard	Duffryn Yard	-	Upper Bank	2/59
6718	8/30	Duffryn Yard	Duffryn Yard	-	Upper Bank	5/58
6719	8/30	Duffryn Yard	Duffryn Yard	-	Upper Bank	7/60
6720	9/30	Duffryn Yard	Duffryn Yard	-	Upper Bank	7/61

No.	Date	First Shed	Last GWR Shed	Wdn GWR	Last BR Shed	Wdn BR
6721	9/30	Stourbridge	Cardiff E Dock	-	Upper Bank	2/59
6722	9/30	Chester	Barry	-	Barry	1/58
6723	10/30	Chester	Barry	-	Upper Bank	2/59
6724	10/30	Swindon	Barry	-	Swansea E Dock	11/63
6725	3/30	Pill	Pill	-	Duffryn Yard	10/59
6726	3/30	Pill	Pill	-	Pill	7/58
6727	4/30	Pill	Pill	-	Pill	4/58
6728	4/30	Pill	Pill	-	Pill	6/60
6729	4/30	Pill	Pill	-	Pill	6/59
6730	5/30	Pill	Pill	-	Pill	9/57
6731	5/30	Pill	Pill	-	Pill	5/58
6732	5/30	Pill	Pill	-	Pill	7/58
6733	5/30	Pill	Barry	-	Barry	5/58
6734	7/30	Pill	Danygraig	-	Paxton Street	2/59
6735	7/30	Pill	Pill	-	Pill	6/59
6736	7/30	Swindon	Barry	-	Swindon	2/59
6737	8/30	Pill	Swindon	-	Swindon	7/57
6738	8/30	Swindon	Barry	-	Upper Bank	10/62
6739	9/30	Swindon	Swindon	-	Pill	6/62
6740	10/30	Swindon	Barry	-	Barry	1/58
6741	10/30	Swindon	Swindon	-	Swansea E Dock	12/63
6742	11/30	Pontypool	Pontypool	-	Swansea E Dock	12/63
6743	11/30	Ebbw Junction	Pill	-	Pill	4/59
6744	11/30	Stafford Road	Cardiff E Dock	-	Cardiff E Dock	7/58
6745	12/30	Tyseley	Barry	-	Pill	6/59
6746	12/30	Tyseley	Barry	-	Barry	8/58
6747	12/30	Tyseley	Barry	-	Cardiff E Dock	7/58
6748	12/30	Stourbridge	Barry	-	Barry	1/58
6749	1/31	Oxley	Duffryn Yard	-	Upper Bank	10/62
6750	6/47	Barry	Barry	-	Pill	1/60
6751	6/47	Cardiff E Dock	Cardiff E Dock	-	Pill	8/60
6752	7/47	Barry	Barry	-	Pill	1/61
6753	8/47	Barry	Barry	-	Upper Bank	1/61
6754	8/47	Barry	Barry	-	Pill	12/62
6755	8/47	Pill	Pill	-	Upper Bank	6/62
6756	8/47	Pill	Pill	-	Pill	7/60
6757	9/47	Severn Tunnel	Severn Tunnel	-	Upper Bank	12/62
6758	9/47	Barry	Barry	-	Pill	6/62
6759	9/47	Pill	Pill	-	Swansea E Dock	4/60
6800	8/36	Penznace	Landore	-	Ebbw Junction	6/64
6801	8/36	Laira	Penzance	-	Penzance	10/60
6802	9/36	Paddington	Reading	-	Pontypool	8/61
6803	9/36	Oxley	Banbury	-	Oxley	9/65
6804	9/36	Bristol SPM	Westbury	-	Barrow Road	8/64
6805	9/36	Cardiff	Cardiff	-	Pontypool	3/61
6806	9/36	Landore	Landore	-	Worcester	10/64
6807	9/36	Worcester	Worcester	-	Worcester	12/63
6808	9/36	Penzance	Penzance	-	Oxley	8/64
6809	11/36	Paddington	Southall	-	Southall	7/63
6810	11/36	Cardiff	Cardiff	-	Llanelly	10/64
6811	11/36	Cardiff	Cardiff	-	Oxley	7/64
6812	12/36	Cardiff	Stafford Road	-	Llanelly	2/65
6813	12/36	Exeter	Newton Abbot	-	Worcester	9/65
6814	12/36	Exeter	Newton Abbot	-	Bristol SPM	12/63
6815	12/36	Oxley	Severn Tunnel	-	Severn Tunnel	11/65
6816	12/36	Oxley	Banbury	-	Worcester	7/65
6817	12/36	Cardiff	Cardiff	-	Worcester	4/65
6818	12/36	Ebbw Junction	Carmarthen	-	Ebbw Junction	4/64
6819	12/36	Pontypool	Birkenhead	-	Worcester	11/65
6820	1/37	Ebbw Junction	Pontypool	-	Cardiff E Dock	7/65

No.	Date	First Shed	Last GWR Shed	Wdn GWR	Last BR Shed	Wdn BR
6821	1/37	Cardiff	Ebbw Junction	-	Llanelly	11/64
6822	1/37	Exeter	Newton Abbot	-	Barrow Road	9/64
6823	1/37	Goodwick	Goodwick	-	Oxley	6/65
6824	1/37	Llanelly	Carmarthen	-	Oxford	4/64
6825	2/37	Exeter	Penzance	-	Bristol SPM	6/64
6826	2/37	Paddington	Southall	-	Cardiff E Dock	5/65
6827	2/37	Truro	Cardiff	-	Oxley	9/65
6828	2/37	Landore	Landore	-	Oxley	7/63
6829	3/37	Truro	Newton Abbot	-	Worcester	11/65
6830	8/37	Ebbw Junction	Bristol SPM	-	Oxley	10/65
6831	8/37	Banbury	Tyseley	-	Oxley	10/65
6832	8/37	Tyseley	Banbury	-	Neath	1/64
6833	8/37	Oxley	Tyseley	-	Oxley	10/65
6834	8/37	Severn Tunnel	Severn Tunnel	-	Tyseley	6/64
6835	9/37	Oxley	Banbury	-	Bristol SPM	5/63
6836	9/37	Bristol SPM	Bristol SPM	-	Worcester	8/65
6837	9/37	Truro	Ebbw Junction	-	Cardiff E Dock	7/65
6838	9/37	Swindon	Penzance	-	Worcester	11/65
6839	9/37	Banbury	Banbury	-	Oxley	5/64
6840	9/37	Ebbw Junction	Pontypool	-	Llanelly	2/65
6841	9/37	Oxley	Banbury	-	Oxford	6/65
6842	9/37	Bristol SPM	Bristol SPM	-	Tyseley	11/64
6843	10/37	Banbury	Tyseley	-	Llanelly	2/64
6844	10/37	Banbury	Stafford Road	-	Llanelly	4/64
6845	10/37	Westbury	Westbury	-	Tyseley	9/64
6846	10/37	Salisbury	Bristol SPM	-	Barrow Road	9/64
6847	10/37	Weymouth	Tyseley	-	Worcester	12/65
6848	10/37	Stafford Road	Stafford Road	-	Oxford	12/65
6849	10/37	Oxley	Banbury	-	Oxford	12/65
6850	10/37	Westbury	Bristol SPM	-	Severn Tunnel	12/64
6851	11/37	Worcester	Worcester	-	Tyseley	8/65
6852	11/37	Weymouth	Bristol SPM	-	Ebbw Junction	1/64
6853	11/37	Tyseley	Tyseley	-	Tyseley	10/65
6854	11/37	Oxley	Banbury	-	Oxley	9/65
6855	11/37	Tyseley	Tyseley	-	Tyseley	10/65
6856	11/37	Stafford Road	Oxley	-	Worcester	11/65
6857	11/37	Landore	Landore	-	Tyseley	10/65
6858	12/37	Tyseley	Tyseley	-	Tyseley	10/65
6859	12/37	Bristol SPM	Chester	-	Severn Tunnel	11/65
6860	2/39	Tyseley	Tyseley	-	Cardiff E Dock	2/65
6861	2/39	Bristol SPM	Bristol SPM	-	Tyseley	10/65
6862	2/39	Oxley	Oxley	-	Tyseley	6/65
6863	2/39	Paddington	Bristol SPM	-	Cardiff E Dock	11/64
6864	2/39	Paddington	Reading	-	Tyseley	10/65
6865	3/39	Paddington	Paddington	-	Bristol SPM	5/62
6866	3/39	Tyseley	Tyseley	-	Tyseley	5/65
6867	3/39	Bristol SPM	Bristol SPM	-	Llanelly	8/64
6868	3/39	Ebbw Junction	Ebbw Junction	-	Oxford	10/65
6869	3/39	Ebbw Junction	Paddington	-	Cardiff E Dock	7/65
6870	3/39	Salisbury	Ebbw Junction	-	Oxley	9/65
6871	3/39	Weymouth	Severn Tunnel	-	Oxley	10/65
6872	3/39	Weymouth	Landore	-	Oxford	12/65
6873	4/39	Westbury	Severn Tunnel	-	Bristol SPM	6/64
6874	4/39	Weymouth	Ebbw Junction	-	Oxford	9/65
6875	4/39	Pontypool	Pontypool	-	Bristol SPM	3/64
6876	4/39	Bristol SPM	Bristol SPM	-	Worcester	11/65
6877	4/39	Worcester	Worcester	-	Worcester	3/65
6878	5/39	Oxley	Birkenhead	-	Worcester	11/64
6879	5/39	Oxley	Oxley	-	Tyseley	10/65
6900	6/40	Paddington	Paddington	-	Barrow Road	10/64

No	Date	First Shed	Last GWR Shed	Wdn GWR	Last BR Shed	Wdn BR
6901	7/40	Stafford Road	Stafford Road	-	Pontypool	6/64
6902	7/40	Paddington	Swindon	-	Swindon	5/61
6903	7/40	Landore	Landore	-	Banbury	9/65
6904	7/40	Banbury	Tyseley	-	Banbury	1/65
6905	7/40	Worcester	Hereford	-	Neath	6/64
6906	11/40	Stafford Road	Banbury	-	Banbury	4/65
6907	11/40	Laira	Laira	-	Banbury	2/65
6908	11/40	Stafford Road	Stafford Road	-	Barrow Road	7/65
6909	11/40	Paddington	Bristol SPM	-	Didcot	6/64
6910	12/40	Paddington	Paddington	-	Oxford	10/65
6911	1/41	Laira	Penzance	-	Banbury	4/65
6912	1/41	Bristol SPM	Bristol SPM	-	Cardiff E Dock	2/64
6913	2/41	Laira	Laira	-	Gloucester	6/64
6914	2/41	Tyseley	Tyseley	-	Cardiff E Dock	4/64
6915	2/41	Stafford Road	Oxley	-	Banbury	2/65
6916	6/41	Bristol SPM	Worcester	-	Banbury	8/65
6917	6/41	Gloucester	Gloucester	-	Banbury	9/65
6918	6/41	Landore	Landore	-	Barrow Road	9/65
6919	6/41	Carmarthen	Carmarthen	-	Bristol SPM	8/63
6920	7/41	Paddington	Hereford	-	Duffryn Yard	12/63
6921	7/41	Hereford	Worcester	-	Oxford	10/65
6922	7/41	Bristol SPM	Bristol SPM	-	Tyseley	4/65
6923	7/41	Didcot	Didcot	-	Oxford	12/65
6924	8/41	Stafford Road	Stafford Road	-	Oxford	10/65
6925	8/41	Paddington	Oxford	-	Oxley	11/64
6926	11/41	Ebbw Junction	Ebbw Junction	-	Banbury	5/65
6927	11/41	Ebbw Junction	Ebbw Junction	-	Oxford	10/65
6928	11/41	Cardiff	Cardiff	-	Didcot	6/65
6929	11/41	Banbury	Banbury	-	Banbury	10/63
6930	11/41	Worcester	Worcester	-	Banbury	10/65
6931	12/41	Truro	Truro	-	Oxford	10/65
6932	12/41	Oxley	Oxley	-	Oxford	12/65
6933	12/41	Oxford	Oxford	-	Oxley	11/64
6934	12/41	Newton Abbot	Newton Abbot	-	Banbury	10/65
6935	12/41	Hereford	Swindon	-	Cardiff E Dock	2/65
6936	7/42	Worcester	Worcester	-	Cardiff E Dock	11/64
6937	7/42	Oxford	Oxford	-	Oxford	12/65
6938	7/42	Worcester	Worcester	-	Didcot	3/65
6939	7/42	Oxley	Oxley	-	Cardiff E Dock	10/63
6940	8/42	Gloucester	Gloucester	-	Gloucester	5/64
6941	8/42	Chester	Chester	-	Pontypool	4/64
6942	8/42	Oxley	Oxley	-	Gloucester	12/64
6943	8/42	Worcester	Hereford	-	Gloucester	12/63
6944	9/42	Bristol SPM	Bristol SPM	-	Severn Tunnel	11/65
6945	9/42	Bristol SPM	Weymouth	-	Cardiff E Dock	9/64
6946	12/42	Cardiff	Cardiff	-	Pontypool	6/64
6947	12/42	Worcester	Worcester	-	Oxford	11/65
6948	12/42	Cardiff	Cardiff	-	Gloucester	12/63
6949+	12/42	Landore	Laira	-	Shrewsbury	5/61
6950	12/42	Worcester	Worcester	-	Cardiff E Dock	6/64
6951	2/43	Worcester	Worcester	-	Tyseley	12/65
6952	2/43	Didcot	Didcot	-	Tyseley	12/65
6953+	2/43	Reading	Paddington	-	Oxford	12/65
6954	3/43	Bristol SPM	Bristol SPM	-	Bristol SPM	5/64
6955	3/43	Bristol SPM	Salisbury	-	Severn Tunnel	2/65
6956	3/43	Oxley	Oxley	-	Oxford	12/65
6957+	4/43	Newton Abbot	Paddington	-	Oxford	10/65
6958	4/43	Bristol SPM	Bath Road	-	Worcester	6/65
6959	3/44	Swindon	Paddington	-	Oxford	12/65
6960	3/44	Paddington	Paddington	-	Oxford	6/64

No.	Date	First Shed	Last GWR Shed	Wdn GWR	Last BR Shed	Wdn BR
6961	3/44	Westbury	Paddington	-	Oxford	9/65
6962	4/44	Bath Road	Paddington	-	Paddington	1/63
6963	4/44	Stafford Road	Shrewsbury	-	Oxford	7/65
6964	5/44	Stafford Road	Stafford Road	-	Banbury	9/65
6965	7/44	Gloucester	Swindon	-	Barrow Road	10/65
6966	5/44	Salisbury	Westbury	-	Oxford	9/64
6967	8/44	Oxley	Oxley	-	Oxford	12/65
6968	9/44	Paddington	Paddington	-	Westbury	9/63
6969	9/44	Severn Tunnel	Cardiff	-	Didcot	2/65
6970	9/44	Oxley	Oxley	-	Oxford	6/64
6971	10/47	Bath Road	Bath Road	-	Tyseley	10/64
6972	10/47	Bath Road	Bath Road	-	Bristol SPM	3/64
6973	10/47	Paddington	Paddington	-	Barrow Road	8/65
6974	10/47	Paddington	Paddington	-	Oxford	5/65
6795	10/47	Oxley	Oxley	-	Neath	12/63
6976	10/47	Shrewsbury	Shrewsbury	-	Banbury	10/65
6977	11/47	Paddington	Paddington	-	Westbury	12/63
6978	11/47	Salisbury	Salisbury	-	Ebbw Junction	7/65
6979	11/47	Banbury	Banbury	-	Banbury	2/65
6980	11/47	Shrewsbury	Shrewsbury	-	Tyseley	10/65
7000	5/46	Newton Abbot	Newton Abbot	-	Worcester	12/63
7001	5/46	Cardiff	Cardiff	-	Oxley	9/63
7002	6/46	Landore	Landore	-	Worcester	3/64
7003	6/46	Landore	Landore	-	Gloucester	8/64
7004	6/46	Gloucester	Gloucester	-	Reading	1/64
7005	6/46	Worcester	Worcester	-	Worcester	9/64
7006	6/46	Shrewsbury	Shrewsbury	-	Paddington	12/63
7007	7/46	Stafford Road	Stafford Road	-	Worcester	2/63
7200	8/34	Llanelly	Newton Abbot	-	Llanelly	7/63
7201	8/34	Llanelly	Cardiff	-	Llanelly	4/65
7202	9/34	Ebbw Junction	Severn Tunnel	-	Barry	6/64
7203	9/34	Ebbw Junction	Ebbw Junction	-	Severn Tunnel	12/63
7204	9/34	Cardiff	Didcot	-	Duffryn Yard	2/64
7205	10/34	Aberdare	Aberdare	-	Severn Tunnel	6/65
7206	10/34	Llanelly	Pontypool	-	Ebbw Junction	7/64
7207	10/34	Severn Tunnel	Oxley	-	Banbury	11/64
7208	10/34	Severn Tunnel	Bristol SPM	-	Barry	4/64
7209	11/34	Severn Tunnel	Severn Tunnel	-	Barry	7/64
7210	11/34	Severn Tunnel	Severn Tunnel	-	Pontypool	4/65
7211	11/34	Ebbw Junction	Swansea E Dock	-	Severn Tunnel	5/64
7212	9/34	Ebbw Junction	Ebbw Junction	-	Aberdare	2/64
7213	9/34	Cardiff	Aberdare	-	Llanelly	9/64
7214	9/34	Swansea E Dock	Didcot	-	Aberdare	12/63
7215	11/34	Pantyffynon	Bristol SPM	-	Swansea E Dock	6/63
7216	11/34	Swindon	Severn Tunnel	-	Duffryn Yard	10/63
7217	9/34	Ebbw Junction	Ebbw Junction	-	Aberbeeg	7/64
7218	11/34	Banbury	Leamington	-	Banbury	9/64
7219	10/34	Cardiff	Cardiff	-	Radyr	1/64
7220	9/35	Pontypool	Newton Abbot	-	Pontypool	9/64
7221	9/35	Pontypool	Aberdare	-	Banbury	11/64
7222	10/35	Cardiff	Oxley	-	Aberdare	1/65
7223	10/35	Cardiff	Severn Tunnel	-	Llanelly	11/64
7224	11/35	Pontypool	Severn Tunnel	-	Ebbw Junction	12/62
7225	10/35	Llanelly	Landore	-	Severn Tunnel	5/64
7226	11/35	Oxley	Oxley	-	Aberdare	11/64
7227	11/35	Banbury	Oxley	-	Pontypool	6/63
7228	12/35	Severn Tunnel	Didcot	-	Barry	7/63
7229	8/35	Severn Tunnel	Severn Tunnel	-	Ebbw Junction	8/64
7230	12/35	Severn Tunnel	Pontypool	-	Pontypool	7/64
7231	12/35	Severn Tunnel	Ebbw Junction	-	Llanelly	10/64

No.	Date	First Shed	Last GWR Shed	Wdn GWR	Last BR Shed	Wdn BR
7232	8/35	Severn Tunnel	Pontypool	-	Llanelly	5/65
7233	9/35	Pontypool	Pontypool	-	Pontypool	9/64
7234	9/35	Oxley	Bristol SPM	-	Aberdare	10/63
7235	1/36	Pontypool	Pontypool	-	Llanelly	4/64
7236	1/36	Oxley	Oxley	-	Banbury	11/63
7237	1/36	Oxley	Bristol SPM	-	Llanelly	6/63
7238	2/36	Banbury	Oxley	-	Barry	4/64
7239	2/36	Banbury	Severn Tunnel	-	Llanelly	10/63
7240	8/37	Ebbw Junction	Oxley	-	Pontypool	9/64
7241	9/37	Ebbw Junction	Ebbw Junction	-	Pontypool	11/62
7242	9/37	Aberdare	Aberdare	-	Llanelly	6/64
7243	1/38	Swindon	Oxley	-	Neath	7/64
7244	3/38	Llanelly	Landore	-	Llanelly	2/65
7245	5/38	Ebbw Junction	Ebbw Junction	-	Radyr	9/64
7246	7/38	Ebbw Junction	Severn Tunnel	-	Pontypool	9/63
7247	9/38	Neath	Ebbw Junction	-	Aberdare	3/63
7248	10/38	Ebbw Junction	Oxley	-	Llanelly	6/65
7249	11/38	Ebbw Junction	Ebbw Junction	-	Llanelly	6/65
7250	2/39	Swansea E Dock	Newton Abbot	-	Radyr	9/64
7251	3/39	Severn Tunnel	Severn Tunnel	-	Aberdare	1/64
7252	4/39	Ebbw Junction	Didcot	-	Severn Tunnel	6/65
7253	12/39	Ebbw Junction	Ebbw Junction	-	Severn Tunnel	4/65
7320	4/25	Cardiff	Reading	-	Gloucester	11/64
7321	4/25	Oxley	Swindon	-	Llanelly	9/62
7400	7/36	Carmarthen	Carmarthen	-	Carmarthen	6/60
7401	8/36	Carmarthen	Carmarthen	-	Hereford	8/59
7402	8/36	Pontypool	Stourbridge	-	Carmarthen	7/62
7403	8/36	Stourbridge	Bala	-	Severn Tunnel	1/64
7404	8/36	Weymouth	Oxford	-	Swansea E Dock	6/64
7405	8/36	Brecon	Oswestry	-	Carmarthen	12/63
7406	8/36	Croes Newydd	Aberystwyth	-	Aberystwyth	3/62
7407	8/36	Didcot	Carmarthen	-	Carmarthen	12/63
7408	8/36	Weymouth	Weymouth	-	Carmarthen	8/62
7409	8/36	Croes Newydd	Bala	-	Croes Newydd	8/61
7410	12/36	Aberystwyth	Oswestry	-	Oswestry	1/61
7411	12/36	Goodwick	Fairford	-	Oxford	5/59
7412	12/36	Goodwick	Fairford	-	Oxford	7/63
7413	12/36	Goodwick	Whitland	-	Stourbridge	9/64
7414	12/36	Stourbridge	Bala	-	Stourbridge	9/64
7415	12/36	Yeovil	Swindon	-	Swindon	2/59
7416	12/36	Gloucester	Worcester	-	Hereford	1/59
7417	1/37	Goodwick	Whitland	-	Machynlleth	9/61
7418	1/37	Chippenham	Swindon	-	Stourbridge	9/64
7419	1/37	Duffryn Yard	Carmarthen	-	Carmarthen	7/60
7420	5/37	Brecon	Ross-on-Wye	-	Tyseley	7/59
7421	5/37	Taunton	Taunton	-	Swindon	11/61
7422	5/37	Laira	Truro	-	Carmarthen	3/62
7423	5/37	Ebbw Junction	Aberdare	-	Aberdare	7/64
7424	5/37	Lydney	Swindon	-	Stourbridge	9/64
7425	6/37	Duffryn Yard	Carmarthen	-	Carmarthen	6/62
7426	6/37	Severn Tunnel	Pontypool	-	Tyseley	7/63
7427	6/37	Laira	Newton Abbot	-	Swansea E Dock	6/64
7428	6/37	Ebbw Junction	Stourbridge	-	Oswestry	10/62
7429	6/37	Ebbw Junction	Severn Tunnel	-	Stourbridge	2/61
7700	1/30	Gloucester	Kidderminster	-	Kidderminster	5/61
7701	2/30	Neath	Neath N&B	-	Neath	2/60
7702	2/30	Duffryn Yard	Leamington	-	Leamington	9/60
7703	2/30	Banbury	Aberbeeg	-	Pill	3/60
7704	2/30	Duffryn Yard	Swansea E Dock	-	Swansea E Dock	12/60
7705	3/30	Oxley	Stourbridge	-	Didcot	8/59

No.	Date	First Shed	Last GWR Shed	Wdn GWR	Last BR Shed	Wdn BR
7706	3/30	Oxley	Duffryn Yard	-	Duffryn Yard	3/60
7707	3/30	Worcester	Hereford	-	Worcester	11/60
7708	3/30	Paddington	Reading	-	Reading	6/60
7709	3/30	Paddington	St. Blazey	-	St. Blazey	8/60
7710	3/30	Paddington	Didcot	-	Didcot	9/58
7711	3/30	Paddington	Bristol SPM	-	St. Blazey	10/56
7712	3/30	Paddington	Ebbw Junction	-	Pontypool	7/60
7713	3/30	Paddington	Paddington	-	Taunton	8/62
7714	5/30	Tyseley	Birkenhead	-	Wrexham	1/59
7715	5/30	St. Blazey	St. Balzey	-	Duffryn Yard	5/63
7716	5/30	St. Blazey	Exeter	-	Exeter	12/59
7717	5/30	Paddington	Merthyr	-	Radyr	3/60
7718	5/30	Paddington	Bristol SPM	-	Llanelly	4/62
7719	6/30	Bristol SPM	Bristol SPM	-	Worcester	9/60
7720	6/30	Tondu	Aberbeeg	-	Aberdare	5/62
7721	6/30	Ebbw Junction	Llantrisant	-	Pill	7/62
7722	6/30	Tondu	Treherbert	-	Stourbridge	10/60
7723	6/30	Ebbw Junction	Gloucester	-	Gloucester LMS	8/60
7724	7/30	Gloucester	Pontypool	-	Pontypool	9/62
7725	12/29	Bristol SPM	Tondu	-	Tondu	8/60
7726	12/29	Bristol SPM	Bristol SPM	-	Abercynon	8/60
7727	12/29	Frome	Salisbury	-	Westbury	1/60
7728	12/29	Westbury	Bristol SPM	-	Bristol SPM	5/60
7729	12/29	Bristol SPM	Bristol SPM	-	Bristol SPM	7/62
7730	12/29	Westbury	Southall	-	Southall	3/59
7731	12/29	Paddington	Southall	-	Southall	4/59
7732	12/29	Paddington	Southall	-	Tondu	10/62
7733	12/29	Paddington	Duffryn Yard	-	Abercynon	5/60
7734	12/29	Paddington	Paddington	-	Paddington	4/59
7735	12/29	Bristol SPM	Tyseley	-	Tyseley	5/59
7736	12/29	Bristol SPM	Ebbw Junction	-	Ebbw Junction	5/62
7737	1/30	Landore	Neath	-	Neath	3/60
7738	1/30	Landore	Paddington	-	Ebbw Junction	2/59
7739	1/30	Swansea E Dock	Neath N&B	-	Neath	11/62
7740	1/30	Duffryn Yard	Aberbeeg	-	Pontypool	12/60
7741	2/30	Gloucester	Gloucester	-	Lydney	12/61
7742	2/30	Swansea E Dock	Neath	-	Neath	7/59
7743	2/30	Swansea E Dock	Neath N&B	-	Neath	8/59
7744	2/30	Duffryn Yard	Duffryn Yard	-	Abercynon	9/62
7745	2/30	Neath	Pantyffynon	-	Pantyffynon	3/61
7746	2/30	Cardiff	Tondu	-	Tondu	10/59
7747	2/30	Cardiff	Goodwick	-	Goodwick	2/61
7748	2/30	Tondu	Aberdare	-	Westbury	4/61
7749	2/30	Westbury	Bath	-	Bristol SPM	9/62
7750	11/30	Paddington	Worcester	-	Lydney	1/59
7751	12/30	Pontypol	Cardiff E Dock	-	Radyr	10/59
7752	11/30	Aberdare	Tondu	-	Tondu	11/59
7753	12/30	Duffryn Yard	Ebbw Junction	-	Tondu	4/62
7754	12/30	Reading	Paddington	-	Wellington	1/59
7755	12/30	Neath	Llanelly	-	Aberbeeg	5/62
7756	12/30	Duffryn Yard	Swansea E Dock	-	Gloucester LMS	6/61
7757	12/30	Danygraig	Neath	-	Neath	9/60
7758	12/30	Tyseley	Tyseley	-	Duffryn Yard	6/60
7759	12/30	Oxley	Oxley	-	Oxley	3/60
7760	12/30	Laira	Paddington	-	Oxford	12/61
7761	12/30	Taunton	Exeter	-	Exeter	1/61
7762	1/31	Bristol SPM	Laira	-	Stourbridge	5/62
7763	1/31	Oxley	Banbury	-	Tyseley	11/59
7764	1/31	Cardiff	Severn Tunnel	-	Barry	5/62
7765	1/31	Brecon	Pantyffynon	-	Llanelly	7/62

No.	Date	First Shed	Last GWR Shed	Wdn GWR	Last BR Shed	Wdn BR
7766	2/31	Merthyr	Merthyr	-	Radyr	11/60
7767	2/31	Llanelly	Neath	-	Neath	3/60
7768	2/31	Pontypool	Ebbw Junction	-	Ebbw Junction	11/59
7769	2/31	Neath N&B	Glyn Neath	-	Neath	8/59
7770	2/31	Gloucester	Tondu	-	Tondu	4/59
7771	3/31	Ebbw Junction	Ebbw Junction	-	Pontypool	11/61
7772	3/31	Aberdare	Merthyr	-	Stourbridge	11/61
7773	3/31	Aberdare	Aberdare	-	Aberdare	12/59
7774	3/31	Aberbeeg	Pill	-	Tredegar	11/59
7775	11/30	Pontypool	Tondu	-	Cardiff	11/60
7776	11/30	Ebbw Junction	Llanelly	-	Llanelly	1/61
7777	11/30	Ebbw Junction	Reading	-	Worcester	11/60
7778	11/30	Paddington	Aberbeeg	-	Tondu	12/59
7779	11/30	Paddington	Bristol SPM	-	Radyr	10/58
7780	11/30	Neath	Weston-S-Mare	-	Weymouth	6/63
7781	11/30	Ebbw Junction	Ebbw Junction	-	Ebbw Junction	7/60
7782	11/30	Bath Road	Bristol SPM	-	Barrow Road	10/64
7783	11/30	Bristol SPM	Bristol SPM	-	Bristol SPM	9/62
7784	11/30	Bristol SPM	Westbury	-	Westbury	3/62
7785	12/30	Swansea E Dock	Pantyffynon	-	Llanelly	5/62
7786	12/30	Neath	Neath	-	Neath	5/62
7787	12/30	Duffryn Yard	Landore	-	Aberbeeg	6/61
7788	12/30	Paddington	Reading	-	Stourbridge	7/62
7789	12/30	Paddington	Abebeeg	-	Severn Tunnel	11/59
7790	12/30	Bristol SPM	Bristol SPM	-	Duffryn Yard	11/62
7791	12/30	Paddington	Paddington	-	Paddington	12/59
7792	12/30	Bristol SPM	Swindon	-	Swindon	11/57
7793	12/30	Bristol SPM	Bristol SPM	-	Severn Tunnel	4/60
7794	12/30	Bristol SPM	Swindon	-	Brecon	10/60
7795	12/30	Bristol SPM	Bristol SPM	-	Bristol SPM	2/58
7796	1/31	Oxley	Oxley	-	Pontypool	2/62
7797	1/31	Tyseley	Oxley	-	Oxley	10/59
7798	1/31	Slough	Tondu	-	Tondu	5/61
7799	1/31	Paddington	Neath N&B	-	Neath	5/62
7800	1/38	Stafford Road	Banbury	-	Shrewsbury	8/64
7801	1/38	Bath Road	Bristol SPM	-	Shrewsbury	7/65
7802	1/38	Paddington	Aberystwyth	-	Shrewsbury	11/65
7803	1/38	Neyland	Aberystwyth	-	Shrewsbury	4/65
7804	2/38	Bath Road	Bristol SPM	-	Severn Tunnel	9/65
7805	3/38	Shrewsbury	Banbury	-	Tyseley	12/64
7806	3/38	Shrewsbury	Banbury	-	Oxley	11/64
7807	3/38	Neyland	Oswestry	-	Machynlleth	11/64
7808	3/38	Paddington	Oswestry	-	Gloucester	12/65
7809	4/38	Westbury	Bath Road	-	Shrewsbury	4/63
7810	12/38	Banbury	Leamington	-	Machynlleth	9/64
7811	12/38	Banbury	Banbury	-	Cardiff E Dock	7/65
7812	1/39	Bath Road	Bath Road	-	Shrewsbury	11/65
7813	1/39	Oxley	Oxley	-	Didcot	5/65
7814	1/39	Westbury	Bath Road	-	Gloucester	9/65
7815	1/39	Gloucester	Gloucester	-	Gloucester	10/64
7816	1/39	Neyland	Neyland	-	Gloucester	11/65
7817	1/39	Croes Newydd	Croes Newydd	-	Reading	6/64
7818	1/39	Worcester	Cheltenham	-	Tyseley	1/65
7819	2/39	Carmarthen	Whitchurch	-	Shrewsbury	11/65
8100	9/38	Leamington	Leamington	-	Leamington	10/62
8101	3/38	Kidderminster	Kidderminster	-	Kidderminster	3/61
8102	6/38	Tyseley	Whitland	-	Worcester	5/64
8103	11/38	Stourbridge	Oswestry	-	Radyr	11/63
8104	1/39	Cardiff	Neath	-	Worcester	12/64
8105	4/39	Bristol SPM	Bristol SPM	-	Worcester	6/57

No.	Date	First Shed	Last GWR Shed	Wdn GWR	Last BR Shed	Wdn BR
8106	6/39	Worcester	Worcester	-	Worcester	12/63
8107	7/39	Croes Newydd	Cardigan	-	Worcester	5/62
8108	9/39	Newton Abbot	Tyseley	-	Tyseley	11/60
8109	11/39	Leamington	Leamington	-	Tyseley	6/65
8700+	2/31	Westbury	Paddington	-	Paddington	10/63
8700	3/34	Tyseley	Tyseley	-	Tyseley	2/62
8701	2/31	Kidderminster	Gloucester	-	Gloucester	3/63
8702	2/31	Swindon	Bristol SPM	-	Hereford	5/64
8703	2/31	Bristol SPM	Bristol SPM	-	Bristol SPM	1/58
8704	2/31	Chester	Stourbridge	-	Stourbridge	2/60
8705	2/31	Chester	Stafford Road	-	Aberbeeg	4/61
8706	3/31	Llanelly	Llanelly	-	Llanelly	7/61
8707	2/31	Llanelly	Paddington	-	Pontypool	7/64
8708	3/31	Llanelly	Pantyffynon	-	Llanelly	5/60
8709	2/31	Laira	Laira	-	Pontypool	9/62
8710	3/31	Severn Tunnel	Ebbw Junction	-	Tondu	3/63
8711	3/31	Ebbw Junction	Ebbw Junction	-	Swindon	3/62
8712	3/31	Pontypool	Tondu	-	Tondu	1/63
8713	3/31	Bristol SPM	Bristol SPM	-	Tyseley	3/62
8714	3/31	Bristol SPM	Bristol SPM	-	Neath	11/64
8715	3/31	Llanelly	Neath	-	Neath	4/62
8716	3/31	Tondu	Pontypool	-	Pontypool	4/64
8717	3/31	Gloucester	Gloucester	-	Abercynon	7/64
8718	3/31	Gloucester	Kidderminster	-	Shrewsbury	7/66
8719	3/31	Newton Abbot	Laira	-	St. Blazey	5/62
8720	3/31	Danygraig	Danygraig	-	Didcot	9/64
8721	3/31	Reading	Tondu	-	Tondu	7/61
8722	4/31	Westbury	Bristol SPM	-	Cardiff	4/61
8723	4/31	Ebbw Junction	Aberbeeg	-	Aberdare	7/64
8724	4/31	Pontypool	Aberbeeg	-	Duffryn Yard	7/62
8725	12/30	Birkenhead	Birkenhead	-	Barrow Road	10/62
8726	12/30	Tyseley	Stafford Road	-	Llanelly	4/61
8727	12/30	Reading	Kidderminster	-	Croes Newydd	4/62
8728	1/31	Paddington	Cardiff	-	Nine Elms	7/63
8729	1/31	Paddington	Banbury	-	Gloucester	12/62
8730	2/31	Bristol SPM	Bristol SPM	-	Abercynon	7/62
8731	2/31	Kidderminster	Gloucester	-	Southall	7/62
8732	2/31	Neath	Pantyffynon	-	Swansea E Dock	4/64
8733	3/31	Gloucester	Swindon	-	Stourbridge	2/62
8734	3/31	Chester	Stafford Road	-	Croes Newydd	3/62
8735	3/31	Aberbeeg	Gloucester	-	Abercynon	1/62
8736	4/31	Merthyr	Merthyr	-	Llanelly	3/62
8737	4/31	Brecon	Bristol SPM	-	Neath	12/62
8738	4/31	Cardiff	Paddington	-	Neyland	3/63
8739	5/31	Ebbw Junction	Aberbeeg	-	Oxley	11/64
8740	5/31	Reading	Cardiff	-	Tondu	2/61
8741	5/31	Swindon	Bristol SPM	-	Bristol SPM	5/62
8742	6/31	Reading	Stourbridge	-	Stourbridge	9/62
8743	6/31	Cardiff	Cardiff E Dock	-	Paddington	6/64
8744	6/31	Westbury	Frome	-	Westbury	10/62
8745	6/31	Worcester	Frome	-	Gloucester	8/65
8746	7/31	Bristol SPM	Bristol SPM	-	Duffryn Yard	12/62
8747	7/31	Bristol SPM	Bristol SPM	-	Neath	5/64
8748	7/31	Ebbw Junction	Bridgend	-	Tondu	9/62
8749	7/31	Whitland	Pantyffynon	-	Gloucester	10/64
8750	9/33	Paddington	Paddington	-	Swansea E Dock	5/62
8751	10/33	Paddington	Paddington	-	Ebbw Junction	12/62
8752	10/33	Paddington	Southall	-	Southall	1/63
8753	10/33	Paddington	Southall	-	Paddington	2/62
8754	10/33	Paddington	Paddington	-	Paddington	11/60

No.	Date	First Shed	Last GWR Shed	Wdn GWR	Last BR Shed	Wdn BR
8755	10/33	Paddington	Pontypool	-	Paddington	12/57
8756	10/33	Paddington	Paddington	-	Paddington	10/62
8757	10/33	Southall	Paddington	-	Paddington	9/62
8758	11/33	Oxford	Southall	-	Southall	1/59
8759	10/33	Paddington	Paddington	-	Paddington	1/63
8760	10/33	Paddington	Paddington	-	Neath	1/62
8761	10/33	Paddington	Paddington	-	Paddington	5/62
8762	10/33	Paddington	Paddington	-	Paddington	8/61
8763	10/33	Paddington	Paddington	-	Paddington	8/62
8764	10/33	Paddington	Soutahll	-	Barry	5/62
8765	12/33	Paddington	Paddington	-	Paddington	9/62
8766	12/33	Paddington	Bristol SPM	-	Cardiff	7/63
8767	12/33	Paddington	Paddington	-	Croes Newydd	7/66
8768	12/33	Paddington	Paddington	-	Didcot	9/64
8769	12/33	Paddington	Paddington	-	Southall	4/61
8770	1/34	Paddington	Paddington	-	Southall	12/62
8771	1/34	Paddington	Paddington	-	Paddington	7/62
8772	1/34	Paddington	Paddington	-	Duffryn Yard	8/61
8773	1/34	Paddington	Paddington	-	Paddington	10/62
8774	1/34	Southall	Southall	-	Cardiff	8/61
8775	1/34	Neath	Neath	-	Neath	12/61
8776	1/34	Pontypool	Aberbeeg	-	Cardiff	12/62
8777	1/34	Ebbw Junction	Tondu	-	Carmarthen	4/61
8778	1/34	Tondu	Ebbw Junction	-	Ebbw Junction	8/60
8779	1/34	Westbury	Chippenham	-	Swindon	2/62
8780	1/34	Lanore	Paddington	-	Radyr	7/62
8781	1/34	Llanelly	Gloucester	-	Ebbw Junction	11/62
8782	1/34	Neath	Neath N&B	-	Neath	11/61
8783	1/34	St. Blazey	St. Blazey	-	Llanelly	6/63
8784	1/34	Tyseley	Tyseley	-	Neath	4/62
8785	1/34	Landore	Pantyffynon	-	Llanelly	12/63
8786	1/34	Ebbw Junction	Ebbw Junction	-	Nine Elms	7/63
8787	1/34	Stourbridge	Banbury	-	Hereford	8/61
8788	1/34	Aberbeeg	Pontypool	-	Neath	5/62
8789	1/34	Landore	Landore	-	Swansea E Dock	6/61
8790	2/34	Bristol SPM	Bristol SPM	-	Bristol SPM	5/62
8791	2/34	Stourbridge	Stourbridge	-	Neath	3/63
8792	2/34	Stafford Road	Stourbridge	-	Stourbridge	2/62
8793	3/34	Bristol SPM	Bristol SPM	-	Worcester	12/64
8794	3/34	Ebbw Junction	Aberbeeg	-	Nine Elms	7/63
8795	3/34	Bath	Bristol SPM	-	Barrow Road	7/65
8796	3/34	Llantrisant	Ebbw Junction	-	Stafford Road	4/61
8797	3/34	Tyseley	Stourbridge	-	Stourbridge	4/62
8798	3/34	Oxley	Oxley	-	Swansea E Dock	6/61
8799	6/34	Pontypool	Severn Tunnel	-	Weymouth	11/62
9300+	2/32	Truro	Southall	-	Gloucester	11/61
9301+	2/32	Penzance	Southall	-	Bristol SPM	8/62
9302+	2/32	Truro	Paddington	-	Didcot	10/62
9303+	2/32	Penzance	Reading	-	Pontypool	4/64
9304+	2/32	Truro	Reading	-	Bristol SPM	9/63
9305+	2/32	Reading	Reading	-	Didcot	11/64
9306+	3/32	Oxley	Paddington	-	Severn Tunnel	4/62
9307+	3/32	Westbury	Reading	-	Shrewsbury	1/63
9308+	3/32	Bristol SPM	Paddington	-	Shrewsbury	9/62
9309+	3/32	Paddington	Reading	-	Reading	9/62
9310+	3/32	Paddington	Paddington	-	Taunton	4/64
9311+	3/32	Swindon	Southall	-	Taunton	10/63
9312+	3/32	Oxley	Oxley	-	Carmarthen	4/62
9313+	3/32	Weymouth	Reading	-	Gloucester	9/63
9314+	3/32	Paddington	Oxley	-	Shrewsbury	9/62

No.	Date	First Shed	Last GWR Shed	Wdn GWR	Last BR Shed	Wdn BR
9315+	3/32	Oxley	Reading	-	Swindon	9/64
9316+	4/32	Westbury	Oxford	-	Bristol SPM	8/62
9317+	4/32	Weymouth	Oxford	-	Severn Tunnel	6/64
9318+	4/32	Weymouth	Reading	-	Didcot	6/64
9319+	4/32	Weymouth	Reading	-	Croes Newydd	9/62
9400	2/47	Chippenham	Swindon	-	Paddington	12/59
9401	2/47	Paddington	Paddington	-	Bromsgrove	7/63
9402	4/47	Paddington	Paddington	-	Reading	9/59
9403	3/47	Paddington	Paddington	-	Reading	6/59
9404	4/47	Paddington	Paddington	-	Paddington	6/65
9405	5/47	Paddington	Paddington	-	Barrow Road	6/65
9406	5/47	Paddington	Paddington	-	Paddington	9/64
9407	5/47	Paddington	Paddington	-	Paddington	7/62
9408	5/47	Oxley	Oxley	-	Llanelly	5/63
9409	5/47	Paddington	Paddington	-	Southall	5/62
9600	2/45	Swindon	Swindon	-	Ebbw Junction	9/65
9601	4/45	Yeovil	Yeovil	-	Barrow Road	12/64
9602	5/45	Goodwick	Goodwick	-	Pontypool	3/65
9603	5/45	Goodwick	Goodwick	-	Barry	12/63
9604	5/45	Bristol SPM	Bristol SPM	-	Swindon	12/62
9605	5/45	Weston-S-Mare	Bristol SPM	-	Westbury	9/65
9606	6/45	Bristol SPM	Bristol SPM	-	Gloucester	11/64
9607	6/45	Shrewsbury	Aberdare	-	Aberdare	3/64
9608	6/45	Tyseley	Tyseley	-	Stourbridge	7/66
9609	6/45	Aberdare	Aberdare	-	Llanelly	10/65
9610	7/45	Tyseley	Tyseley	-	Croes Newydd	9/66
9611	8/45	Oxford	Oxford	-	Radyr	4/65
9612	8/45	Westbury	Westbury	-	Westbury	12/63
9613	9/45	Stourbridge	Stourbridge	-	Stourbridge	10/65
9614	9/45	Brecon	Brecon	-	Stourbridge	7/66
9615	9/45	Westbury	Yeovil	-	Radyr	7/65
9616	9/45	Llantrisant	Ebbw Junction	-	Severn Tunnel	8/65
9617	10/45	Duffryn Yard	Duffryn Yard	-	Neath	6/65
9618	10/45	Merthyr	Merthyr	-	Merthyr	12/63
9619	10/45	Hereford	Hereford	-	Severn Tunnel	7/65
9620	10/45	Bristol SPM	Bristol SPM	-	Gloucester	7/64
9621	10/45	Stafford Road	Stafford Road	-	Llanelly	10/64
9622	11/45	Merthyr	Merthyr	-	Radyr	7/65
9623	11/45	Taunton	Newton Abbot	-	Barrow Road	7/65
9624	11/45	Wellington	Wellington	-	Stourbridge	1/65
9625	11/45	Swansea E Dock	Swansea E Dock	-	Cardiff E Dock	6/65
9626	11/45	Bristol SPM	Bristol SPM	-	Worcester	12/65
9627	11/45	Neath N&B	Neath N&B	-	Neath	7/62
9628	12/45	Weymouth	Salisbury	-	Westbury	3/63
9629	12/45	Cardiff	Cardiff	-	Pontypool	10/64
9630	12/45	Wellington	Wellington	-	Croes Newydd	9/66
9631	12/45	Barry	Barry	-	Llanelly	6/65
9632	12/45	Ebbw Junction	Ebbw Junction	-	Abercynon	11/64
9633	1/46	Newton Abbot	Newton Abbot	-	Duffryn Yard	10/63
9634	1/46	Duffryn Yard	Duffryn Yard	-	Neath	5/64
9635	1/46	Tyseley	Tyseley	-	Taunton	6/64
9636	2/46	Stourbridge	Stourbridge	-	Wellington	10/63
9637	2/46	Ebbw Junction	Ebbw Junction	-	Neath	9/64
9638	2/46	Merthyr	Abercynon	-	Merthyr	12/63
9639	2/46	Wellington	Wellington	-	Croes Newydd	9/65
9640	3/46	Slough	Slough	-	Oxley	7/66
9641	3/46	Southall	Southall	-	Croes Newydd	10/66
9642	4/46	Westbury	Weymouth	-	Paddington	11/64
9643	4/46	Merthyr	Merthyr	-	Merthyr	5/62
9644	4/46	Ebbw Junction	Ebbw Junction	-	Radyr	6/65

No.	Date	First Shed	Last GWR Shed	Wdn GWR	Last BR Shed	Wdn BR
9645	4/46	Swansea E Dock	Swansea E Dock	-	Goodwick	10/63
9646	5/46	Newton Abbot	Exeter	-	Stourbridge	5/65
9647	5/46	Exeter	Exeter	-	Templecombe	6/65
9648	5/46	Cardiff	Cardiff	-	Llanelly	7/64
9649	6/46	Tondu	Bridgend	-	Ebbw Junction	7/65
9650	6/46	Pontypool	Pontypool	-	Pontypool	12/64
9651	6/46	Birkenhead	Birkenhead	-	Cardiff E Dock	7/65
9652	11/46	Milford Haven	Milford Haven	-	Llanelly	1/63
9653	11/46	Slough	Slough	-	Oxford	7/65
9654	11/46	Oxford	Oxford	-	Oxford	10/64
9655	11/46	St. Blazey	St. Blazey	-	Pontypool	5/64
9656	11/46	Croes Newydd	Croes Newydd	-	Severn Tunnel	11/65
9657	11/46	Shrewsbury	Shrewsbury	-	Shrewsbury	4/66
9658	11/46	Paddington	Paddington	-	Oxley	10/66
9659	11/46	Paddington	Paddington	-	Southall	6/65
9660	12/46	Tondu	Tondu	-	Neath	11/64
9661	12/46	Paddington	Paddington	-	Oxley	11/64
9701	9/33	Paddington	Paddington	-	Paddington	1/61
9702	9/33	Paddington	Paddington	-	Paddington	5/62
9703	9/33	Paddington	Paddington	-	Paddington	12/61
9704	10/33	Paddington	Paddington	-	Paddington	12/63
9705	10/33	Paddington	Paddington	-	Paddington	10/61
9706	12/33	Paddington	Paddington	-	Paddington	11/64
9707	11/33	Paddington	Paddington	-	Taunton	9/64
9708	12/33	Paddington	Paddington	-	Paddington	1/59
9709	12/33	Paddington	Paddington	-	Paddington	5/62
9710	12/33	Paddington	Paddington	-	Paddington	10/64
9711	6/34	Laira	Laira	-	Barrow Road	7/65
9712	6/34	Ebbw Junction	Aberdare	-	Aberdare	9/62
9713	6/34	Severn Tunnel	Cardiff	-	Nine Elms	7/63
9714	6/34	Stafford Road	Oxley	-	Goodwick	11/61
9715	6/34	Shrewsbury	Oxley	-	Nine Elms	7/63
9716	6/34	Taunton	Laira	-	Neath	6/65
9717	6/34	Newton Abbot	Newton Abbot	-	Hereford	12/62
9718	6/34	Exeter	Taunton	-	Taunton	5/62
9719	7/34	Stafford Road	Shrewsbury	-	Stourbridge	7/62
9720	11/34	Yeovil	Chippenham	-	Swindon	11/61
9721	11/34	Bristol SPM	Chippenham	-	Swindon	6/62
9722	12/34	Oxford	Reading	-	Slough	7/62
9723	12/34	Aberbeeg	Aberbeeg	-	Abercynon	7/62
9724	12/34	Tyseley	Tyseley	-	Stourbridge	1/66
9725	12/34	Paddington	Paddington	-	Paddington	12/62
9726	12/34	Paddington	Paddington	-	Didcot	6/65
9727	12/34	Stafford Road	Gloucester	-	Barry	11/62
9728	12/34	Stourbridge	Chester	-	Abercynon	5/62
9729	12/34	Bristol SPM	Bristol SPM	-	Westbury	10/64
9730	1/35	Stafford Road	Oxley	-	Pontypool	5/64
9731	1/35	Reading	Ebbw Junction	-	Aberdare	5/63
9732	1/35	Westbury	Bristol SPM	-	Swansea E Dock	4/64
9733	1/35	Bristol SPM	Tyseley	-	Stourbridge	9/65
9734	1/35	Neath	Neath	-	Neath	7/64
9735	1/35	Duffryn Yard	Duffryn Yard	-	Duffryn Yard	3/61
9736	1/35	Duffryn Yard	Duffryn Yard	-	Duffryn Yard	7/61
9737	1/35	Duffryn Yard	Duffryn Yard	-	Duffryn Yard	12/60
9738	1/35	Landore	Landore	-	Tondu	1/62
9739	1/35	Stafford Road	Oxley	-	Oxley	7/61
9740	1/35	Banbury	Leamington	-	Swindon	2/62
9741	1/35	Stafford Road	Stourbridge	-	Wellington	8/62
9742	1/35	Leamington	Oxley	-	Neath	9/64
9743	1/35	Llanelly	Llanelly	-	Neath	4/64

No.	Date	First Shed	Last GWR Shed	Wdn GWR	Last BR Shed	Wdn BR
9744	1/35	Neath	Swansea E Dock	-	Swansea E Dock	1/63
9745	2/35	Severn Tunnel	Severn Tunnel	-	Severn Tunnel	6/61
9746	2/35	Cardiff	Llantrisant	-	Swansea E Dock	3/64
9747	2/35	Banbury	Oxley	-	Aberdare	1/63
9748	2/35	Stafford Road	Tyseley	-	Swansea E Dock	1/64
9749	2/35	Southall	Reading	-	Reading	11/60
9750	5/35	Taunton	Neath	-	Neath	5/62
9751	5/35	Taunton	Paddington	-	Paddington	6/61
9752	5/35	Stafford Road	Oxley	-	Swansea E Dock	12/63
9753	5/35	Tyseley	Tyseley	-	Tyseley	5/65
9754	6/35	Paddington	Paddington	-	Yeovil Town	6/65
9755	6/35	Southall	Southall	-	Paddington	5/63
9756	6/35	Neath	Glyn Neath	-	Weymouth	11/62
9757	6/35	Southall	Taunton	-	Swansea E Dock	8/62
9758	6/35	Reading	Paddington	-	Paddington	5/62
9759	6/35	Yeovil	Cardiff	-	Abercynon	10/62
9760	9/35	Swansea E Dock	Goodwick	-	Duffryn Yard	12/63
9761	9/35	Llanelly	Landore	-	Swansea E Dock	10/62
9762	9/35	Paddington	Frome	-	Westbury	5/61
9763	9/35	Paddington	Reading	-	Croes Newydd	9/63
9764	9/35	Bristol SPM	Bristol SPM	-	Yeovil Town	8/63
9765	9/35	St. Blazey	Laira	-	Exeter	12/61
9766	10/35	Duffryn Yard	Duffryn Yard	-	Neath	7/64
9767	10/35	Stafford Road	Stourbridge	-	Stourbridge	6/61
9768	10/35	Tyseley	Oxley	-	Oxley	12/64
9769	10/35	Chester	Oxley	-	Westbury	3/63
9770	3/36	Taunton	Laira	-	Bath LMS	12/63
9771	3/36	Bristol SPM	Yeovil	-	Bristol SPM	5/61
9772	3/36	Bristol SPM	Swindon	-	Swindon	1/59
9773	3/36	Bristol SPM	Swindon	-	Llanelly	12/65
9774	3/36	Banbury	Chester	-	Tyseley	11/66
9775	3/36	Landore	Landore	-	Barry	11/62
9776	3/36	Landore	Gloucester	-	Oxley	4/66
9777	3/36	Swansea E Dock	Landore	-	Neath	5/64
9778	4/36	Severn Tunnel	Cardiff	-	Radyr	11/64
9779	4/36	Milford Haven	Neath N&B	-	Neath	2/64
9780	6/36	Tondu	Llantrisant	-	Radyr	7/65
9781	6/36	Duffryn Yard	Slough	-	Didcot	5/61
9782	7/36	Stafford Road	Banbury	-	Stourbridge	10/64
9783	7/36	Duffryn Yard	Neath	-	Neath N&B	5/62
9784	7/36	Bristol SPM	Paddington	-	Paddington	5/63
9785	5/36	Neath N&B	Duffryn Yard	-	Duffryn Yard	9/62
9786	5/36	Neath	Neath N&B	-	Neath	5/64
9787	5/36	Llanelly	Llanelly	-	Neath	9/64
9788	5/36	Neath	Llanelly	-	Neath	4/64
9789	5/36	Paddington	Slough	-	Oxford	12/65
9790	5/36	Chippenham	Swindon	-	Westbury	9/65
9791	5/36	Paddington	Reading	-	Didcot	1/64
9792	5/36	Brecon	Neath N&B	-	Neath	3/64
9793	6/36	Stafford Road	Tyseley	-	Croes Newydd	8/63
9794	6/36	Stourbridge	Chester	-	Barry	9/64
9795	9/36	Westbury	Swindon	-	Swindon	11/60
9796	9/36	Ebbw Junction	Cardiff	-	Pontypool	2/65
9797	9/36	Ebbw Junction	Pontypool	-	Pontypool	9/62
9798	9/36	Tyseley	Tyseley	-	Barry	10/64
9799	9/36	Duffryn Yard	Duffryn Yard	-	Duffryn Yard	10/63

No.	Date	First Shed	Last BR Shed	Wdn BR
1500	6/49	Paddington	Paddington	12/63
1501	7/49	Paddington	Southall	1/61
1502	7/49	Paddington	Didcot	1/61
1503	8/49	Paddington	Paddington	12/63
1504	8/49	Paddington	Paddington	5/63
1505	8/49	Paddington	Paddington	5/62
1506	9/49	Pill	Paddington	12/63
1507	9/49	Pill	Paddington	12/63
1508	9/49	Pill	Cardiff	9/62
1509	9/49	Pill	Ebbw Junction	8/59
1600	10/49	Barry	Barry	3/59
1601	10/49	Duffryn Yard	Whitland	8/60
1602	10/49	Duffryn Yard	Oswestry	9/60
1603	10/49	Machynlleth	Oswestry	6/59
1604	10/49	Oswestry	Oswestry	7/60
1605	11/49	Southall	Gloucester	2/62
1606	11/49	Danygraig	Llanelly	9/61
1607	11/49	Llanelly	Llanelly	8/65
1608	11/49	Newton Abbot	Gloucester	9/63
1609	11/49	Llanelly	Llanelly	7/62
1610	11/49	Cardiff E Dock	Abercynon	12/59
1611	11/49	Whitland	Llanelly	10/59
1612	11/49	Lydney	Radyr	7/65
1613	12/49	Carmarthen	Radyr	3/65
1614	12/49	Llanelly	Ebbw Junction	2/64
1615	12/49	Barry	Llanelly	6/61
1616	12/49	Lydney	Lydney	10/59
1617	12/49	Oxford	Hereford	11/63
1618	12/49	Llanelly	Croes Newydd	5/62
1619	12/49	Wellington	Croes Newydd	5/63
1620	6/50	Abercynon	Abercynon	6/60
1621	6/50	Birkenhead	Swindon	1/63
1622	6/50	Paxton Street	Slough	6/64
1623	6/50	Lydney	Llanelly	6/65
1624	6/50	Croes Newydd	St. Balzey	2/62
1625	8/50	Lydney	Gloucester	6/60
1626	8/50	St. Blazey	Lydney	8/62
1627	8/50	Lydney	Oxford	6/64
1628	8/50	Whitland	Croes Newydd	9/66
1629	9/50	Cathays	Worcester	6/60
1630	1/51	Lydney	Oxford	6/64
1631	1/51	Lydney	Hereford	11/64
1632	1/51	Lydney	Croes Newydd	4/65
1633	1/51	Llanelly	Llanelly	10/62
1634	2/51	Danygraig	Swindon	6/61
1635	2/51	Oswestry	Croes Newydd	10/59
1636	2/51	Oswestry	Didcot	6/64
1637	2/51	Whitland	Whitland	6/60
1638	3/51	Llanelly	Croes Newydd	8/66
1639	3/51	Lydney	Worcester	11/64
1640	3/51	Swansea E Dock	Swindon	7/61
1641	3/51	Swansea E Dock	Abercynon	11/64
1642	4/51	Lydney	Lydney	1/62
1643	4/51	Danygraig	Llanelly	10/65
1644	4/51	Neath	Llanelly	10/59
1645	4/51	Neath	Neath	10/62
1646	5/51	Swansea E Dock	Dingwall	12/62
1647	5/51	Swindon	Worcester	4/61
1648	5/51	Swindon	Neyland	5/63
1649	5/51	Bristol SPM	Dingwall	12/62

No.	Date	First Shed	Last BR Shed	Wdn BR
1650	11/54	Laira	Gloucester	2/64
1651	12/54	Llanelly	Llanelly	10/65
1652	12/54	Swansea E Dock	Swansea E Dock	1/60
1653	12/54	Ebbw Junction	Ebbw Junction	12/62
1654	12/54	Burry Port	Slough	6/64
1655	1/55	Burry Port	Llanelly	7/65
1656	1/55	Ebbw Junction	Ebbw Junction	6/64
1657	1/55	Hereford	Hereford	11/64
1658	2/55	Swindon	Swindon	11/64
1659	2/55	Carmarthen	Croes Newydd	10/60
1660	2/55	Croes Newydd	Croes Newydd	2/66
1661	3/55	Kidderminster	Worcester	7/64
1662	3/55	Worcester	Swindon	12/63
1663	3/55	Wellington	Croes Newydd	1/65
1664	3/55	St. Blazey	Swindon	11/64
1665	4/55	Llanelly	Llanelly	7/64
1666	4/55	Burry Port	Gloucester	2/64
1667	5/55	Hereford	Hereford	11/64
1668	5/55	Taunton	Oswestry	1/65
1669	5/55	Bristol SPM	Llanelly	10/65
3218	1/48	Banbury	Templecombe	5/65
3219	1/48	Worcester	Reading	12/63
3400	12/55	Cardiff E Dock	Radyr	11/64
3401	1/56	Cardiff E Dock	Radyr	11/64
3402	2/56	Cardiff E Dock	Radyr	11/64
3403	2/56	Cardiff E Dock	Radyr	9/64
3404	3/56	Cardiff E Dock	Radyr	7/62
3405	4/56	Cardiff E Dock	Radyr	11/64
3406	5/56	Cardiff E Dock	Radyr	11/64
3407	8/56	Cardiff E Dock	Radyr	10/62
3408	8/56	Cardiff E Dock	Radyr	9/62
3409	10/56	Cardiff E Dock	Radyr	10/64
4160	9/48	Barry	Severn Tunnel	6/65
4161	9/48	Barry	Worcester	11/65
4162	9/48	Treherbert	Leamington	7/60
4163	9/48	Barry	Cheltenham	10/62
4164	9/48	Duffryn Yard	Severn Tunnel	1/60
4165	10/48	Tyseley	Oxley	10/65
4166	10/48	Tyseley	Radyr	6/64
4167	10/48	Newton Abbot	Tyseley	5/64
4168	11/48	Neyland	Tyseley	10/65
4169	11/48	Neath	Radyr	5/65
4170	10/49	Tyseley	Tyseley	9/60
4171	10/49	Leamington	Leamington	10/64
4172	10/49	Tyseley	Stourbridge	1/65
4173	10/49	Stourbridge	Stourbridge	12/64
4174	11/49	Cheltenham	Pontypool	6/64
4175	11/49	Kidderminster	Banbury	10/65
4176	11/49	Exeter	Oxley	10/65
4177	11/49	Treherbert	Radyr	5/65
4178	11/49	Carmarthen	Oxley	9/65
4179	12/49	Newton Abbot	Tyseley	2/65
6760	11/48	Pill	Duffryn Yard	12/63
6761	11/48	Duffryn Yard	Paxton Street	1/61
6762	11/48	Tondu	Pill	3/63
6763	11/48	Danygraig	Swansea E Dock	12/63
6764	11/48	Ebbw Junction	Swansea E Dock	12/63
6765	12/48	Cardiff E Dock	Swansea E Dock	5/64
6766	1/49	Danygraig	Duffryn Yard	8/60
6767	1/49	Cardiff E Dock	Upper Bank	12/62

No.	Date	First Shed	Last BR Shed	Wdn BR
6768	1/49	Duffryn Yard	Duffryn Yard	1/64
6769	1/49	Barry	Bristol SPM	12/63
6770	10/50	Cardiff E Dock	Upper Bank	10/62
6771	10/50	Cardiff E Dock	Barry	3/58
6772	11/50	Pill	Swansea E Dock	12/63
6773	11/50	Cardiff E Dock	Paxton Street	10/59
6774	11/50	Barry	Upper Bank	12/59
6775	11/50	Barry	Pill	8/60
6776	11/50	Duffryn Yard	Upper Bank	1/61
6777	12/50	Duffryn Yard	Swansea E Dock	5/64
6778	12/50	Cardiff E Dock	Upper Bank	5/62
6779	12/50	Cardiff E Dock	Gurnos	6/59
6981	2/48	Weston-S-Mare	Bristol SPM	3/64
6982	1/48	Westbury	Barrow Road	8/64
6983	2/48	Paddington	Oxford	8/65
6984	2/48	Hereford	Barrow Road	12/65
6985	2/48	Oxley	Gloucester	9/64
6986	3/48	Tyseley	Didcot	4/65
6987	3/48	Gloucester	Cardiff E Dock	9/64
6988	3/48	Weymouth	Westbury	9/64
6989	3/48	Hereford	Gloucester	6/64
6990	4/48	Paddington	Barrow Road	12/65
6991	11/48	Westbury	Oxford	12/65
6992	11/48	Gloucester	Worcester	6/64
6993	12/48	Weymouth	Oxford	12/65
6994	12/48	Exeter	Oxley	11/64
6995	12/48	Taunton	Worcester	3/65
6996	1/49	Reading	Didcot	10/64
6997	1/49	Bath Road	Barrow Road	11/64
6998	1/49	Cardiff	Oxford	12/65
6999	2/49	Cardiff	Oxford	12/65
7008	5/48	Oxford	Paddington	9/64
7009	5/48	Landore	Gloucester	3/63
7010	6/48	Oxford	Reading	3/64
7011	6/48	Bath Road	Oxley	2/65
7012	6/48	Landore	Oxley	11/64
7013	7/48	Paddington	Gloucester (as 4082)	9/64
7014	7/48	Bath Road	Tyseley	2/65
7015	7/48	Swindon	Paddington	4/63
7016	8/48	Cardiff	Cardiff E Dock	11/62
7017	8/48	Paddington	Paddington	2/63
7018	5/49	Landore	Paddington	9/63
7019	5/49	Bath Road	Oxley	2/65
7020	5/49	Cardiff	Southall	9/64
7021	6/49	Landore	Paddington	9/63
7022	6/49	Cardiff	Gloucester	6/65
7023	6/49	Cardiff	Oxley	2/65
7024	6/49	Paddington	Oxley	2/65
7025	8/49	Paddington	Worcester	9/64
7026	8/49	Stafford Road	Tyseley	10/64
7027	8/49	Laira	Reading	12/63
7028	5/50	Landore	Llanelly	12/63
7029	5/50	Newton Abbot	Gloucester	12/65
7030	6/50	Paddington	Paddington	2/63
7031	6/50	Laira	Worcester	7/63
7032	6/50	Paddington	Paddington	9/64
7033	7/50	Paddington	Paddington	1/63
7034	8/50	Bath Road	Gloucester	6/65
7035	8/50	Shrewsbury	Paddington	6/64

No.	Date	First Shed	Last BR Shed	Wdn BR
7036	8/50	Paddington	Paddington	9/63
7037	8/50	Swindon	Swindon	3/63
7430	8/48	Stourbridge	Stourbridge	12/63
7431	8/48	Trawsfynnydd	Croes Newydd	8/64
7432	8/48	Stourbridge	Stourbridge	9/64
7433	8/48	Croes Newydd	Croes Newydd	2/61
7434	8/48	Oswestry	Oswestry	10/62
7435	9/48	Stourbridge	Aberdare	7/64
7436	9/48	Oxford	Bristol SPM	6/64
7437	9/48	Worcester	Llanelly	3/65
7438	10/48	Danygraig	Tyseley	2/59
7439	10/48	Llanelly	Llanelly	4/65
7440	1/50	Croes Newydd	Carmarthen	10/62
7441	1/50	Slough	Stourbridge	12/63
7442	2/50	Slough	Carmarthen	12/63
7443	2/50	Croes Newydd	Stourbridge	9/64
7444	2/50	Carmarthen	Llanelly	7/64
7445	3/50	Cathays	Carmarthen	3/64
7446	3/50	St. Blazey	Llanelly	7/64
7447	3/50	Croes Newydd	Stourbridge	4/59
7448	4/50	Stourbridge	Pontypool	4/63
7449	4/50	Stourbridge	Stourbridge	6/63
7820	11/50	Oswestry	Shrewsbury	11/65
7821	11/50	Oswestry	Shrewsbury	11/65
7822	12/50	Oswestry	Shrewsbury	11/65
7823	12/50	Chester	Tyseley	7/64
7824	12/50	Cheltenham	Oxley	11/64
7825	12/50	Croes Newydd	Reading	5/64
7826	12/50	Croes Newydd	Cardiff E Dock	4/65
7827	12/50	Chester	Shrewsbury	10/65
7828	12/50	Neath	Shrewsbury	10/65
7829	12/50	Neath	Gloucester	12/65
7900	4/49	Bath Road	Oxford	12/64
7901	3/49	Bath Road	Bristol SPM	2/64
7902	3/49	Paddington	Paddington	6/64
7903	4/49	Paddington	Cardiff E Dock	6/64
7904	4/49	Paddington	Oxford	12/65
7905	4/49	Laira	Banbury	5/64
7906	12/49	Bristol SPM	Oxford	3/65
7907	1/50	Bristol SPM	Oxford	12/65
7908	1/50	Bristol SPM	Tyseley	10/65
7909	1/50	Laira	Oxford	11/65
7910	1/50	Southall	Oxford	2/65
7911	2/50	Paddington	Oxford	12/63
7912	3/50	Stafford Road	Banbury	10/65
7913	3/50	Tyseley	Severn Tunnel	3/65
7914	3/50	Swindon	Barrow Road	12/65
7915	3/50	Stafford Road	Tyseley	10/65
7916	4/50	St. Blazey	Cardiff E Dock	12/64
7917	4/50	Bath Road	Oxford	8/65
7918	4/50	Tyseley	Tyseley	2/65
7919	5/50	Reading	Oxford	12/65
7920	9/50	Gloucester	Worcester	6/65
7921	9/50	Chester	Paddington	12/63
7922	9/50	Carmarthen	Oxford	12/65
7923	9/50	Swindon	Cardiff E Dock	6/65
7924	9/50	Westbury	Oxford	12/65
7925	10/50	Penzance	Oxford	12/65
7926	10/50	Gloucester	Gloucester	12/64
7927	10/50	Reading	Oxford	12/65

No.	Date	First Shed	Last BR Shed	Wdn BR
7928	10/50	Worcester	Worcester	3/65
7929	11/50	Tyseley	Tyseley	8/65
8400	8/49	Tyseley	Bromsgrove	9/64
8401	8/49	Cardiff	Bromsgrove	9/64
8402	9/49	Pill	Gloucester	11/64
8403	9/49	Newton Abbot	Barrow Road	6/65
8404	9/49	Laira	Bromsgrove	11/61
8405	12/49	Tyseley	Bromsgrove	9/64
8406	12/49	Ebbw Junction	Bromsgrove	1/61
8407	12/49	Tyseley	Duffryn Yard	10/62
8408	1/50	Danygraig	Swansea E Dock	9/59
8409	1/50	Penzance	Bromsgrove	8/64
8410	1/50	Tyseley	Aberbeeg	1/60
8411	1/50	Stafford Road	Stourbridge	6/60
8412	1/50	Truro	Worcester	7/59
8413	1/50	Bristol SPM	Southall	1/61
8414	1/50	Cardiff E Dock	Swansea E Dock	4/64
8415	3/50	Tyseley	Worcester	6/65
8416	3/50	Cardiff E Dock	Duffryn Yard	10/62
8417	3/50	Oxley	Tyseley	3/59
8418	3/50	Stourbridge	Bromsgrove	8/64
8419	4/50	Stourbridge	Barry	1/60
8420	7/50	Neath	Paddington	6/65
8421	7/50	Exeter	Gloucester	12/59
8422	10/50	St. Blazey	Southall	7/62
8423	10/50	Chester	Swansea E Dock	12/59
8424	12/50	Llanelly	Oxford	1/61
8425	2/51	Laira	Cardiff E Dock	11/63
8426	1/51	St. Blazey	Southall	11/63
8427	2/51	Worcester	Worcester	9/60
8428	2/51	Oxley	Oxley	10/62
8429	3/51	Cardiff E Dock	Aberbeeg	1/60
8430	1/53	Reading	Reading	5/63
8431	1/53	Swansea E Dock	Neath	8/64
8432	2/53	Paddington	Oxford	7/59
8433	3/53	Paddington	Paddington	6/65
8434	2/53	Paddington	Paddington	6/59
8435	4/53	Didcot	Paddington	2/62
8436	5/53	Aberbeeg	Bath LMS	6/65
8437	5/53	Stourbridge	Aberbeeg	11/64
8438	6/53	Stourbridge	Radyr	10/62
8439	6/53	Cardiff	Neath	10/62
8440	3/54	Ebbw Junction	Aberbeeg	7/62
8441	4/54	Cardiff E Dock	Cardiff	12/61
8442	3/54	Danygraig	Neath	6/59
8443	3/54	Swansea E Dock	Swansea E Dock	6/59
8444	4/54	Aberdare	Aberbeeg	7/63
8445	6/54	Aberdare	Cardiff E Dock	9/62
8446	6/54	Cardiff	Barry	9/64
8447	7/54	Cardiff	Cardiff	6/59
8448	6/54	Tondu	Tondu	8/59
8449	6/54	Oxley	Shrewsbury	9/62
8450	8/49	Ebbw Junction	Barry	6/59
8451	9/49	Cardiff E Dock	Southall	11/61
8452	10/49	Tyseley	Cardiff E Dock	4/64
8453	11/49	Ebbw Junction	Tondu	10/62
8454	12/49	Leamington	Duffryn Yard	1/61
8455	1/50	Barry	Radyr	1/60
8456	2/50	Taunton	Southall	10/63
8457	3/50	Cardiff E Dock	Cardiff	1/61

No.	Date	First Shed	Last BR Shed	Wdn BR
8458	4/50	Cardiff E Dock	Paddington	8/63
8459	5/50	Banbury	Paddington	6/65
8460	6/50	Barry	Cardiff	1/61
8461	6/50	Barry	Pontypool	11/53
8462	7/50	Stafford Road	Bath Road	8/59
8463	9/50	Tyseley	Landore	1/60
8464	10/50	Duffryn Yard	Reading	12/63
8465	11/50	Duffryn Yard	Southall	11/63
8466	11/50	Duffryn Yard	Southall	7/64
8467	2/51	Llanelly	Llanelly	2/62
8468	2/51	Tyseley	Cardiff	5/60
8469	3/51	Barry	Radyr	11/64
8470	5/51	Barry	Radyr	1/62
8471	5/51	Rhymney	Barrow Road	6/65
8472	7/51	Swindon	Paddington	3/63
8473	8/51	Penzance	Penzance	1/61
8474	10/51	Llanelly	Llanelly	5/65
8475	11/51	Llanelly	Radyr	9/64
8476	3/52	Danygraig	Swansea E Dock	1/61
8477	3/52	Llanelly	Llanelly	7/62
8478	5/52	Cardiff E Dock	Radyr	1/63
8479	6/52	Westbury	Radyr	10/64
8480	3/52	Worcester	Neath	7/64
8481	3/52	Radyr	Paddington	6/65
8482	4/52	Radyr	Duffryn Yard	9/62
8483	5/52	Danygraig	Swansea E Dock	2/62
8484	5/52	Cardiff E Dock	Radyr	6/65
8485	6/52	Truro	St. Blazey	6/59
8486	6/52	Truro	Bath LMS	6/65
8487	7/52	Gloucester	Paddington	11/63
8488	7/52	Gloucester	Llanelly	5/65
8489	7/52	Cardiff E Dock	Aberbeeg	2/62
8490	7/52	Duffryn Yard	Duffryn Yard	10/62
8491	8/52	Bristol SPM	Gloucester	7/63
8492	8/52	Bristol SPM	Bristol SPM	6/59
8493	9/52	Ebbw Junction	Aberbeeg	11/64
8494	9/52	Ebbw Junction	Didcot	1/62
8495	10/52	Ferndale	Radyr	11/64
8496	10/52	Worcester	Reading	7/63
8497	11/52	Tondu	Radyr	7/64
8498	11/52	Tondu	Southall	6/65
8499	11/52	Tondu	Aberbeeg	6/62
9410	2/50	Barry	Paddington	7/62
9411	2/50	Barry	Paddington	6/65
9412	3/50	Barry	Neath	3/63
9413	3/50	Didcot	Southall	11/63
9414	4/50	Slough	Slough	8/60
9415	5/50	Slough	Paddington	6/65
9416	5/50	Oxford	Aberbeeg	1/62
9417	5/50	Didcot	Southall	6/59
9418	5/50	Paddington	Southall	6/65
9419	6/50	Paddington	Paddington	3/63
9420	6/50	Reading	Paddington	3/64
9421	7/50	Slough	Slough	2/62
9422	7/50	Paddington	Paddington	12/63
9423	7/50	Reading	Paddington	3/63
9424	8/50	Slough	Ebbw Junction	11/62
9425	9/50	Chester	Aberbeeg	11/63
9426	9/50	Banbury	Radyr	5/65
9427	9/50	Stourbridge	Southall	6/59

No.	Date	First Shed	Last BR Shed	Wdn BR
9428	9/50	Stafford Road	Stafford Road	6/60
9429	10/50	Worcester	Llanelly	12/63
9430	10/50	Neath	Barrow Road	6/65
9431	11/50	Duffryn Yard	Swansea E Dock	4/64
9432	11/50	Tyseley	Tyseley	11/59
9433	12/50	Laira	Laira	7/62
9434	1/51	Truro	Penzance	6/60
9435	1/51	Stafford Road	Paddington	9/64
9436	1/51	Neath	Landore	7/60
9437	1/51	Duffryn Yard	Radyr	6/65
9438	2/51	Banbury	Gloucester	6/59
9439	2/51	Exeter	Exeter	6/59
9440	2/51	Newton Abbot	Paddington	7/63
9441	3/51	Worcester	Swansea E Dock	11/63
9442	3/51	Neath	Neath	7/64
9443	3/51	Llanelly	Cardiff	6/59
9444	4/51	Duffryn Yard	Aberbeeg	3/63
9445	4/51	Gloucester	Gloucester	1/60
9446	5/51	Neath	Radyr	5/65
9447	5/51	Neath	Reading	1/61
9448	5/51	Neath	Neath	7/62
9449	6/51	Banbury	Banbury	6/60
9450	6/51	Stourbridge	Reading	6/64
9451	7/51	Neath	Tondu	7/62
9452	7/51	Llanelly	Llanelly	5/65
9453	7/51	Bristol SPM	Gloucester	11/64
9454	9/51	Duffryn Yard	Duffryn Yard	1/62
9455	9/51	Neath	Paddington	4/63
9456	9/51	Duffryn Yard	Radyr	4/64
9457	10/51	Danygraig	Radyr	7/64
9458	10/51	Ebbw Junction	Aberbeeg	1/61
9459	11/51	Ebbw Junction	Aberbeeg	9/59
9460	11/51	Cardiff	Aberbeeg	2/62
9461	12/51	Duffryn Yard	Radyr	5/65
9462	12/51	Newton Abbot	Newton Abbot	11/60
9463	1/52	Penzance	Southall	6/65
9464	1/52	Gloucester	Radyr	6/65
9465	1/52	Llanelly	Llanelly	2/62
9466	2/52	Gloucester	Radyr	7/64
9467	2/52	Laira	Laira	5/62
9468	2/52	Llanelly	Ebbw Junction	8/60
9469	3/52	Llanelly	Paddington	3/62
9470	3/52	Radyr	Paddington	9/64
9471	4/52	Gloucester	Gloucester	9/64
9472	4/52	Llanelly	Llanelly	5/65
9473	5/52	Neath	Neath	7/64
9474	5/52	Llanelly	Oxley	11/61
9475	6/52	Worcester	Radyr	5/65
9476	6/52	Swindon	Swindon	6/62
9477	7/52	Stourbridge	Southall	6/65
9478	7/52	Neath	Neath	7/62
9479	7/52	Llanelly	Paddington	7/63
9480	8/52	Worcester	Radyr	4/65
9481	8/52	Bristol SPM	Bristol SPM	1/61
9482	9/52	Ebbw Junction	Aberbeeg	11/63
9483	10/52	Duffryn Yard	Duffryn Yard	7/63
9484	10/52	Landore	Swansea E Dock	4/64
9485	10/52	Danygraig	Llanelly	7/64
9486	11/52	Neath	Worcester	7/62
9487	11/52	Duffryn Yard	Exeter	7/62

No.	Date	First Shed	Last BR Shed	Wdn BR
9488	12/52	Bristol SPM	Radyr	4/65
9489	1/53	Swansea E Dock	Swansea E Dock	4/64
9490	2/54	Ebbw Junction	Worcester	12/64
9491	3/54	Danygraig	Swansea E Dock	6/59
9492	5/54	Whitland	Gloucester	6/59
9493	6/54	Cardiff	Bromsgrove	9/64
9494	8/54	Cardiff E Dock	Aberbeeg	11/64
9495	10/54	Bristol SPM	Southall	6/65
9496	10/54	Stafford Road	Stafford Road	12/59
9497	12/54	Exeter	Exeter	5/62
9498	3/55	Tyseley	Paddington	9/64
9499	7/55	Ebbw Junction	Aberbeeg	9/59
9662	4/48	Ebbw Junction	Ebbw Junction	9/65
9663	4/48	Taunton	Taunton	9/64
9664	4/48	Ebbw Junction	Ebbw Junction	5/64
9665	4/48	Bristol SPM	Hereford	2/63
9666	4/48	Neath	Ebbw Junction	9/65
9667	5/48	Ebbw Junction	Radyr	6/65
9668	5/48	Newton Abbot	Abercynon	12/63
9669	5/48	Croes Newydd	Croes Newydd	1/66
9670	5/48	Taunton	Yeovil Town	6/65
9671	6/48	Laira	Cardiff E Dock	3/65
9672	6/48	Shrewsbury	Gloucester	12/65
9673	2/49	Laira	St. Blazey	5/60
9674	3/49	Tondu	Westbury	4/64
9675	3/49	Merthyr	Ebbw Junction	10/65
9676	3/49	Cathays	Cardiff E Dock	6/65
9677	3/49	Cardiff E Dock	Llanelly	11/64
9678	4/49	Newton Abbot	Cardiff E Dock	6/65
9679	4/49	Cardiff E Dock	Merthyr	11/64
9680	4/49	Tyseley	Gloucester	12/65
9681	5/49	Tondu	Cardiff E Dock	8/65
9682	5/49	Tyseley	Radyr	8/65
15101	4/48	Paddington	Bescot	8/67
15102	4/48	Paddington	Bescot	7/67
15103	5/48	Paddington	Bescot	8/67
15104	5/48	Paddington	Bescot	7/67
15105	7/48	Paddington	Bescot	8/67
15106	7/48	Paddington	Bescot	8/67
15107	11/49	Bristol SPM	Bristol SPM	6/58
18000	5/50	Paddington	Paddington	12/60
18100	4/52	Paddington	Paddington	1/58

NOTES ON Nos. 6700-49 (New in 1930-1):

Nos. 6700-9/25-32 went into traffic immediately at the sheds listed.
Nos. 6710-4/33-5/7 went into traffic immediately at the sheds listed but were put into store in 8/30 (No. 6737 in 9/30); they later enterred traffic as follows:-

6710	7/32 Bristol SPM	6712	Westbury	6733	7/32 Swindon
6711	7/32 Swindon	6713/4	Swansea E Dock	6734/5/7	8/32 Swindon

Nos. 6715-24/36/8-49 went into store and entered traffic at the depots given at the following dates:-

6715 8/32; 6716-9 7/32; 6720-3 8/32; 6724 7/32; 6736/8-49 8/32

WITHDRAWN AND RE-INSTATED:

5775 1/63 from Pontypool; re-instated 1/63 to Pontypool,
6145 6/65 from Didcot; re-instated 6/65 to Oxford,
6638 4/63 from Radyr; re-instated 4/63 to Radyr,
6826 8/64 from Cardiff E Dock; re-instated 9/64 to Cardiff E Dock,
6836 3/65 from Worcester; re-instated 4/65 to Worcester,
9726 12/64 from Didcot; re-instated 1/65 to Southall.

Diesel Railcars

The AEC railcars were built to the following Lot numbers (carriage series):-

1516	No. 1	1547	Nos. 8, 10-2/7	1636	No. 34
1522	Nos. 2-7	1564	No. 18	1637	Nos. 35-8
1546	Nos. 9, 13-6	1635	Nos. 19-33		

No.	Date	First Shed	Last GWR Shed	Wdn GWR	Last BR Shed	Wdn BR
1	2/34	Southall	Reading	-	Oxford	8/55
2	7/34	Tyseley	Ebbw Junction	-	Ebbw Junction	2/54
3	7/34	Cardiff	Landore	-	Oxford	3/55
4	9/34	Tyseley	Landore	-	Ebbw Junction	7/58
5	7/35	Oxford	Worcester	-	Worcester	12/57
6	8/35	Worcester	Worcester	-	Worcester	4/58
7	7/35	Worcester	Worcester	-	Stourbridge	1/59
8	3/36	Stourbridge	Stourbridge	-	Stourbridge	1/59
9	2/36	Oxford	Oxford	5/46	-	-
10	2/36	Banbury	Oxford	-	Reading	4/56
11	2/36	Weymouth	Oxford	-	Landore	11/56
12	2/36	Oxford	Reading	-	Oxford	6/57
13	3/36	Neath	Carmarthen	-	Ebbw Junction	8/60
14	3/36	Pontypool	Worcester	-	Storubridge	8/60
15	4/36	Cheltenham	Carmarthen	-	Stourbridge	1/59
16	4/36	Ebbw Junction	Landore	-	Reading	10/57
17	4/36	Southall	Southall	-	Leamington	1/59
18	4/37	Reading	Llantrisant	-	Reading	5/57
19	7/40	Ebbw Junction	Reading	-	Cheltenham	2/60
20	6/40	Ebbw Junction	Yeovil	-	Worcester	10/62
21	7/40	Bristol	Yeovil	-	Southall	8/62
22	9/40	Ebbw Junction	Worcester	-	Worcester	10/62
23	9/40	Ebbw Junction	Ebbw Junction	-	Worcester	10/62
24	9/40	Bristol	Yeovil	-	Worcester	10/62
25	9/40	Worcester	Cheltenham	-	Bristol	8/62
26	9/40	Cheltenham	Leamington	-	Worcester	10/62
27	11/40	Ebbw Junction	Worcester	-	Southall	9/60
28	12/40	Ebbw Junction	Yeovil	-	Bristol	9/60
29	1/41	Ebbw Junction	Leamington	-	Reading	8/62
30	1/41	Ebbw Junction	Pontypool	-	Southall	8/62
31	2/41	Llanelly	Worcester	-	Southall	8/62
32	2/41	Llanelly	Worcester	-	Worcester	10/62
33	3/41	Worcester	Stourbridge	-	Reading	8/62
34	9/41	Southall	Southall	-	Southall	9/60
35	11/41	Stourbridge	Reading	-	Bristol	4/57
36	11/41	Stourbridge	Reading	-	Bristol	4/57
37	2/42	Stourbridge	Yeovil	-	Bristol	9/49
38	2/42	Stourbridge	Yeovil	-	Reading	8/62

NOTE:

Nos. 21-7 were put into store at Cherry Orchard (near Cardiff) when completed.
Nos. 19-38 went into traffic as follows:-

19	12/40	23	9/41	27	5/41	31	5/41	35	12/41
20	6/40	24	6/41	28	1/41	32	5/41	36	12/41
21	6/41	25	8/41	29	5/41	33	5/41	37	4/42
22	9/41	26	3/42	30	5/41	34	10/41	38	4/42

Sheds of 2-8-0 Purchased 1925

These locomotives, purchased in bulk in May 1925, were built by the following manufacturers (the GWR numbers are those first allotted in 1925):-

1st GWR Nos.	ROD Nos.	Builder	Order No.	1st GWR Nos.	ROD Nos.	Builder	Order No.
3020	1648	RS 3732	E106	3058	1897	NBL 21874	L693
3021	1660	RS 3704	E104	3059	1940	NBL 21917	L693
3022	1698	RS 3746	E107	3060-9	1944/7/	NBL 22202/5/	L703
3023	1704	NW 1247	L926		50/1/8/	8/9/16/8/9/	
3024	1725	NW 1281	L1550		60/1/4/	22/4/5	
3025	1635	RS 3755	E109		6/7		
3026/7	1726/30	NW 1282/6	L1550	3070	1977	NBL 22035	L704
3028/9	1644/6	RS 3752/4	E108	3071	1919	NBL 21896	L693
3030/1	1731/2	NW 1287/8	L1550	3072-7	1982/3/	NBL 22040/1/	L704
3032	1708	NW 1251	L926		7/9/90/	5/7/8/50	
3033	1830	NBL 21797	L689		2		
3034-7	1733-6	RS 3760-3	E109	3078	1935	NBL 21912	L693
3038	1613	K 5199	479	3079	1623	K 5209	479
3039	1740	RS 3767	E109	3080/1	1996/7	NBL 22054/5	L704
3040	2146	NBL 22236	L714	3082-4	2069/	NBL 22160/2/	L712
3041	1742	RS 3769	E109		71/2	3	
3042-4	2154-6	NBL 22244-6	L714	3085	1681	RS 3725	E105
3045	1744	RS 3771	E109	3086-90	2105-9	NBL 22195-9	L713
3046-50	1790-4	NBL 22083-7	L707	3091	1826	NBL 21793	L689
3051	1985	NBL 22043	L704	3092-7	2113/4/	NBL 22203/4/	L713
3052/3	1795/6	NBL 22088/9	L707		7/8/20/	7/8/10/2	
3054/5	1810/8	NBL 21777/85	L689		2		
3056	1838	NBL 21805	L689	3098	1709	NW 1252	L926
3057	1852	NBL 21829	L692	3099	2124	NBL 22214	L713

Most of these engines never worked during 1925 (sheds in brackets indicate non-working state) and in October of that year all were taken out of traffic. They were sorted into two groups, renumbered as shown in the last column and returned to traffic in 1926-7 as shown in the main lists.

1st GWR No.	Date	1st 1925 Shed	Last 1925 Shed	2nd No.
3020	10/18	(Paddington)	(Paddington)	3056
3021	11/17	(Bristol SPM)	(Bristol SPM)	3050
3022	3/19	Cardiff	Cardiff	3025
3023	-/17	(Paddington)	(Paddington)	3027
3024	-/19	(Oxley)	(Oxley)	3028
3025	-/19	Carmarthen	Carmarthen	3057
3026	9/19	(Oxley)	(Oxley)	3026
3027	-/19	(Oxley)	(Oxley)	3069
3028	5/19	(Newton Abbot)	(Newton Abbot)	3038
3029	6/19	(Landore)	(Landore)	3071
3030	-/19	(Oxley)	(Oxley)	3061
3031	-/19	(Oxley)	(Oxley)	3075
3032	-/18	Paddington	Paddington	3032
3033	11/17	(Reading)	(Reading)	3049
3034	8/19	(Carmarthen)	(Carmarthen)	3084
3035	9/19	(Oxley)	(Oxley)	3085
3036	9/19	(Oxley)	(Oxley)	3086
3037	9/19	(Oxley)	(Oxley)	3087
3038	5/18	(Neath)	Carmarthen	3035
3039	10/19	(Llanelly)	(Llanelly)	3090

1st GWR No.	Date	1st 1925 Shed	Last 1925 Shed	2nd No.
3040	6/19	(Aberdare)	Severn Tunnel	3091
3041	11/19	(Landore)	(Landore)	3093
3042	7/19	Ebbw Junction	Ebbw Junction	3040
3043	7/19	(Neath)	(Neath)	3043
3044	7/19	(Paddington)	(Paddington)	3044
3045	11/19	(Newton Abbot)	Exeter	3097
3046	3/19	(Oxley)	(Oxley)	3022
3047	3/19	(Oxley)	(Oxley)	3098
3048	3/19	(Oxley)	(Oxley)	3048
3049	3/19	(Oxley)	(Oxley)	3021
3050	3/19	(Oxley)	(Oxley)	3020
3051	12/18	(Paddington)	(Paddington)	3082
3052	4/19	(Oxley)	(Oxley)	3024
3053	4/19	(Oxley)	(Oxley)	3029
3054	9/17	(Ebbw Junction)	(Ebbw Junction)	3030
3055	10/17	Swindon	Swindon	3055
3056	12/17	Bristol SPM	Bristol SPM	3031
3057	-/18	(Paddington)	(Paddington)	3033
3058	2/18	(Reading)	(Reading)	3034
3059	8/18	(Oxley)	(Worcester)	3059
3060	10/18	(Oxley)	(Oxley)	3060
3061	10/18	(Gloucester)	(Neath)	3089
3062	10/18	(Oxley)	(Oxley)	3062
3063	10/18	(Oxley)	Ebbw Junction	3063
3064	12/18	(Oxley)	(Oxley)	3064
3065	12/18	(Oxley)	(Oxley)	3023
3066	1/19	(Oxley)	(Hereford)	3066
3067	1/19	(Oxley)	(Oxley)	3067
3068	2/19	(Oxley)	(Oxley)	3068
3069	2/19	(Oxley)	(Hereford)	3036
3070	10/18	(Oxley)	(Oxley)	3070
3071	5/18	(Cardiff)	(Cardiff)	3037
3072	11/18	(Oxley)	(Ebbw Junction)	3072
3073	11/18	(Oxley)	(Oxley)	3073
3074	12/18	(Oxley)	(Oxley)	3074
3075	12/18	(Oxley)	(Oxley)	3054
3076	12/18	(Oxley)	(Oxley)	3076
3077	1/19	(Neath)	Carmarthen	3077
3078	7/18	(Paddington)	Carmarthen	3078
3079	8/18	(Cardiff)	Cardiff	3079
3080	2/19	(Oxley)	(Oxley)	3080
3081	2/19	(Oxley)	(Oxley)	3081
3082	7/19	(Neath)	Neath	3058
3083	8/19	(Llanelly)	(Llanelly)	3083
3084	8/19	(Swindon)	Swindon	3051
3085	6/18	(Newton Abbot)	Exeter	3065
3086	7/19	(Bristol SPM)	Bristol SPM	3039
3087	7/19	Bristol SPM	Bristol SPM	3052
3088	7/19	(Ebbw Junction)	(Pontypool)	3088
3089	7/19	(Ebbw Junction)	(Ebbw Junction)	3094
3090	7/19	(Ebbw Junction)	(Pontypool)	3041
3091	11/17	(Paddington)	(Paddington)	3042
3092	7/19	(Paddington)	(Paddington)	3092
3093	7/19	(Swindon)	(Swindon)	3045
3094	7/19	(Bristol SPM)	Bristol SPM	3046
3095	7/19	(Aberdare)	Aberdare	3095
3096	8/19	(Llanelly)	(Llanelly)	3096
3097	8/19	(Bristol SPM)	(Bristol SPM)	3053
3098	-/18	(Reading)	(Reading)	3047
3099	8/19	(Ebbw Junction)	(Ebbw Junction)	3099

Locomotives Lent to GWR 1939-47

LMS 0-6-0 Diesel-Electric Shunter

LMS No.	To GWR	First GWR Shed	Last GWR Shed	Left GWR
7069	9/39	+	+	3/40
7076	3/40	+	+	9/41

+ Not shown in GWR records but used at Swansea Docks.

LMS 2F 0-6-0

LMS No.	To GWR	First GWR Shed	Last GWR Shed	Left GWR
3023	10/39	Westbury	Weymouth	9/45
3027	1/40	Oxford	Oxford	11/45
3038	2/40	Didcot	Oswestry	11/45
3039	12/39	Bristol SPM	Oswestry	11/45
3047	10/39	Bristol SPM	Oswestry	11/45
3048	12/39	Westbury	Westbury	9/45
3071	3/40	Bristol SPM	Bristol SPM	9/45
3078	1/40	Bristol SPM	Bristol SPM	9/45
3081	1/40	Southall	Southall	9/45
3085	12/39	Bristol SPM	Oswestry	11/45
3090	3/40	Bristol SPM	Westbury	9/45
3094	2/40	Bristol SPM	Bristol SPM	9/45
3096	1/40	Westbury	Westbury	9/45
3103	2/40	Bristol SPM	Bristol SPM	9/45
3108	11/39	Didcot	Didcot	9/45
3109	10/39	Paddington	Oxford	9/45
3113	2/40	Paddington	Southall	11/45
3119	10/39	Southall	Tyseley	10/45
3121	11/39	Paddington	Didcot	9/45
3126	12/39	Southall	Machynlleth	11/45
3127	11/39	Paddington	Duffryn Yard	10/45
3196	12/39	Reading	Oswestry	11/45
3372	10/39	Paddington	Oxford	11/45
3473	2/40	Bristol SPM	Swindon	9/45
3485	11/39	Paddington	Oxford	10/45
3492	4/40	Bristol SPM	Swindon	10/45
3516	11/39	Paddington	Leamington	11/45
3517	10/39	Westbury	Bristol SPM	9/45
3526	12/39	Westbury	Duffryn Yard	10/45
3536	12/39	Bristol SPM	Moat Lane	8/45
3543	11/39	Westbury	Bristol SPM	10/45
3545	11/39	Paddington	Southall	9/45
3564	10/39	Oxford	Didcot	9/45
3603	12/39	Westbury	Bristol SPM	10/45
3616	3/40	Reading	Duffryn Yard	10/45
3688	1/40	Paddington	Southall	9/45
3689	11/39	Westbury	Oswestry	11/45
3696	1/40	Slough	Slough	9/45
3725	2/40	Paddington	Southall	9/45
3739	2/40	Bristol SPM	Llanidloes	11/45

NOTE:

LMS Nos. 3437, 3445 and 3758 have been listed as on loan to the GWR, but they do not appear in the company's records of locomotive allocations.

LNER J25 0-6-0

LNER No.	To GWR	First GWR Shed	Last GWR Shed	Left GWR
29	11/39	Stourbridge	Stourbridge	1/46
257	11/39	Shrewsbury	Shrewsbury	8/46
536	11/39	Shrewsbury	Shrewsbury	8/46
1725	3/40	Tyseley	Tyseley	8/46
1963	1/40	Croes Newydd	Croes Newydd	8/46
1964	3/40	Gloucester	Gloucester	3/43
1967	3/40	Croes Newydd	Croes Newydd	11/46
1969	10/39	Tyseley	Tyseley	8/46
1970	2/40	Croes Newydd	Stourbridge	1/46
1973	1/40	Stourbridge	Stourbridge	3/43
1981	1/40	Gloucester	Gloucester	11/46
1982	10/39	Tyseley	Tyseley	8/46
1983	3/40	Croes Newydd	Tyseley	3/43
1986	10/39	Worcester	Kidderminster	1/46
1989	11/39	Chester	Worcester	1/46
1992	1/40	Chester	Chester	8/46
1994	2/40	Hereford	Kidderminster	1/46
2000	1/40	Stourbridge	Stourbridge	3/43
2040	1/40	Worcester	Kidderminster	1/46
2043	2/40	Croes Newydd	Stourbridge	12/45
2047	10/39	Tyseley	Tyseley	3/43
2051	3/40	Worcester	Kidderminster	11/46
2053	3/40	Hereford	Kidderminster	1/46
2058	2/40	Worcester	Worcester	3/43
2059	2/40	Croes Newydd	Shrewsbury	8/46
2061	10/39	Worcester	Worcester	1/46
2065	10/39	Worcester	Worcester	3/43
2069	10/39	Tyseley	Banbury	3/43
2071	11/39	Gloucester	Shrewsbury	11/46
2072	2/40	Hereford	Evesham	11/46
2073	11/39	Gloucester	Evesham	1/46
2075	10/39	Worcester	Hereford	1/46
2076	11/39	Gloucester	Gloucester	11/46
2126	3/40	Stourbridge	Stourbridge	8/46
2134	3/40	Chester	Croes Newydd	11/46
2135	11/39	Gloucester	Gloucester	3/43
2136	1/40	Gloucester	Worcester	11/46
2138	1/40	Worcester	Worcester	11/46
2141	2/40	Worcester	Worcester	3/43
2142	2/40	Worcester	Shrewsbury	11/46

NOTE:

LNER Nos. 257, 536, 1725, 1963/7/9/81/2/92, 2051/9/71/2/6, 2126/34/6/8/42 were altered to their new LNER numbers (5713, 5671/2, 5647/9/50/6/7/66, 5704/10, 5696/7, 5700/15/20/2/3/7 respectively) while on loan to the GWR.

LMS(WD) 8F 2-8-0

LMS No.	To GWR	First GWR Shed	Last GWR Shed	Left GWR
8226	10/40	Severn Tunnel	Severn Tunnel	9/41
8237	10/40	Severn Tunnel	Severn Tunnel	9/41

LMS No.	To GWR	First GWR Shed	Last GWR Shed	Left GWR
8238	10/40	Severn Tunnel	Severn Tunnel	1/41
8239	10/40	Neath	Neath	7/41
8240	10/40	Llanelly	Llanelly	9/41
8241	10/40	Ebbw Junction	Ebbw Junction	1/41
8242	10/40	Ebbw Junction	Severn Tunnel	1/41
8243	11/40	Neath	Neath	1/41
8244	11/40	Llanelly	Llanelly	9/41
8252	12/40	Ebbw Junction	Severn Tunnel	1/41
8253	12/40	Ebbw Junction	Severn Tunnel	1/41
8257	12/40	Ebbw Junction	Cardiff	7/41
8261	12/40	Ebbw Junction	Severn Tunnel	7/41
8286	9/40	Ebbw Junction	Ebbw Junction	9/41
8287	9/40	Neath	Neath	7/41
8288	9/40	Llanelly	Severn Tunnel	9/41
8289	9/40	Ebbw Junction	Ebbw Junction	7/41
8290	9/40	Ebbw Junction	Ebbw Junction	9/41
8291	9/40	Ebbw Junction	Ebbw Junction	1/41
8292	10/40	Ebbw Junction	Llanelly	9/41
8293	10/40	Ebbw Junction	Cardiff	10/41
8294	10/40	Cardiff	Cardiff	9/41
8295	10/40	Ebbw Junction	Cardiff	9/41
8296	11/40	Landore	Severn Tunnel	9/41
8300	12/40	Ebbw Junction	Ebbw Junction	9/41

LNER 04 2-8-0

LNER No.	To GWR	First GWR Shed	Last GWR Shed	Left GWR
5391	12/40	Bristol SPM	Bristol SPM	7/42
5393	11/40	Chester	Chester	2/43
6205	11/40	Southall	Southall	7/42
6209	11/40	Bristol SPM	Bristol SPM	8/42
6220	11/40	Oxley	Oxley	2/43
6231	11/40	Banbury	Oxley	2/43
6232	11/40	Tyseley	Tyseley	4/41
6236	11/40	Oxley	Oxley	4/41
6245	11/40	Bristol SPM	Bristol SPM	2/43
6258	12/40	Ebbw Junction	Oxley	7/42
6265	12/40	Ebbw Junction	Oxley	2/43
6275	11/40	Paddington	Southall	1/43
6277	11/40	Oxley	Chester	2/43
6295	11/40	Oxley	Shrewsbury	2/43
6321	11/40	Oxley	Oxley	2/43
6336	11/40	Oxley	Shrewsbury	2/43
6350	11/40	Oxley	Tyseley	2/43
6365	11/40	Oxley	Tyseley	2/43
6524	11/40	Oxley	Oxley	1/43
6540	11/40	Oxley	Oxley	7/42
6562	11/40	Paddington	Southall	1/43
6566	11/40	Paddington	Southall	2/43
6573	11/40	Paddington	Southall	7/42
6586	11/40	Oxley	Banbury	7/42
6594	12/40	Bristol SPM	Bristol SPM	2/43
6595	11/40	Oxley	Oxley	2/43
6621	11/40	Oxley	Oxley	7/42
6629	12/40	Bristol SPM	Bristol SPM	7/42
6633	11/40	Oxley	Oxley	1/43
6639	11/40	Oxley	Oxley	7/42

SR Locomotives

SR No.	To GWR	First GWR Shed	Last GWR Shed	Left GWR	SR Class	Type
82	12/41	+	+	9/43	B4	0-4-0T
87	12/41	+	+	9/43	B4	0-4-0T
478	11/41	Paddington	Paddington	1/42	H15	4-6-0
496	11/41	Southall	Southall	7/43	S15	4-6-0
497	11/41	Southall	Southall	5/43	S15	4-6-0
498	11/41	Paddington	Paddington	5/43	S15	4-6-0
499	11/41	Paddington	Paddington	3/43	S15	4-6-0
2089	12/41	Gloucester	Bath Road	7/43	I3	4-4-2T
2091	11/41	Gloucester	Bath Road	7/43	I3	4-4-2T
2327	11/41	Paddington	Paddington	7/43	N15X	4-6-0
2328	11/41	Paddington	Paddington	12/42	N15X	4-6-0
2329	11/41	Newton Abbot	Exeter	7/43	N15X	4-6-0
2330	1/42	Paddington	Paddington	7/43	N15X	4-6-0
2331	11/41	Swindon	Swindon	7/43	N15X	4-6-0
2332	11/41	Gloucester	Gloucester	7/43	N15X	4-6-0

\+ Transferred to Admiralty use at Trecwn (Fishguard) 1/42 but not taken off GWR records until 9/43.

USA 2-8-0

USA No.	To GWR	First GWR Shed	Last GWR Shed	Left GWR
1601	2/43	Ebbw Junction	Ebbw Junction	10/44
1602	3/43	Pontypool	Pontypool	9/44
1603	2/43	Pontypool	Pontypool	9/44
1604	1/43	Ebbw Junction	Ebbw Junction	9/44
1605	1/43	Stourbridge	Stourbridge	9/44
1606	3/43	Swindon	Swindon	9/44
1607	1/43	Cardiff	Cardiff	9/44
1608	2/43	Oxley	Croes Newydd	9/44
1609	1/43	Cardiff	Pontypool	9/44
1610	2/43	Croes Newydd	Croes Newydd	9/44
1611	2/43	Stourbridge	Stourbridge	9/44
1612	2/43	Oxley	Oxley	9/44
1613	2/43	Oxley	Oxley	9/44
1614	2/43	Oxley	Oxley	9/44
1615	3/43	Oxley	Shrewsbury	10/44
1616	2/43	Pontypool	Pontypool	10/44
1617	1/43	Oxley	Shrewsbury	9/44
1618	2/43	Oxley	Stourbridge	9/44
1619	2/43	Ebbw Junction	Pontypool	10/44
1620	2/43	Oxley	Shrewsbury	9/44
1621	2/43	Stourbridge	Stourbridge	9/44
1622	2/43	Oxley	Stourbridge	9/44
1623	2/43	Pontypool	Pontypool	9/44
1624	1/43	Ebbw Junction	Ebbw Junction	9/44
1628	11/43	Worcester	Worcester	9/44
1632	11/43	Bristol SPM	Bristol SPM	9/44
1639	8/43	Severn Tunnel	Severn Tunnel	9/44
1641	9/43	Neath	Neath	9/44
1642	9/43	Llanelly	Llanelly	9/44
1643	9/43	Gloucester	Gloucester	9/44
1644	9/43	Gloucester	Gloucester	9/44
1645	9/43	Gloucester	Gloucester	9/44
1646	10/43	Chester	Stourbridge	9/44

USA No.	To GWR	First GWR Shed	Last GWR Shed	Left GWR
1647	9/43	Oxley	Banbury	9/44
1648	9/43	Oxley	Oxley	9/44
1649	9/43	Oxley	Banbury	9/44
1651	10/43	Ebbw Junction	Ebbw Junction	9/44
1654	10/43	Banbury	Banbury	9/44
1655	10/43	Cardiff	Cardiff	9/44
1656	10/43	Oxford	Oxford	9/44
1658	10/43	Banbury	Banbury	9/44
1659	10/43	Paddington	Paddington	9/44
1660	12/43	Worcester	Worcester	9/44
1661	1/44	Hereford	Hereford	9/44
1662	10/43	Paddington	Paddington	9/44
1663	10/43	Oxford	Oxford	9/44
1664	10/43	Llanelly	Llanelly	9/44
1665	11/43	Worcester	Worcester	9/44
1681	11/43	Bristol SPM	Bristol SPM	9/44
1682	11/43	Bristol SPM	Bristol SPM	9/44
1683	11/43	Bristol SPM	Bristol SPM	9/44
1684	11/43	Worcester	Worcester	9/44
1687	12/43	Bristol SPM	Bristol SPM	9/44
1688	10/43	Paddington	Paddington	10/44
1689	10/43	Didcot	Didcot	9/44
1749	11/43	Bristol SPM	Bristol SPM	9/44
1757	11/43	Bristol SPM	Bristol SPM	9/44
1835	2/43	Pontypool	Pontypool	9/44
1836	NOT	RECORDED ON GWR	(USED ON LNER)	
1841	6/43	Oxley	Oxley	9/44
1877	6/43	Paddington	Paddington	9/44
1880	6/43	Oxford	Oxford	9/44
1881	5/43	Aberdare	Ebbw Junction	9/44
1883	5/43	Ebbw Junction	Ebbw Junction	9/44
1884	5/43	Ebbw Junction	Ebbw Junction	9/44
1891	6/43	Oxford	Oxford	9/44
1892	6/43	Aberdare	Ebbw Junction	9/44
1893	6/43	Aberdare	Severn Tunnel	9/44
1894	6/43	Pontypool	Pontypool	9/44
1895	6/43	Ebbw Junction	Ebbw Junction	9/44
1896	6/43	Cardiff	Cardiff	9/44
1897	5/43	Oxford	Oxford	9/44
1898	6/43	Severn Tunnel	Severn Tunnel	9/44
1899	6/43	Aberdare	Ebbw Junction	9/44
1900	6/43	Shrewsbury	Shrewsbury	9/44
1901	7/43	Shrewsbury	Shrewsbury	9/44
1902	7/43	Oxley	Oxley	9/44
1909	8/43	Cardiff	Cardiff	9/44
1910	8/43	Oxley	Tyseley	9/44
1913	8/43	Bristol SPM	Bristol SPM	9/44
1914	8/43	Swindon	Swindon	9/44
1915	9/43	Didcot	Didcot	9/44
2096	6/43	Paddington	Paddington	9/44
2098	6/43	Oxford	Oxford	9/44
2100	6/43	Paddington	Paddington	9/44
2102	5/43	Oxford	Oxford	9/44
2103	6/43	Oxford	Oxford	9/44
2109	7/43	Severn Tunnel	Severn Tunnel	9/44
2110	6/43	Oxley	Oxley	9/44
2112	6/43	Pontypool	Pontypool	9/44
2116	8/43	Oxley	Tyseley	9/44
2118	7/43	Ebbw Junction	Ebbw Junction	9/44
2122	8/43	Bristol SPM	Bristol SPM	9/44

USA No.	To GWR	First GWR Shed	Last GWR Shed	Left GWR
2129	7/43	Paddington	Paddington	9/44
2130	7/43	Swindon	Swindon	9/44
2131	7/43	Paddington	Paddington	9/44
2132	8/43	Banbury	Banbury	9/44
2133	6/43	Oxley	Oxley	9/44
2134	8/43	Banbury	Banbury	9/44
2135	7/43	Oxley	Oxley	9/44
2136	7/43	Oxley	Oxley	9/44
2137	7/43	Tyseley	Tyseley	9/44
2138	7/43	Oxley	Oxley	9/44
2139	6/43	Ebbw Junction	Ebbw Junction	9/44
2140	7/43	Oxley	Oxley	9/44
2141	8/43	Swindon	Swindon	9/44
2142	6/43	Paddington	Didcot	9/44
2143	6/43	Swindon	Swindon	9/44
2144	7/43	Cardiff	Cardiff	9/44
2145	6/43	Paddington	Paddington	10/44
2147	8/43	Banbury	Banbury	9/44
2148	6/43	Paddington	Paddington	9/44
2149	7/43	Cardiff	Cardiff	9/44
2150	6/43	Pontypool	Pontypool	9/44
2151	6/43	Paddington	Paddington	9/44
2159	9/43	Chester	Chester	9/44
2164	10/43	Bristol SPM	Bristol SPM	9/44
2165	10/43	Landore	Llanelly	9/44
2167	11/43	Bristol SPM	Bristol SPM	9/44
2267	10/43	Landore	Llanelly	9/44
2269	9/43	Neath	Neath	9/44
2270	9/43	Worcester	Gloucester	9/44
2279	1/44	Hereford	Worcester	9/44
2280	9/43	Oxford	Oxford	9/44
2290	10/43	Llanelly	Llanelly	9/44
2294	8/43	Cardiff	Cardiff	9/44
2312	8/43	Ebbw Junction	Ebbw Junction	9/44
2313	7/43	Severn Tunnel	Severn Tunnel	9/44
2314	6/43	Oxley	Oxley	9/44
2315	10/43	Neath	Neath	9/44
2318	7/43	Paddington	Paddington	9/44
2319	9/43	Ebbw Junction	Ebbw Junction	9/44
2323	7/43	Ebbw Junction	Ebbw Junction	10/44
2324	8/43	Ebbw Junction	Ebbw Junction	9/44
2326	7/43	Tyseley	Tyseley	10/44
2327	7/43	Tyseley	Tyseley	9/44
2338	2/44	Worcester	Worcester	9/44
2339	9/43	Neath	Neath	9/44
2349	9/43	Neath	Neath	9/44
2350	9/43	Llanelly	Llanelly	9/44
2351	9/43	Neath	Neath	9/44
2352	9/43	Didcot	Didcot	9/44
2353	3/44	Gloucester	Gloucester	9/44
2354	8/43	Tyseley	Tyseley	9/44
2357	11/43	Worcester	Worcester	9/44
2358	9/43	Neath	Neath	9/44
2359	9/43	Llanelly	Llanelly	9/44
2360	9/43	Oxley	Oxley	9/44
2368	8/43	Severn Tunnel	Severn Tunnel	10/44
2369	8/43	Severn Tunnel	Severn Tunnel	9/44
2375	9/43	Shrewsbury	Shrewsbury	9/44
2377	9/43	Didcot	Didcot	9/44
2403	7/43	Ebbw Junction	Ebbw Junction	9/44

USA No.	To GWR	First GWR Shed	Last GWR Shed	Left GWR
2404	7/43	Swindon	Swindon	9/44
2405	8/43	Severn Tunnel	Severn Tunnel	9/44
2407	9/43	Aberdare	Ebbw Junction	9/44
2408	9/43	Bristol SPM	Bristol SPM	9/44
2410	9/43	Neath	Neath	9/44
2422	8/43	Westbury	Swindon	9/44
2423	9/43	Paddington	Paddington	9/44
2424	9/43	Worcester	Gloucester	9/44
2430	9/43	Didcot	Didcot	9/44
2431	10/43	Banbury	Banbury	9/44
2432	10/43	Paddington	Paddington	9/44
2433	8/43	Bristol SPM	Bristol SPM	9/44
2434	8/43	Westbury	Swindon	9/44
2435	1/44	Hereford	Worcester	9/44
2438	8/43	Aberdare	Ebbw Junction	9/44
2439	9/43	Tyseley	Tyseley	9/44
2440	9/43	Oxley	Banbury	9/44
2441	9/43	Banbury	Banbury	9/44
2442	1/44	Hereford	Worcester	9/44
2443	2/44	Hereford	Worcester	9/44
2448	11/43	Tyseley	Tyseley	9/44
2450	11/43	Bristol SPM	Bristol SPM	9/44

NOTE:

No. 1836 is shown in some sources as being lent to the GWR in 2/43 and transferred to the LNER in 3/43, but it is not shown in the co-pany's records of locomotive allocations.

LMS 8F 2-8-0 (Swindon built)

These locomotives were built at Swindon to following Lot numbers:-

8400-9 Lot 348
8410-29 Lot 349
8430-9 Lot 351
8440-79 Lot 353

LMS No.	Date	First GWR Shed	Last GWR Shed	Left GWR
8400	5/43	Swindon	Bristol SPM	11/46
8401	7/43	Swindon	Bristol SPM	12/46
8402	7/43	Swindon	Bristol SPM	11/46
8403	7/43	Bristol SPM	Bristol SPM	11/46
8404	7/43	Didcot	Paddington	3/47
8405	8/43	Didcot	Paddington	12/46
8406	8/43	Didcot	Didcot	1/47
8407	8/43	Newton Abbot	Newton Abbot	12/46
8408	9/43	Newton Abbot	Newton Abbot	4/47
8409	9/43	Newton Abbot	Newton Abbot	3/47
8410	9/43	Bristol SPM	Laira	3/47
8411	9/43	Bristol SPM	Bristol SPM	2/47
8412	10/43	Bristol SPM	Bristol SPM	3/47
8413	10/43	Paddington	Paddington	3/47
8414	10/43	Laira	Laira	1/47
8415	10/43	Bristol SPM	Bristol SPM	12/46
8416	11/43	Oxley	Oxley	4/47
8417	11/43	Oxley	Oxley	11/46
8418	11/43	Oxley	Oxley	2/47
8419	11/43	Oxley	Oxley	3/47

LMS No.	Date	First GWR Shed	Last GWR Shed	Left GWR
8420	11/43	Exeter	Exeter	4/47
8421	12/43	Bristol SPM	Bristol SPM	4/47
8422	12/43	Oxley	Croes Newydd	5/47
8423	12/43	Paddington	Paddington	5/47
8424	12/43	Bristol SPM	Bristol SPM	4/47
8425	12/43	Exeter	Newton Abbot	4/47
8426	12/43	Oxley	Oxley	5/47
8427	1/44	Laira	Laira	8/47
8428	1/44	Oxley	Chester	8/47
8429	1/44	Oxley	Oxley	4/47
8430	3/44	Oxley	Oxley	4/47
8431	3/44	Newton Abbot	Gloucester	3/47
8432	3/44	Exeter	Newton Abbot	4/47
8433	3/44	Taunton	Gloucester	5/47
8434	3/44	Taunton	Laira	4/47
8435	4/44	Penzance	Laira	4/47
8436	4/44	Croes Newydd	Croes Newydd	4/47
8437	5/44	Chester	Chester	4/47
8438	5/44	Paddington	Paddington	5/47
8439	5/44	Birkenhead	Birkenhead	3/47
8440	6/44	Exeter	Taunton	4/47
8441	6/44	Bristol SPM	Bristol SPM	4/47
8442	6/44	Paddington	Paddington	4/47
8443	6/44	Paddington	Paddington	3/47
8444	6/44	Oxley	Oxley	4/47
8445	7/44	Didcot	Didcot	5/47
8446	7/44	Reading	Reading	4/47
8447	7/44	Oxley	Oxley	5/47
8448	7/44	Bristol SPM	Bristol SPM	5/47
8449	8/44	Reading	Reading	4/47
8450	8/44	Reading	Reading	4/47
8451	9/44	Worcester	Gloucester	8/47
8452	9/44	Gloucester	Gloucester	4/47
8453	9/44	Ebbw Junction	Oxley	4/47
8454	9/44	Ebbw Junction	Oxley	5/47
8455	10/44	Ebbw Junction	Oxley	4/47
8456	10/44	Cardiff	Cardiff	4/47
8457	10/44	Oxley	Oxley	6/47
8458	11/44	Oxley	Oxley	5/47
8459	11/44	Oxley	Oxley	4/47
8460	12/44	Oxley	Oxley	5/47
8461	12/44	Tyseley	Tyseley	4/47
8462	12/44	Ebbw Junction	Ebbw Junction	5/47
8463	1/45	Cardiff	Cardiff	4/47
8464	1/45	Cardiff	Cardiff	8/47
8465	2/45	Cardiff	Cardiff	5/47
8466	2/45	Severn Tunnel	Severn Tunnel	9/47
8467	3/45	Severn Tunnel	Severn Tunnel	10/47
8468	3/45	Pontypool	Pontypool	5/47
8469	3/45	Pontypool	Pontypool	4/47
8470	4/45	Ebbw Junction	Ebbw Junction	5/47
8471	4/45	Cardiff	Cardiff	5/47
8472	5/45	Cardiff	Cardiff	5/47
8473	5/45	Pontypool	Pontypool	5/47
8474	5/45	Oxley	Oxley	5/47
8475	6/45	Tyseley	Tyseley	5/47
8476	6/45	Reading	Reading	5/47
8477	6/45	Paddington	Paddington	10/47
8478	6/45	Oxley	Oxley	9/47
8479	7/45	Reading	Reading	5/47

LNER Y9 0-4-0ST

LNER No.	To GWR	First GWR Shed	Last GWR Shed	Left GWR
10090	6/43	Neyland	+	9/43
10093	6/43	Neyland	+	9/43

\+ Transferred off GWR stock to the Admiralty 11/9/1943. Recorded as despatched from Neyland on 30/8/1945 for Glasgow.

USA 0-6-0T

USA No.	To GWR	First GWR Shed	Last GWR Shed	Left GWR
1307	6/44	Duffryn Yard	Duffryn Yard	9/44
1945	6/44	Barry	Barry	9/44
1946	6/44	Barry	Barry	9/44
1948	6/44	Cardiff E Dock	Cardiff E Dock	9/44
1976	6/44	Cardiff E Dock	Barry	9/44
1979	6/44	Radyr	Radyr	9/44
1981	6/44	Barry	Barry	9/44
1983	6/44	Radyr	Radyr	9/44
1985	6/44	Barry	Barry	9/44
1987	6/44	+	+	9/44
4381	6/44	Ebbw Junction	Ebbw Junction	8/44
4383	6/44	Ebbw Junction	Ebbw Junction	8/44
4384	6/44	Pill	Ebbw Junction	8/44
4385	6/44	Ebbw Junction	Ebbw Junction	7/44
4388	6/44	Ebbw Junction	Ebbw Junction	8/44
4389	6/44	Pill	Ebbw Junction	8/44
4390	6/44	Pill	Pill	8/44
4392	6/44	Ebbw Junction	Ebbw Junction	8/44
4393	6/44	Ebbw Junction	Ebbw Junction	8/44
4395	6/44	Pill	Ebbw Junction	8/44
4397	6/44	Ebbw Junction	Ebbw Junction	8/44
4398	6/44	Ebbw Junction	Ebbw Junction	7/44
4399	6/44	Danygraig	Danygraig	9/44
4401	6/44	Danygraig	Danygraig	9/44

\+ Not shown in GWR records of locomotive allocations, but shown elsewhere as Duffryn Yard.

WD 2-8-0 (Loans 1944-5)

WD No.	To GWR	First GWR Shed	Last GWR Shed	Left GWR
7011	9/44	Oxley	Oxley	1/45
7040	11/44	Reading	Reading	12/44
7041	11/44	Reading	Reading	1/45
7042	9/44	Oxley	Oxley	1/45
7043	9/44	Worcester	Worcester	1/45
7044	9/44	Oxley	Oxley	1/45
7045	9/44	Croes Newydd	Oxley	1/45
7047	9/44	Oxley	Oxley	1/45
7048	9/44	Oxley	Oxley	1/45
7049	9/44	Banbury	Banbury	1/45
7087	9/44	Ebbw Junction	Ebbw Junction	3/45
7088	9/44	Severn Tunnel	Severn Tunnel	1/45
7089	9/44	Severn Tunnel	Severn Tunnel	3/45

WD No.	To GWR	First GWR Shed	Last GWR Shed	Left GWR
7091	9/44	Severn Tunnel	Ebbw Junction	2/45
7094	9/44	Paddington	Paddington	1/45
7111	9/44	Pontypool	Pontypool	3/45
7113	9/44	Ebbw Junction	Ebbw Junction	2/45
7114	9/44	Cardiff	Cardiff	12/44
7116	9/44	Cardiff	Cardiff	1/45
7135	9/44	Severn Tunnel	Ebbw Junction	2/45
7136	9/44	Ebbw Junction	Ebbw Junction	2/45
7137	9/44	Pontypool	Pontypool	12/44
7138	9/44	Swindon	Swindon	1/45
7140	9/44	Stourbridge	Stourbridge	1/45
7141	9/44	Stourbridge	Stourbridge	3/45
7142	9/44	Pontypool	Pontypool	1/45
7144	9/44	Tyseley	Tyseley	2/45
7146	9/44	Shrewsbury	Shrewsbury	2/45
7148	9/44	Worcester	Worcester	2/45
7150	9/44	Oxley	Stourbridge	1/45
7215	11/44	Oxford	Oxford	2/45
7216	11/44	Oxford	Oxford	2/45
7217	11/44	Paddington	Paddington	2/45
7218	11/44	Paddington	Paddington	2/45
7219	11/44	Didcot	Didcot	3/45
7220	11/44	Reading	Reading	2/45
7229	9/44	Worcester	Worcester	2/45
7231	9/44	Paddington	Paddington	1/45
7233	9/44	Worcester	Worcester	2/45
7240	9/44	Pontypool	Pontypool	2/45
7243	11/44	Worcester	Worcester	12/44
7244	9/44	Banbury	Banbury	1/45
7258	9/44	Tyseley	Tyseley	2/45
7259	9/44	Neath	Neath	2/45
7261	9/44	Bristol SPM	Bristol SPM	11/44
7262	9/44	Banbury	Banbury	3/45
7354	9/44	Banbury	Banbury	2/45
7366	9/44	Oxley	Oxley	2/45
7377	9/44	Paddington	Paddington	12/44
7383	9/44	Ebbw Junction	Ebbw Junction	2/45
7385	9/44	Bristol SPM	Bristol SPM	2/45
7386	9/44	Ebbw Junction	Ebbw Junction	12/44
7399	9/44	Bristol SPM	Bristol SPM	12/44
7400	9/44	Worcester	Worcester	12/44
7401	9/44	Neath	Neath	12/44
7403	9/44	Oxley	Oxley	12/44
7404	9/44	Worcester	Worcester	12/44
7408	9/44	Cardiff	Cardiff	12/44
7409	9/44	Swindon	Swindon	12/44
7412	9/44	Westbury	Westbury	12/44
7413	9/44	Neath	Neath	12/44
7414	9/44	Ebbw Junction	Ebbw Junction	12/44
7415	9/44	Ebbw Junction	Ebbw Junction	2/45
7416	9/44	Banbury	Banbury	2/45
7418	9/44	Worcester	Worcester	12/44
7419	9/44	Stourbridge	Stourbridge	12/44
7420	9/44	Banbury	Banbury	2/45
7421	9/44	Shrewsbury	Shrewsbury	2/45
7450	9/44	Ebbw Junction	Ebbw Junction	2/45
7451	9/44	Pontypool	Pontypool	1/45
7453	9/44	Bristol SPM	Bristol SPM	12/44
7454	9/44	Neath	Neath	11/44
7455	9/44	Bristol SPM	Bristol SPM	11/44

WD No.	To GWR	First GWR Shed	Last GWR Shed	Left GWR
7456	9/44	Bristol SPM	Bristol SPM	12/44
7457	9/44	Swindon	Swindon	11/44
7458	9/44	Worcester	Worcester	11/44
7459	9/44	Neath	Neath	11/44
7460	9/44	Neath	Neath	11/44
7461	9/44	Worcester	Worcester	11/44
7465	9/44	Paddington	Paddington	11/44
7489	9/44	Reading	Reading	11/44
7490	9/44	Didcot	Reading	11/44
7491	9/44	Oxford	Oxford	11/44
7492	9/44	Paddington	Paddington	12/44
7493	9/44	Southall	Southall	12/44
7494	9/44	Reading	Reading	11/44
7495	9/44	Reading	Reading	11/44
7496	9/44	Oxford	Oxford	12/44
7497	9/44	Reading	Reading	11/44

WD 2-8-0 (Loans 1946-7)

WD No.	To GWR	First GWR Shed	Last GWR Shed	Left GWR
70801	11/46	Bristol SPM	Bristol SPM	-
70808	5/47	Oxley	Croes Newydd	-
70836	11/46	Bristol SPM	Bristol SPM	-
70843	3/47	Didcot	Didcot	-
70876	12/46	Bristol SPM	Bristol SPM	-
77000	7/47	Cardiff	Cardiff	-
77001	1/47	Newton Abbot	Newton Abbot	-
77005	4/47	Paddington	Paddington	-
77012	10/46	Oxley	Banbury	-
77014	5/47	Oxley	Banbury	-
77015	2/47	Paddington	Paddington	-
77026	3/47	Oxley	Banbury	-
77028	3/47	Oxley	Oxley	-
77040	10/46	Oxley	Oxley	-
77049	10/46	Oxley	Oxley	-
77053	4/47	Cardiff	Cardiff	-
77058	1/47	Oxford	Oxford	-
77064	5/47	Oxley	Oxley	-
77077	5/47	Taunton	Taunton	-
77079	3/47	Oxley	Oxley	-
77097	5/47	Tyseley	Tyseley	-
77099	5/47	Pontypool	Pontypool	-
77102	5/47	Ebbw Junction	Ebbw Junction	-
77106	4/47	Cardiff	Cardiff	-
77115	5/47	Oxley	Chester	-
77116	11/46	Bristol SPM	Bristol SPM	-
77123	4/47	Cardiff	Cardiff	-
77130	4/47	Paddington	Paddington	-
77142	3/47	Bristol SPM	Bristol SPM	-
77151	9/46	Oxley	Oxley	-
77161	2/47	Laira	Laira	-
77165	10/46	Oxley	Oxley	-
77184	4/47	Paddington	Paddington	-
77196	3/47	Laira	Laira	-
77200	11/46	Bristol SPM	Bristol SPM	-
77202	3/47	Oxley	Oxley	-
77203	4/47	Gloucester	Gloucester	-

WD No.	To GWR	First GWR Shed	Last GWR Shed	Left GWR
77210	4/47	Newton Abbot	Newton Abbot	-
77214	3/47	Newton Abbot	Newton Abbot	-
77234	5/47	Oxley	Oxley	-
77241	4/47	Newton Abbot	Newton Abbot	-
77247	4/47	Bristol SPM	Bristol SPM	-
77255	4/47	Laira	Laira	-
77257	5/47	Oxley	Chester	-
77288	3/47	Exeter	Cardiff	-
77289	3/47	Bristol SPM	Bristol SPM	-
77291	12/47	+	+	-
77294	4/47	Laira	Laira	-
77297	3/47	Oxley	Banbury	-
77310	4/47	Llanelly	Llanelly	-
77325	4/47	Laira	Laira	-
77326	12/46	Bristol SPM	Bristol SPM	-
77368	10/47	+	+	-
77378	4/47	Reading	Reading	-
77380	4/47	Cardiff	Cardiff	-
77388	3/47	Exeter	Cardiff	-
77393	12/47	+	+	-
77407	5/47	+	+	-
77408	4/47	Oxley	Banbury	-
77421	4/47	Laira	Laira	-
77429	5/47	Llanelly	Llanelly	-
77443	6/47	Pontypool	Pontypool	-
77489	4/47	Llanelly	Llanelly	-
77508	11/46	Bristol SPM	Bristol SPM	-
78510	6/47	Ebbw Junction	Ebbw Junction	-
78512	6/47	Banbury	Banbury	-
78521	7/47	Paddington	Paddington	-
78522	4/47	Reading	Reading	-
78542	11/47	+	+	-
78543	10/47	Cardiff	Cardiff	-
78604	3/47	Banbury	Banbury	-
78622	3/47	Paddington	Paddington	-
78671	4/47	Laira	Laira	-
78695	5/47	Oxley	Oxley	-
78714	5/47	Oxley	Croes Newydd	-
78717	4/47	Laira	Laira	-
79219	5/47	Oxley	Oxley	-
79224	4/47	Oxley	Tyseley	-
79225	4/47	Oxley	Oxley	-
79226	5/47	Birkenhead	Birkenhead	-
79232	4/47	Gloucester	Gloucester	-
79234	3/47	Paddington	Paddington	-
79235	7/47	Pontypool	Pontypool	-
79261	11/46	Bristol SPM	Bristol SPM	-
79274	10/47	Severn Tunnel	Severn Tunnel	-
79278	11/47	+	+	-
79301	11/46	Bristol SPM	Bristol SPM	-
79303	8/47	Didcot	Didcot	-
79309	12/46	Bristol SPM	Bristol SPM	-

\+ Still in Swindon Works on 31/12/1947 (except No. 77407 which was at Stafford Road Works).

All of these locomotives passed on loan to British Railways on the Nationalisation of the GWR.